The Complete Guide To High Fidelity

"Just tried experiment with a diaphragm having an embossing point and held against paraffin paper moving rapidly. The speaking vibrations are indented nicely and there is no doubt that I shall be able to store up and reproduce automatically at any future time the human voice perfectly."

—from the daily log of
Thomas A. Edison
July 18, 1877

Martin Clifford was born in Harlem in Manhattan and is a graduate of the City University of New York. Subsequently he worked as an engineer for Sperry Gyroscope Co. with emphasis on the areas of voltage regulated power supplies, radar, and automatic direction finders for aircraft. He also acquired various licenses—radio amateur (ex W2CDV), first-class radio-telephone (commercial), and a license to teach vocational subjects from the University of the State of New York.

When the demand for engineers grew slack he became a teacher in private vocational schools and specialized in electronic communications. During this time he became interested in writing and gradually became a full-time free lance writer and has had numerous articles published in newspapers, trade, and consumer magazines. His articles have been syndicated in as many as 200 college newspapers.

Most of his writing efforts are devoted to books. He is the author of many books as well as a number of booklets on various topics in electronics, and several correspondence courses in electronics, radio, television, and drafting. He is the author of *The Complete Guide to Car Audio*.

The Complete Guide To High Fidelity

by
Martin Clifford

Howard W. Sams & Co., Inc.
4300 WEST 62ND ST. INDIANAPOLIS, INDIANA 46268 USA

Copyright © 1982 by Howard W. Sams & Co., Inc.
Indianapolis, IN 46268

FIRST EDITION
FIRST PRINTING—1982

All rights reserved. No part of this book shall be
reproduced, stored in a retrieval system, or transmitted by
any means, electronic, mechanical, photocopying,
recording, or otherwise, without written permission from
the publisher. No patent liability is assumed with respect to
the use of the information contained herein. While every
precaution has been taken in the preparation of this book,
the publisher assumes no responsibility for errors or
omissions. Neither is any liability assumed for damages
resulting from the use of the information contained herein.

International Standard Book Number: 0-672-21892-5
Library of Congress Catalog Card Number: 82-50014

Edited by: *Bob Manville*
Illustrated by: *Jill E. Martin*

Printed in the United States of America.

Foreword

As a manufacturer of quality home high fidelity and car audio equipment, the Luxman division of Alpine Electronics of America dedicates its resources toward the design and production of a product line which is synonymous with the best that you, our customers, desire.

However, our best efforts and value to you can be negated, and our efforts downgraded, if the consumer does not know how to shop for audio equipment, does not understand electronic specifications, is not aware of benefit/features and does not know how to plan a quality high fidelity system.

The value to you can be further complicated by the fact that the sound quality from high fidelity components depends on how the components match, how the system is connected and placed, and of course, the basic quality and technology of components purchased.

We at Luxman, in harmony with our dealers, seek to have our components perform to their utmost limits. We feel that our responsibility to you does not end with the sale. We want more.

We want you to have the musical enjoyment we know our components can supply. That is why we have sponsored and have encouraged the publication of this book. This book is not a Luxman catalog. We do not quote prices or model numbers nor have we engaged in any promoting of our products, although it would be tempting for us to do so.

What we do want is an educated consumer. This book is written in nontechnical language, so you will be able to read and understand the meaning of specifications and features. You will know what to look for and how to do comparison shopping. This book will also help you plan your high fidelity system, component by component.

We urge you to select your high fidelity specialist/dealer carefully. If you do, with his help and the help of this book, you should be able to have a high fidelity system that will give you the utmost in pleasure, satisfaction, and value.

REESE HAGGOTT
Executive Vice President
Luxman Division
Alpine Electronics of America

Preface

There is a myth to the effect that a high-fidelity system will bring the concert hall into your home. It's not true and it would be unfortunate if it were. Much of the world's great music, past and present, wasn't composed to be played in that 19th century invention, the concert hall. Bach wrote his organ works as an integral part of church liturgy. Brass bands are for outdoors and much modern music is composed especially for the recording studio and electronic reproduction.

In the recording studio, music is often recorded "dry," that is, the microphones are placed as close to the performers' instruments as possible. Controlled amounts of reverberant sound, sometimes called "wet" sound, are then added later to give the recording a sense of liveliness. So what you hear, via a prerecorded tape or a phono record, is often not *natural* in the sense that it is a true reproduction of a combo, or a band, a vocalist, or an orchestra. Even those works which were intended for concert hall performance, such as the classical and romantic symphonies, aren't intended for playing over a hi-fi system with concert hall volume. The cannon in Tchaikovsky's 1812 Overture would probably shatter your window panes.

When music is recorded on tape or phono records the dynamic range—the sonic separation between the softest and loudest sounds—is compressed. This compression is handled by a recording engineer, so what you do hear isn't what the composer intended, but a squeezed together version to accommodate the technical demands of tape and records. Fortunately, efforts are now being made to widen the dynamic range of recorded music.

There is always some modification of sound by audio equipment. Even the room in which you do your listening is an integral part of your hi-fi system, for the sound coming out of the speakers must interface with that room. What you will hear in the way of bass or treble will depend on the acoustic qualities of that room. You, the listener, are low man or woman on the hi-fi totem pole. You get what's left over, for the acoustic demands of your listening room take precedence.

You can console yourself with the fact that no two concert halls are alike, that no two orchestras play the same compositions the same way, and that what you hear in a concert hall will also depend on where you sit. So whether you listen to music through a hi-fi system in your home or in a concert hall or at a disco or at an outdoor band always involves some measure of compromise. Concert hall listening has the advantage of being *live* listening; hi-fi music in your home supplies diversity and convenience. Both live and in-home listening are desirable, relaxing, and enjoyable.

There are two types of music lovers: active and passive. The passive type supplies a pair of ears and a sense of musical enjoyment. But the passive listener doesn't become involved; doesn't participate in the music making process.

The other kind of listener—the active—is a joiner. He, or she, sings or plays an instrument, studies music, and somehow becomes part of it. They have an inner drive that makes them want to be part of the musical scene.

Fortunately, with a hi-fi system in your home you can be either active or passive. The passive listener has the whole world of recorded music at his fingertips. But for the participant, whole new worlds are opened. With one or more microphones and a tape deck the active hi-fi-er can sing or play along with any orchestral group, become the soloist, and be in the musical spotlight. With tape you can change yourself from a singing soloist into a duet, or into a quartet. You can play an instrument and include your voice at the same time or add your voice later.

The idea that most of us belong in the passive class is beginning to change. You can now use a hi-fi system as an art form. You can join in, modify, add, do what you wish, and in the process learn a lot about music and possibly a lot more about yourself.

This book has several purposes. It will give you an overview of hi-fi and let you make a selection of the kind of hi-fi system your budget permits. But it will also let you plan for the future, show you how you can expand your hi-fi to let you ultimately have the kind you want. This book also explains the technical language of specifications, so you can buy audio components with more confidence, and, more intelligently and capably, compare one component with another. You will also find a description of the features of each hi-fi component, some of which you may find highly appealing, others less so. But knowing about them will also help you in making your choice of components.

You can use your hi-fi system in a number of ways: to relax, to enjoy, and to escape from or to exclude the real world, even if only temporarily. This will give you a chance to "rewind," to "charge your batteries" and to meet the world again, refreshed and strengthened. Or, you can use your hi-fi system to further your own musical abilities. You can record your own efforts either for the pure enjoyment of it, or as a way of being able to listen to yourself for critical evaluation. But in any event, owning and using a hi-fi system can be encompassed in just a single word.

Enjoy!

Martin Clifford

Acknowledgments

My thanks go to Reese Haggott, VP, Marketing, Luxman, for his encouragement in the production of this book and to George Savage, Marketing Services Manager, Luxman, for his considerable assistance.

There were many other manufacturers who supplied essential data and I just hope I haven't forgotten any of them. They include Acousti-phase Corp., AKG Acoustics, Inc., Audio-Technica U.S. Inc., B•I•C-Avnet, Bose Corp., dbx, Inc., Discwasher®, Electro-Voice, Inc., Epicure Products, Inc., Jumetite Laboratories, KLH, Mobile Fidelity Sound Lab., Nakamichi USA Corp., North American Philips Corp., Sony Corp. of America, TDK® Electronics Corp., Thorens Corp., and Yamaha Electronics Corp. USA. My appreciation also to Sound (Canada) magazine.

And a vote of thanks to Steve Garey who read the manuscript and made valuable suggestions.

MARTIN CLIFFORD

Contents

CHAPTER 1

HI-FI SYSTEMS 13
    The Compact System—The Component System—Prepackaged Systems—Dealer Systems—The Basic System—Mono and Stereo—The Stereo Receiver—Adding a Record Player—Adding a Cassette Deck—Adding Headphones—Adding an Open Reel Deck—Adding an Equalizer—Audio and Hi-Fi Systems—Installing Your Hi-Fi System—The Do's and Don't's of Components and Hi-Fi Systems—Specs—Features—Hertz—Sine Waves—Harmonics—Timbre

CHAPTER 2

HI-FI TUNERS AND RECEIVERS 53
    What Is a Receiver?—What Is a Tuner?—Cost Difference—The Tuner-Amplifier—The Casseiver—The AM Signal—The AM Tuner—The AM Band—The AM Antenna—Ground—Frequency Modulation—Signal Separation—The FM Antenna—FM Reception Range—Antenna Gain—Multipath Reception—The Superheterodyne—The FM Tuner or Receiver—Tuning the Tuner or Receiver—Tuning Meters—Controls—Features—Tuner Versus Receiver Connections—What Decibels Are All About—Specifications

CHAPTER 3

PREAMPLIFIERS 105
    Preamplifier Inputs—Input Signal Strength—Phono Equalization—Equalization Accuracy—Nonlinear Versus Linear Inputs—Preamp

Features—Dubbing—Tape Monitoring—Controls—Octaves—Filters—Head Amplifiers—Preamp Specs—Negative Feedback—Duo-Beta Negative Feedback

CHAPTER 4

INTEGRATED AMPLIFIERS AND POWER AMPLIFIERS 123
Voltage Versus Power—AC Line Power—Importance of the Power Supply—Importance of the Load—Adding More Speakers—Short Circuits—Protector Circuits—Cooling—Clipping—Output Capacitor-Less Circuits—Amplifier Efficiency—Amplifier Classes—Audio Output Power—Impedance Matching—Biamplification—Complementary Symmetry—Negative Feedback—Integrated Amplifier Controls—Power Amplifier Controls—Power Output Indicators—Speaker Switching—Subsonic Filter—Power Amplifier Specs—Equalizers

CHAPTER 5

SPEAKERS AND HEADPHONES 157
What Do We Want?—Dynamic Speaker—Speaker Systems—Electrostatic Speakers—Labyrinth Speakers—Transmission Line System—Heil Driver—Piezoelectric Tweeters—Directionality—Efficiency—Minimum and Maximum Speaker Power—Enclosures—Cone Materials—Speaker Size Versus Power—Ferrofluids—Distortion—Time Delay Distortion—Acoustics—Standing Waves—Speaker Specifications—Connecting Speakers—Headphones

CHAPTER 6

RECORD PLAYERS, PHONO CARTRIDGES, AND RECORDS 199
Record Player Speeds—Automatic Record Player—Record Changer—Turntables—Drive Motors—Drive System—The Problem of Record Player Specs—Record Warp—Vacuum Disc Stabilizer—Tonearms—Record Player Maintenance—The Phono Cartridge—Cartridge Requirements—Damping—The Stylus—Effective Tip Mass—Cartridge Specs—How Records Are Made—Virgin Vinyl—Record Cleaning—Styli Care—Record Player Care—Speed Control—Record Storage

CHAPTER 7

DIGITAL VERSUS ANALOG AUDIO 237
    Analog Versus Digital—The Binary System—Binary to Decimal Conversion—Converting Analog to Digital—The Analog Disc—The Digital Disc—Mechanical Tracking—Variable Capacitance Tracking—Optical Tracking—Analog Recording—Digital Recording—Digital Recording on Tape—Pulse Code Modulation—The Digital Record—Digital and Sound Fidelity—The Compact Disc—Hybrid Recordings—The Analog Phono Record—Direct to Disc Recording—Dynamic Range and Noise

CHAPTER 8

TAPE DECKS AND MAGNETIC TAPES 265
    Tape Players Versus Tape Decks—Types of Decks—Cassette Tape—How a Tape Recorder Works—Testing the Deck—Tape Heads—Pressure Pads—The Cassette Shell—Erasing—Switching—The Bias Oscillator—Bias Requirements—Tape Types—Equalization—Time Constant—Low Frequency Boost—Equalization and Bias—Switchable Bias—Sensitivity—Print Through—Erasability—Dynamic Range—How To Select Cassettes—Open Reel Versus Cassette Tape—Cassette Recording—Precautions—Taping Off the Air—Tape Editing—The Bulk Eraser—Superimposing—Mixers—Tape Deck Features—Dolby®—Twin Speeds—Memory—Specifications

CHAPTER 9

MICROPHONES 311
    Basic Microphones—Microphone Ruggedness—Microphone Sensitivity—Polar Patterns—Proximity Effect—Microphone Specifications—Phase—Dry Versus Reverberant Sound—Microphone Sound Response—The Lavalier Microphone—Using Microphones—Cables

GLOSSARY 331
INDEX 363

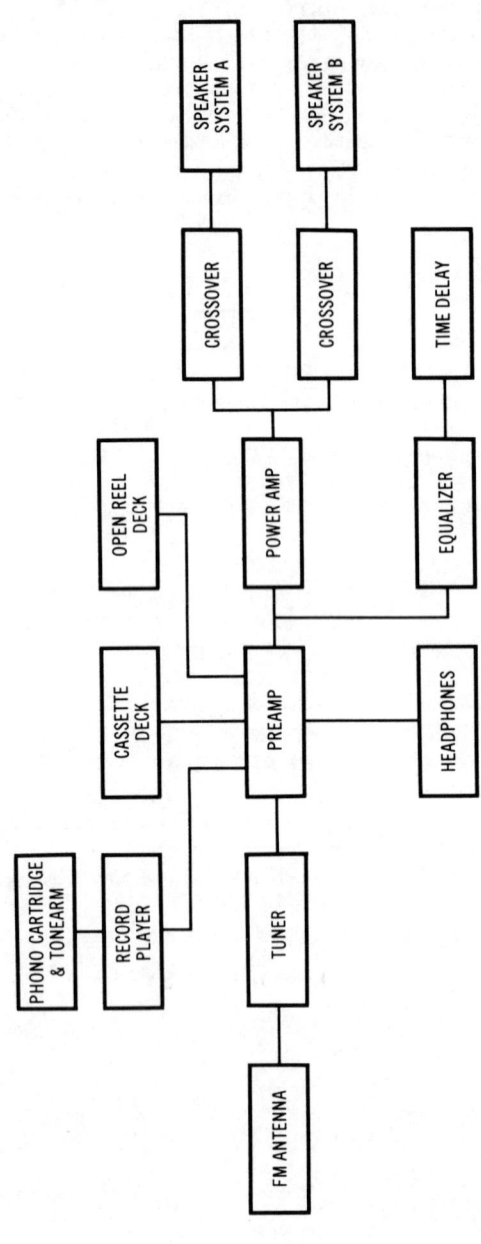

1

Hi-Fi Systems

A high-fidelity (hi-fi) system is probably your most economical form of in-home entertainment. Aside from low operating costs, a good system is superb for letting tired muscles and stretched nerves relax, requires little physical cooperation on your part, and can supply hours of muted background music.

THE COMPACT SYSTEM

There are various paths you can take to on-demand music. The easiest, but not the best, is the compact.

In a compact system you get all the components that make up an audio system—the receiver, record player, tape recorder/player, speakers; all wired, assembled in a cabinet, ready for you to use. All you need to do is insert the ac plug into your nearest power outlet, tack an fm antenna along the baseboard, make your choice of broadcast, disc, or tape, and you are all set. It is no more difficult than connecting a toaster or electric iron. But what you will get will be sound minus its adjective—quality.

THE COMPONENT SYSTEM

The component system is a diametrically opposite approach to the compact; it demands more expertise on your part, quite possibly requires more money, and assumes you are willing to spend some time connecting the different components. You must make sure these components will work together and are all of comparable quality. By taking this approach it is assumed that you are willing to read specifications (specs) even if many of those specs are written in "engineerese." Specs are important because the fidelity of a system is that of the poorest component in that system. Being able to read and understand specs will enable you to avoid intermixing low and high quality components. An expensive pair of speakers, for example, will never make up for a low quality amplifier, a concept that applies to all the other components.

PREPACKAGED SYSTEMS

Because planning a hi-fi system takes thought and effort, manufacturers of components supply prepackaged systems, following the merchandising techniques used by suppliers of compacts. In such an arrangement, all the components have been chosen for you, so you have an assurance of quality plus the knowledge that the components will work as an ensemble.

In a prepackaged system a rack is often supplied to house the components. Many of the racks are attractive pieces of furniture, have tempered glass doors, space for records, and are mounted on casters. A caster-equipped rack is easy to move and makes it easy to get to the rear of the system for the wiring of patch cords, cables that connect one component to another.

MODULAR SYSTEMS

Not all prepackaged systems come supplied with a rack, and the phrase "modular system" is sometimes used when the rack is omitted. If you buy a modular system you must either buy a rack subsequently, or arrange for the components to fit on shelves, whose depth should be that of the widest component. Not uncommonly, a component will be 14-inches deep so this calls for rather wide shelving. The disadvantage of shelving is that it becomes inconvenient to reach the rear of the components making interconnecting wiring somewhat difficult. The shelves and their brackets must be sturdy enough to support the weight of the components.

DEALER SYSTEMS

Dealers of hi-fi equipment will often offer complete systems whose total cost is less than the sum of the individual components. While this can represent a buying bargain it does mean you must take what the dealer offers. You lose your buying independence and can no longer shop for components that have all the features and specs you want.

THE BASIC SYSTEM

You can get started in hi-fi with just an am/fm receiver and a pair of speakers. Not all fm stations broadcast in stereo so some sets are for mono only. However, this also means that you lose the pleasure of listening to stereo sound.

You can get musical entertainment from the basic hi-fi system but it puts you in the unenviable position of waiting for handouts from broadcast stations. You don't get it for free, either, for you take what some disc jockey thinks you should hear. You also become a captive audience for advertising interruptions and if these come at the wrong time and upset your listening mood, that's the price you pay.

The am/fm monophonic system isn't suitable for hi-fi. It is mostly used by am/fm receivers in the home to supply voice and music broadcasts, and also in portables.

To make sure you have the enjoyment of stereo the receiver should be identified as am/fm/fm stereo or am/fm/mpx. The abbreviation for multiplex, mpx, indicates that the receiver has a stereo capability.

MONO AND STEREO

Mono is an abbreviation for monophonic or single channel sound; stereo is stereophonic or two channel sound. In a monophonic recording the entire orchestra is regarded as a single unit. Reproduction of mono sound requires just a single speaker, although more than one can be used.

In stereophonic recording separate microphones are used to pick up the different sounds from the left and right of the orchestra. We refer to these as left and right channels of sound for they are recorded individually, reproduced individually by the in-home hi-fi system and finally emerge from separate loudspeakers positioned to the left and right in your listening room.

THE STEREO RECEIVER

A receiver is an integrated unit and is actually three components in one — a tuner, a preamplifier (preamp) or voltage amplifier, and a power amplifier (power amp). You can, of course, buy each of these components separately.

The tuner selects the one station you want to listen to, amplifies the signal and sends the audio section of that signal along to the preamplifier. Here the amplification process continues and the audio, now magnified in the order of a million times or more, is sent to the power amplifier. This section or unit delivers the audio signal in the form of audio electrical energy to two or more speakers.

This setup, then, is our true basic hi-fi system. It consists of an am/fm stereo receiver and a pair of speakers. Sometimes a receiver is referred to as a tuner/amplifier by the manufacturer, an indication that these integrated components should be regarded as individual units. However it may be called, a tuner/amplifier is a receiver.

ADDING A RECORD PLAYER

The trouble with the simple system consisting of a receiver and a pair of speakers is that is just what it is——simple. Finding the music you want means chasing up and down the am and fm bands, and by the time you do get what you want there's always an advertising break-in, before, after, and during the program. You can get musical independence with a record player or changer.

Before you buy a record player make sure you know in advance just what you're going to get. A record player consists of a turntable (the mechanism for rotating the phonograph (phono) records), a base for housing the turntable, and a tonearm with a phono cartridge sitting on one end. In some cases the phono cartridge is extra; or the base, or the tonearm. Record players and and changers also come equipped with a dust cover, usually plastic, and while you might consider this an essential part of the component—and it is—it is sometimes regarded as an optional extra.

But once you get a record player or changer you can make your hi-fi system more productive. The output signal from the phono cartridge is an audio voltage, but this voltage is delivered to the preamp, then to the power amp and from there to the speakers. But since these units are part of the receiver for which you have already paid and which you already have, including a record player makes great economic sense. A control on the front of the receiver will let you select the sound source you want, either broadcasts or records.

ADDING A CASSETTE DECK

A next logical step, on your way to a more capable hi-fi system, would be to add a cassette deck. Like the record player it takes advantage of your existing receiver's preamps and power amps, and your speakers, making them do extra duty at no extra cost. A cassette deck is so-called since it does not contain preamplifiers and power amplifiers, depending on the receiver to supply these. You can, of course, get a cassette player/recorder in contrast to a cassette deck, and this cassette player/recorder will have its own preamps and power amps.

Cassette decks that are made part of an in-home hi-fi system can record and playback, but there are some that are playback types only.

The cassette deck is an important juncture in the growth of a hi-fi system for it can change you from a passive to an active participant. You can now record, not relying solely on playback. Like the record player or changer it is connected to the preamp in your hi-fi system.

While a record player and a cassette deck are both sound sources, there is one important, fundamental difference between the two. All phono records are prerecorded. What is recorded is what you get and is what you hear. With a cassette deck you can use prerecorded cassette tapes but you also have the option of buying and using blank cassettes, tapes on which you can record. You can use cassettes to record your phono records, keeping these in mint condition, or you can build a cassette music library by recording am or fm broadcasts. If your cassette deck has an input jack for a microphone, you can record your voice, your musical instrument, the voices of your family, or any other sounds within reach of the microphone.

ADDING HEADPHONES

You now have a hi-fi system that is reasonably complete, but you might also consider the addition of headphones, a good idea for a number of reasons. Headphones eliminate all or most of the acoustics of your listening room, bringing you closer to the music to let you hear what it really sounds like. Headphones give you a clue as to how well your hi-fi system is performing and are a more stringent test of that performance than speakers.

Usually the jack or jacks for headphones are on the front panel of the receiver or on the preamplifier if you are using this as a separate component. If there is only one such jack, and you want his and her listening you can get an adapter which will let you do just that.

ADDING AN OPEN REEL DECK

The advantage of a cassette deck is the convenience of cassettes which are easy to insert. With an open reel deck you must thread the tape. Further, since the tape reels are much larger than cassettes, ranging from 5 to 10½ inches, they require more handling, plus the fact that you must thread the tape. Open reel decks are much more difficult to mount than cassette decks and do not fit into a rack unless the rack is specifically made to accommodate them.

All cassette decks operate at a single speed, 1⅞ inches per second (ips), while some open reel decks have a three-speed capability, 3¾ ips, 7½ ips, and 15 ips. You can get a wider frequency response with open reel tapes and generally there are more recording tricks you can do with open reel machines.

Like the cassette deck or record player, the open reel deck takes advantage of the preamps, power amps, and speakers of an existing hi-fi system.

Your hi-fi system can include a record player, or a record player and a cassette deck, or a record player, cassette deck, and open reel deck, or any combination of these three components. Generally, though, most hi-fi systems have a record player and cassette deck.

ADDING AN EQUALIZER

Your listening room is a definite part of your hi-fi system and what you hear depends on what that room does to the sound. It isn't flattering, but the hi-fi listener gets what is left over after the room does its bit in changing the sound. If the room is highly sound absorptive, if it has many cushions, padded furniture, drapes, or cloth wall covering, you may find the sound deficient in treble tones. If, on the contrary, the room is "hard" (if it has many reflective surfaces) you may find the music too brilliant, the treble too strong or shrill, in relation to midrange and bass tones.

You can compensate somewhat by adjusting the bass and treble tone controls on the receiver; a few receivers are also equipped with a midrange tone control. But these controls are coarse, cover a rather wide frequency range, and you may find that adjusting them doesn't let you have the kind of sound you prefer. A better arrangement is to use an equalizer. There are two basic types of equalizers: graphic and parametric, but the graphic is easier to use and generally is more a part of the in-home hi-fi system.

While the graphic equalizer could be called a tone control, there is really no comparison. The equalizer has independent slide adjustments for left and right channel sound, something that isn't true of tone controls. Further, the equalizer divides the audio band, from 20 hertz to 20 kilohertz, into much smaller segments, so you do get finer control of the sound. How much control you get depends on the equalizers. Some divide the audio spectrum into five bands; others use seven or ten or more.

AUDIO AND HI-FI SYSTEMS

The difference between an audio system and a hi-fi system is one of quality. An audio system is a sound reproducer and so is a hi-fi system, but in the latter accuracy of reproduction is top priority.

System 1

The simplest sound system consists of a receiver that will pick up am broadcasts only with the sound delivered to a speaker. The speaker may be a single driver, the name commonly used for speakers, or it can be a speaker system consisting of a low audio frequency reproducer, the woofer, a midrange driver, and a tweeter speaker for treble tones. Sound reproduction is limited to a rather narrow audio band of only 5 kHz (5,000 hertz or 5,000 cycles per second).

Although a separate antenna can be used, as indicated in Fig. 1–1, more commonly the antenna is a loopstick contained within the receiver.

Plans are under way to use stereophonic (stereo) reproduction on the am broadcast bands, but because it uses just one speaker and because it isn't equipped for stereo reproduction, System 1 is monophonic (mono) or single channel sound only.

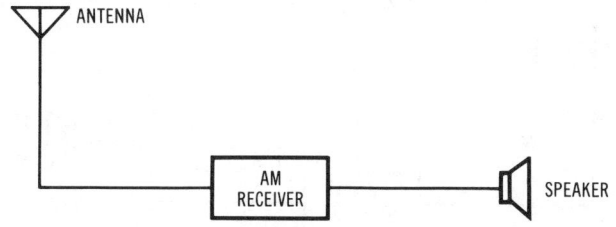

Fig. 1-1. System 1. A monophonic receiver often uses just a single speaker. The antenna is built into the receiver but there is generally a provision on the rear of the set for connecting an outside antenna.

System 2

System 2 is practically the same as System 1 with the exception of the additional speaker. The fact that another speaker has been included does not make it stereo; the sound coming out of both speakers will be the same. Its advantage is that the sound will be spread over a greater area and may help minimize "hole-in-the-wall" effect, a condition in which you are highly conscious that the sound is coming directly from the speaker.

Every speaker imposes a load on the amplifier which supplies it with audio power. In Fig. 1–2 the audio power amplifier in the receiver must be capable of delivering or meeting the audio power demands of the second speaker. If the audio power amplifier cannot cope with this demand the result will be an increase in the level of distortion. In that case it might be preferable to eliminate the additional speaker.

System 3

The system shown in Fig. 1–3 closely resembles that of Fig. 1–2 except that a stereo fm receiver is used. The receiver is limited to fm broadcasts and since many (but not all) fm stations do transmit in stereo, a minimum of two speakers is essential.

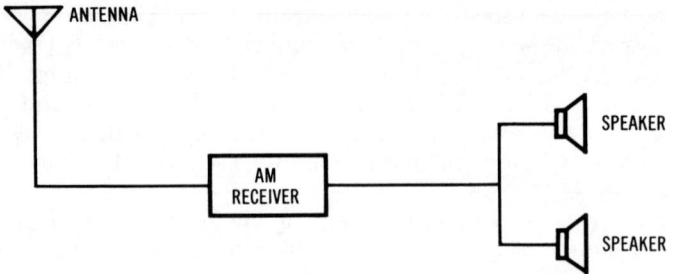

Fig. 1-2. System 2. A monophonic am receiver using two speakers.

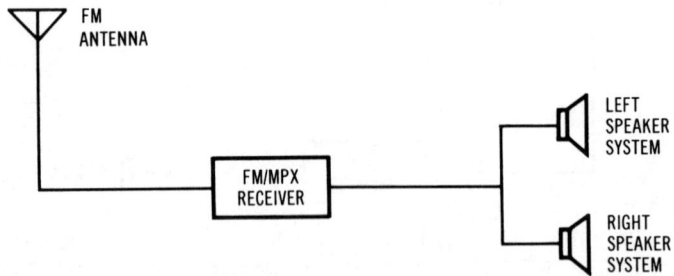

Fig. 1-3. System 3. A fm-only receiver requires a pair of speaker systems.

The receiver can be identified in various ways. It can be called fm/fm stereo or fm/fm mpx or more simply fm/mpx. All of these designations indicate that the receiver is capable of handling either mono or stereo fm broadcasts. The mpx, or multiplex, is a circuit in the receiver capable of decoding or separating stereo channels. Monophonic sound is sometimes referred to as single channel; stereo as two channel sound.

While single speaker reproducers are commonly used for Systems 1 and 2, System 3 and those that follow use two speaker systems, or more.

In System 3 (Fig. 1-3) the antenna that is used is cut especially for fm reception. It may be an outside antenna made for fm or it can be a folded doublet type consisting of a short length of transmission line and generally packed in with the receiver at the time of purchase.

The left speaker system reproduces the left-of-center orchestral sound; the right speaker handles right-of-center orchestral sound.

System 4

The receiver shown in Fig. 1-4 has a considerable advantage over those described earlier. It is a two-band rather than a single-band type

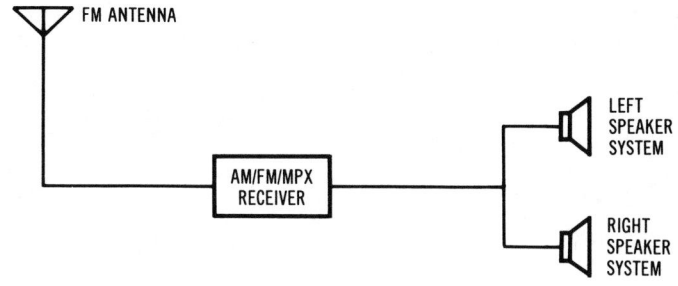

Fig. 1-4. System 4. An am/fm/mpx receiver uses a minimum of two speakers, one for each channel of sound.

and can pick up both am and fm broadcasts. Since the am and fm sections in the receiver use a common power supply and audio power amplifier it is an economical way of having two sound sources, am and fm.

Other than the fact that the receiver shown in Fig. 1–1 is an am receiver and that in Fig. 1–4 is an am/fm/mpx receiver, there is another important difference. Since the receiver in Fig. 1–4 is a stereo type this means that the audio amplifiers in that receiver must be capable of handling stereo signals. To do this the audio amplifiers in the stereo receiver are a pair of amplifiers instead of just a single unit. There is one amplifier for handling left channel sound and another amplifier for strengthening right channel sound.

System 5

In some instances, as indicated in Fig. 1–5, a hi-fi system may use more than just a single pair of speakers. The speakers may be positioned left/right in the rear of the listening room, or they can be placed in another room. This implies, of course, that the receiver is capable of meeting the audio power requirements of the four speakers.

The first set of speakers is sometimes called "A," the second set "B." On the front panel of the receiver you will find a control with positions marked "A," "B," and "A + B." This control lets you select one set of speakers, or the other, or both to work simultaneously.

System 6

System 6, shown in Fig. 1–6, is almost the same as System 4, in Fig. 1–4. It is shown here to emphasize the fact that the speakers used aren't single drivers, but are in fact speaker systems. Speaker systems, described in greater detail in Chapter 5, can consist of two or more drivers.

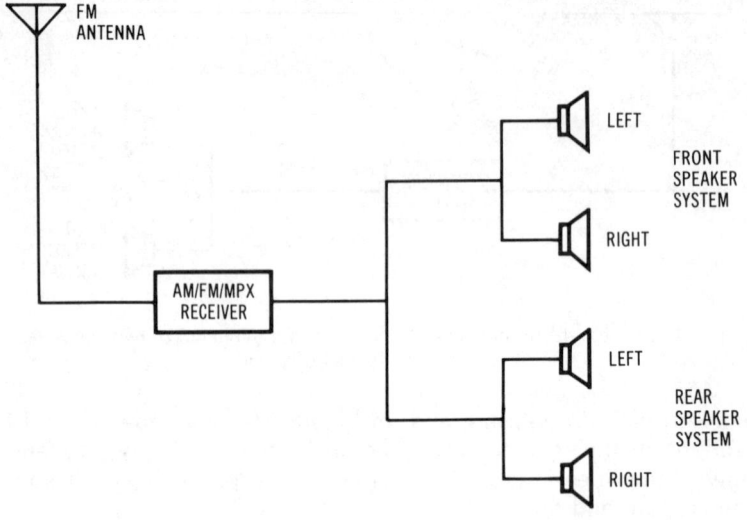

Fig. 1-5. System 5. Hi-fi setup using four speaker systems.

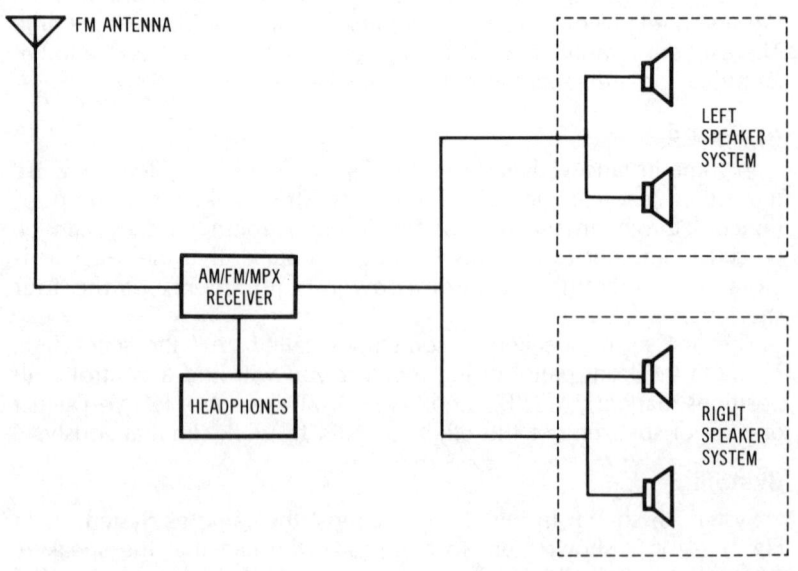

Fig. 1-6. System 6. This is a two-speaker system since the woofer and midrange/tweeter speakers are housed in a single enclosure.

The difference between the two systems is that a pair of headphones have been included in System 6. (Headphones are described in detail in Chapter 5.) Most receivers have one or more jacks for headphones.

System 7

The problem with all the systems described up to this point is that they rely exclusively on am or fm broadcasts. Adding a record player (Fig. 1–7) to the system is a logical next step since it lets you have the kind of music you want when you want it. A control on the front of the receiver lets you select either am or fm broadcasting or record play. Record players are described in detail in Chapter 6.

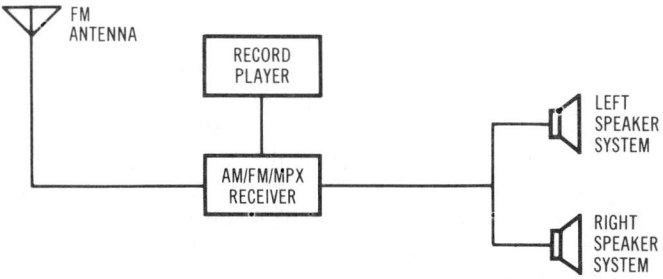

Fig. 1-7. System 7. The hi-fi system now includes a record player.

System 8

One of the advantages of a component system is that you can generally make that system more encompassing. Thus, in Fig. 1–8, a cassette deck has been added.

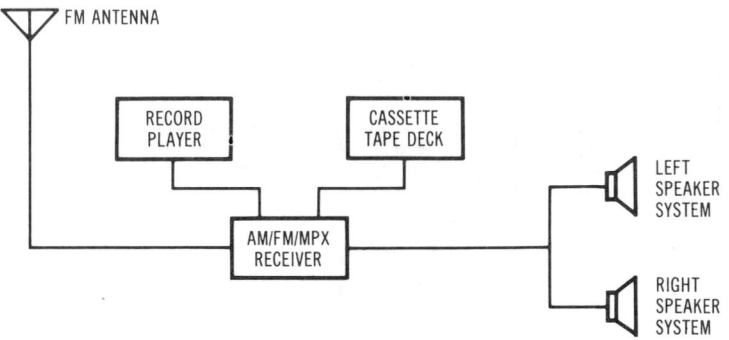

Fig. 1-8. System 8. Adding a tape deck (a tape recorder/player) supplies the system with another sound source.

23

Making a cassette deck a part of the hi-fi system is helpful in several ways. Since the cassette deck has both a record and a playback facility, you can use it to dub your favorite records, that is, to transcribe the music in your records onto tape. In that way you can keep those records as *masters* and, by minimizing wear, have them available for a longer time. You can also record am or fm broadcasts with a cassette deck and use it for playback of prerecorded cassettes.

System 9

At one time an open reel deck was favored by audiophiles over other decks such as cassette or 8-track cartridge, but because of loading convenience and highly improved quality in both tapes and decks, cassette decks have become the favored form of handling tape. Open reel decks, though, can supply longer playing times and can give greater sound fidelity.

Another advantage of the open reel deck (Fig. 1-9) when used in conjunction with a cassette deck, as in System 9, is that you can dub or transfer material from one deck to the other. In this System you also have the option of transferring the music on your records to either cassette or open reel or both. Also, by running an A-B test, that is, by making an aural comparison between the sound supplied by the cassette deck versus that supplied by the open reel deck, you have a basis for making a good evaluation of the sound capabilities of the cassette deck.

System 10

The next step toward a more complete hi-fi system is shown in Fig. 1-10. A pair of headphones have been included. Headphones were also used earlier in System 6. Thus, even though some of the hi-fi systems do not show headphones they can be added at any time.

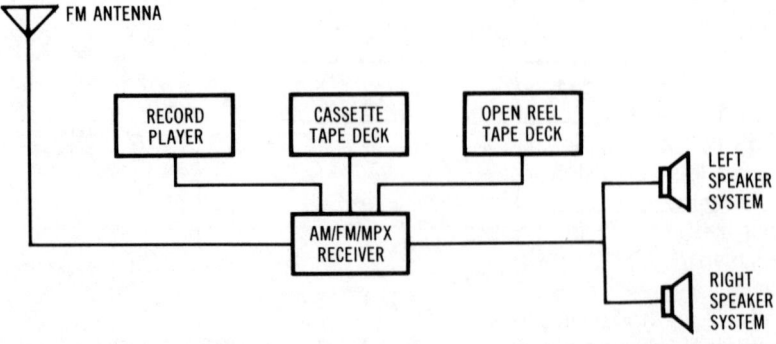

Fig. 1-9. System 9. An open reel deck has been included in this system.

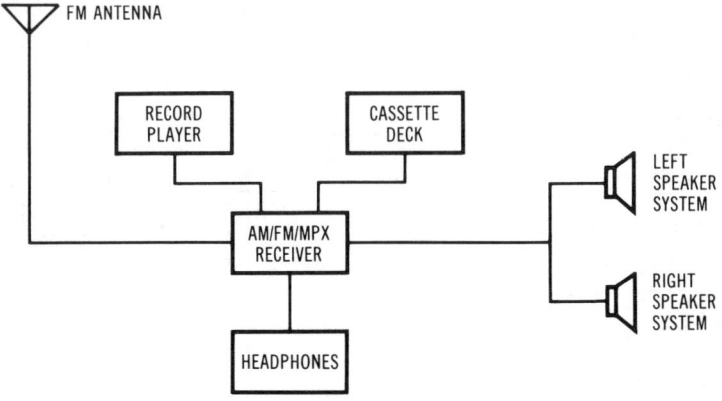

Fig. 1-10. System 10. One or more pairs of headphones can be added to the component hi-fi system.

The advantage of headphones is that they exclude the influence of the listening room on the sound. Headphones bring you closer to the sound and give you a basis of comparison between speaker and headphone listening. On the front panel of the receiver you will find one or more headphone jacks for accommodating the plug (or plugs) of headphones. Headphones usually supply a more critical listening test of the performance of a hi-fi system than do speakers.

System 11

When you buy a receiver you will often find an instruction book packed in with the component to give you some idea of its system capabilities, as shown in Fig. 1–11. A further advantage is that the book shows how the connections are to be made. An illustration of this kind indicates the system capabilities of the receiver, which is important since they can vary considerably from one receiver to the next.

By examining Fig. 1–11 you can see that two separate speaker systems can be accommodated, one set marked "A Speakers" and the other set "B Speakers." A control on the front panel of this receiver lets you listen to either set, or to both at the same time.

Examine the upper right-hand side of the picture and you will see that it is possible to use any of four different antennas. You can connect an outdoor antenna for am reception, but usually there is no need for this except in areas that are remote from am stations. The rather long horizontal tube just below the antenna connections is the internally connected am antenna. This antenna is held in place by a spring clip at the left side and a swivel at the right. The antenna, a

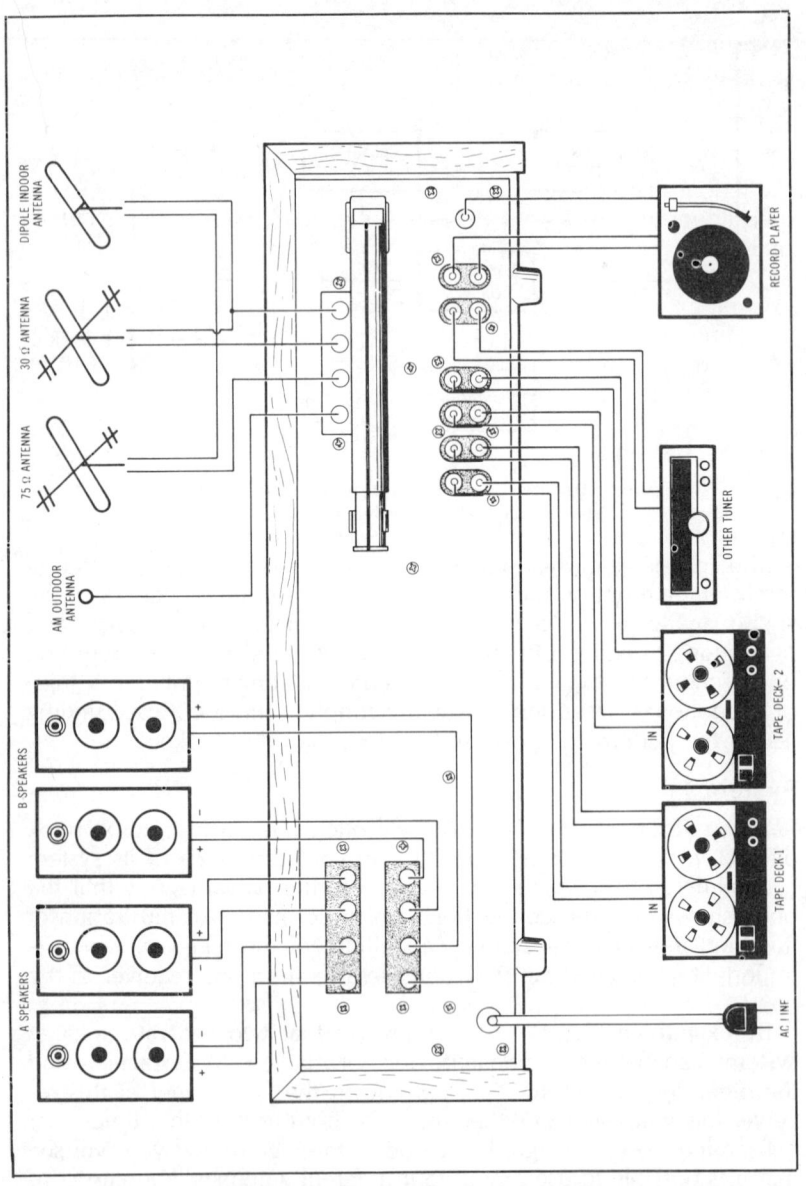

Fig. 1-11. System 11. Although three types of antennas for fm reception are shown, these are options, not requirements since only one fm antenna is needed. One or both of the open reel decks can be replaced by a cassette deck. The connections are the same. Two pairs of speaker systems are shown but only one system need be used.

loopstick, can be pulled away from the spring clip and adjusted for best am reception. This means you may need to move the receiver a few inches away from the rear wall.

The receiver also lets you connect any of three different antennas for fm reception. If you are unable to use an outdoor fm antenna, use the indoor dipole antenna shown at the extreme right. If you can use an outdoor fm antenna, you can disregard the dipole indoor antenna and not bother connecting it to the receiver. The fm antennas shown as 75 and 300 Ω (300 ohms) can be identical antennas, with one antenna made for a 75-ohm connecting line and the other for a 300-ohm line. Although two antennas are shown here, 75 ohms and 300 ohms, only one or the other is required. The following chapter supplies more detailed information about antennas.

Two tape decks are shown at the lower left in Fig. 1-11, marked tape deck 1 and tape deck 2. These are shown as open reel tape decks, but they could also be a pair of cassette decks, or, more likely, a cassette deck and an open reel type.

Finally, there is also a means for connecting a record player, as shown at the lower right.

System 12

As mentioned earlier, a receiver is a three-component integrated system. Instead of using a receiver, you could try an am/fm tuner followed by an integrated amplifier (see Fig. 1-12). An integrated amplifier is a two component unit consisting of a preamplifier followed by a power amplifier.

Note how similar System 12 is to System 4 shown earlier. The difference is that System 12 uses more of a component approach.

System 12 has advantages, and, as you might expect, some disadvantages. The advantage of using a separate component system is that you have the chance to choose units having the capabilities you want. Thus, you might want an amplifier having higher audio power output than that offered by a particular receiver. The individual units could have more features or they might have better spec ratings. Further, if you wanted to "trade up" at some future date, you might decide to keep either the tuner or the integrated amplifier, simply updating just one of the components.

The advantage of buying a receiver is that the tuner, preamplifier, and power amplifier in that receiver have been designed by the manufacturer to work together. Further, you get the receiver as a pre-connected package, that is, the tuner is connected to the preamp and the preamp, in turn is already connected to the power amp. In System 12 it will be your responsibility to connect the tuner to the integrated amp. This doesn't require any great expertise, and is rather easy to do, but it is still an extra connection.

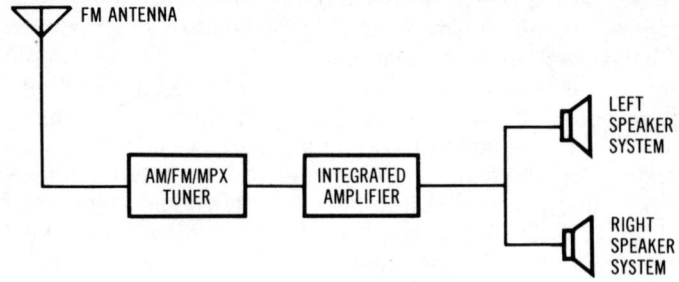

Fig. 1-12. System 12. The integrated amplifier (amp) contains a preamplifier (preamp) and a power amplifier (power amp).

When you buy individual components the burden of selection is on you. You should be a little more *savvy*, understand more about specs and features, be a little more knowledgeable, and, of course, willing to spend more time in "looking around" and shopping. This has a plus side to it, for the more you know about your components, the more confidence you will have in them and the more you will enjoy your hi-fi system.

System 13

Although we are now using an am/fm/mpx tuner followed by an integrated amplifier we can follow the approach used earlier and start adding other components. In Fig. 1-13 we have added a record player and this is shown connected to an input on the integrated amplifier.

The am/fm tuner and the integrated amplifier will have separate

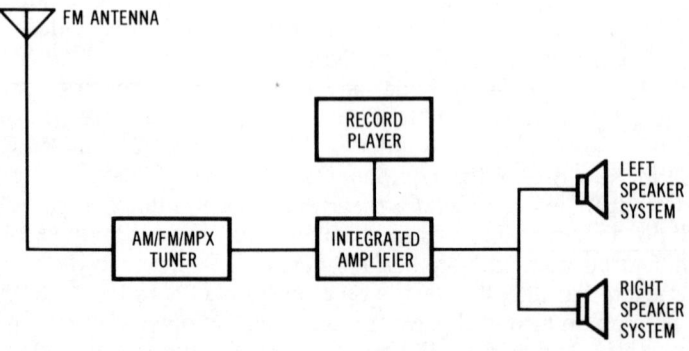

Fig. 1-13. System 13. In this system a tuner and integrated amplifier are used instead of a receiver.

on/off power switches, and so will the record player. This doesn't mean you will need to operate three switches since you will not have the record player on at the same time as the tuner. You can turn the system on, then, just by using two switches—no great difficulty.

System 14

System 14, shown in Fig. 1-14, indicates a cassette deck connected to an input on the integrated amplifier. With this arrangement the hi-fi system has three sound sources: am/fm broadcasts, records, and tapes. In the event you do not have enough wall outlets, the record player and cassette deck can be connected to unswitched outlets on the rear of the integrated amplifier, assuming the amplifier is provided with such outlets. Many are usually so equipped.

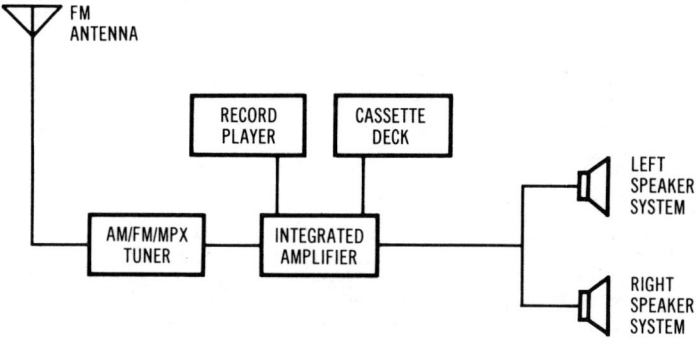

Fig. 1-14. System 14. A cassette deck has been added to the hi-fi system.

System 15

Two components have been added to the system shown in Fig. 1-14 to make the more complete system in Fig. 1-15. We have now included an open reel deck and a pair of headphones. While this is a good arrangement it is by no means the end of the line.

System 16

You can modify System 15 as shown in Fig. 1-16. Here, in System 16, we now have two pairs of headphones and two pairs of speaker systems. Extra headphones are useful if more than one person wants to use them, but it does mean they will both have to listen to the same musical program. If the amplifier has provision for only one pair of headphones you can always buy an adapter that will let you use two. And, if you select two completely different types of headphones you will be able to make an aural comparison to decide which you personally prefer.

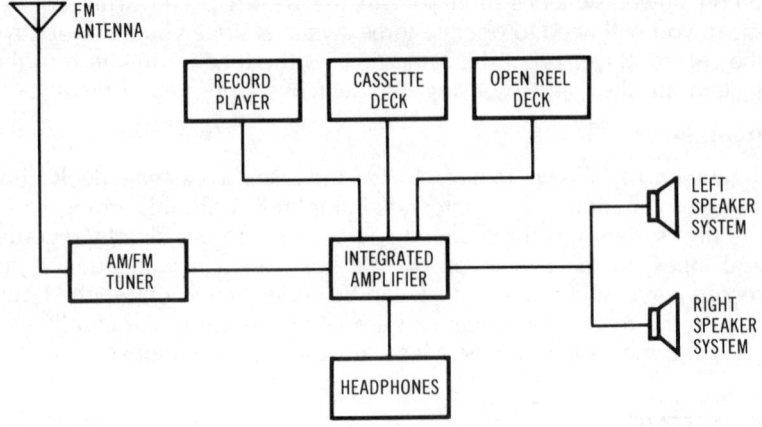

Fig. 1-15. System 15. System now includes two tape decks and a pair of headphones.

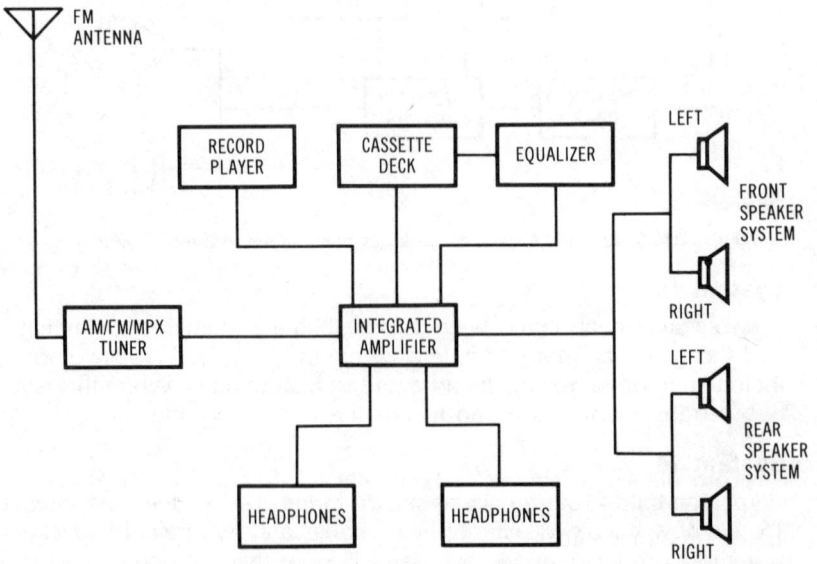

Fig. 1-16. System 16. This system includes an equalizer and an extra pair of headphones.

The speaker systems can be arranged in various ways. You can use one pair of speakers in the rear of your listening room, or in a nearby or adjoining room.

System 16 also shows the inclusion of a new component, an equalizer. An equalizer is a tone control system, but unlike the tone controls on your integrated amplifier, it can do a much better job since it lets you boost or weaken small segments of the audio band. Equalizers are described in greater detail in Chapter 4.

System 17

System 17, shown in Fig. 1-17, is the same as System 16, but with more components added. The notable new unit is covered by the block marked "sound processor." Sound processors could be a dynamic range expander, a noise reducing unit, or some similar component.

System 18

System 18, shown in Fig. 1-18, looks much simpler than the preceding system, but it is an important step toward an all component system. In System 18 the integrated amplifier has been separated into its integral units—a preamplifier and a power amplifier.

Again, there are advantages and disadvantages. With an all component system you may be able to have more features than by using an integrated amplifier, you may have more controls, possibly better specs. But it does mean more work in making rear apron connections and it does mean you must make certain that these components, taken individually, will work well together.

System 19

When using an all component hi-fi system, as shown in Fig. 1-19, you may wish to have more than just one pair of speaker systems. If the power amplifier has the capability, you could connect an additional speaker system pair. We can now follow the same procedure as that shown in earlier systems and begin to add more components.

System 20

In adding more components a logical first step would be to include a record player, as shown in Fig. 1-20. The record player is connected to the preamplifier, but other than the separation of the integrated amplifier into a preamp and power amp, the system is very much the same as that shown earlier in Fig. 1-13.

System 20, though, is much more flexible. Thus, you may be able to use the output of the preamp to drive an extra power amplifier, something you may want to do if you are interested in adding more speaker systems. However, some integrated amplifiers do have a provision for letting you connect a second power amplifier to the preamplifier in that component.

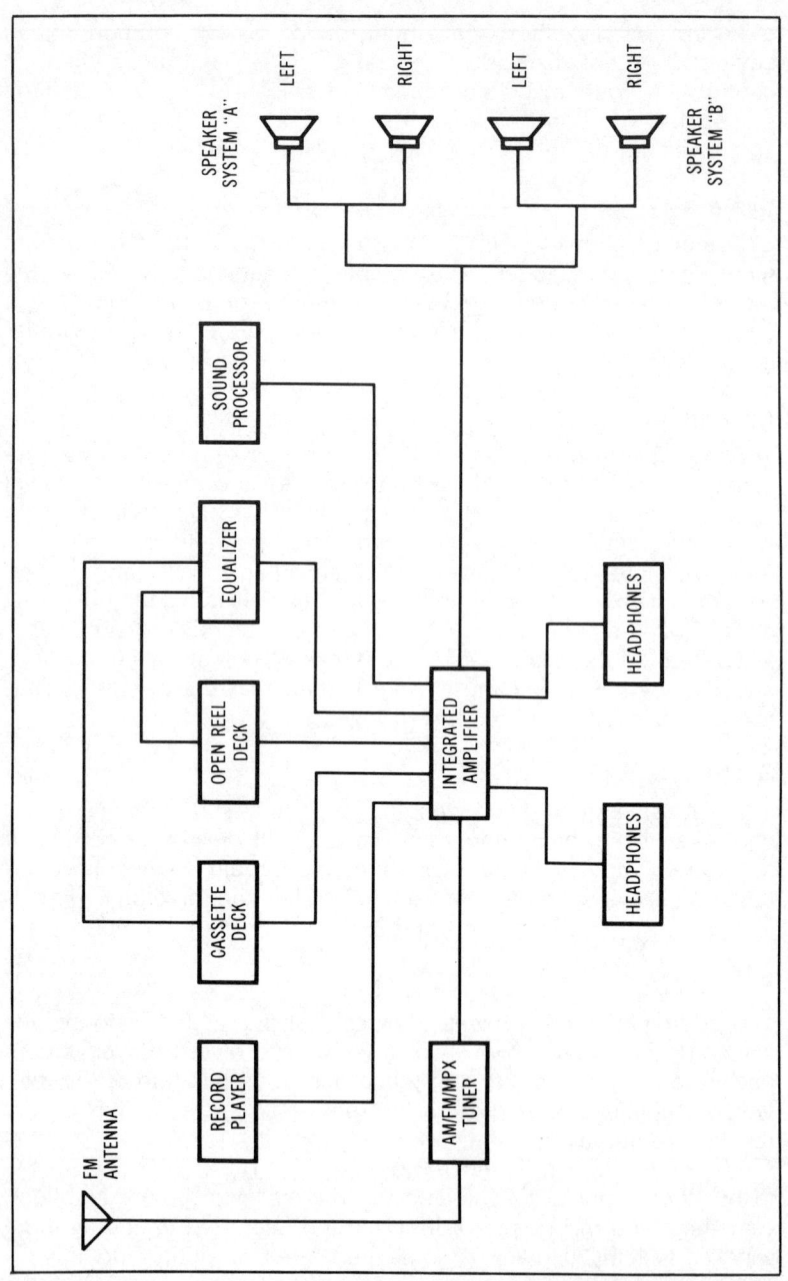

Fig. 1-17. System 17. This arrangement includes a sound processor.

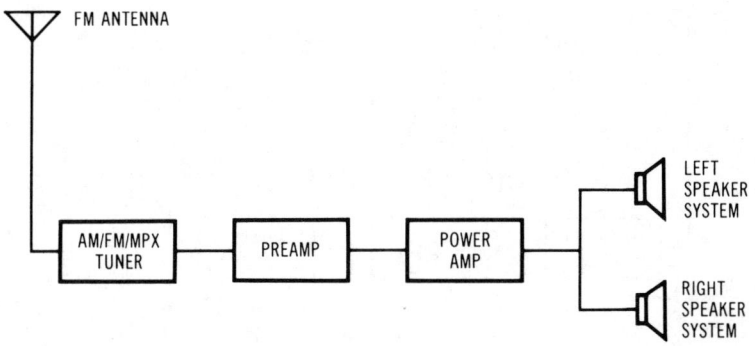

Fig. 1-18. System 18. Instead of an integrated amp, this system uses separate pre- and power-amps.

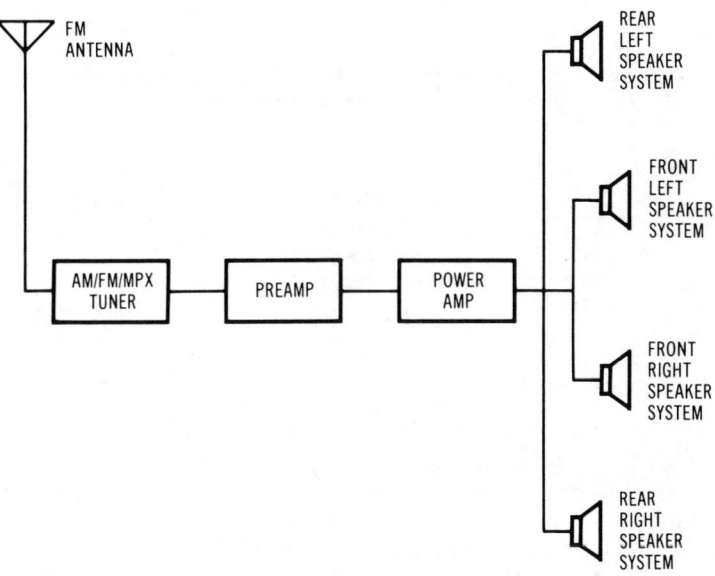

Fig. 1-19. System 19. Hi-fi component setup with front and rear speakers.

System 21

A logical next step, as shown in Fig. 1–21, would be to add a cassette deck to the system. The same advantages apply as in the comparable system shown earlier in Fig. 1–14. You can dub your phono records onto cassettes, using these for playing and keeping the

33

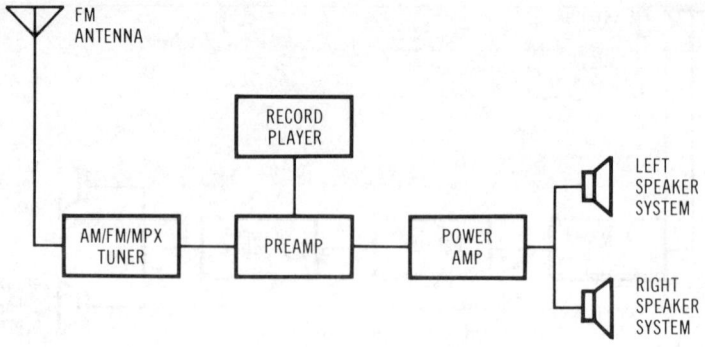

Fig. 1-20. System 20. A record player is added to the all-component system.

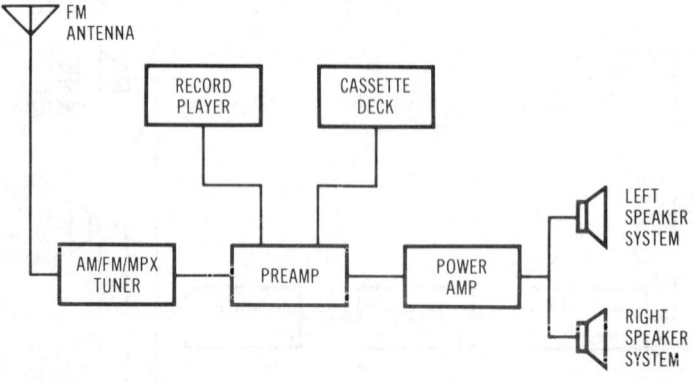

Fig. 1-21. System 21. The next possible step in the growth of the all-component system is the addition of a cassette deck.

records as masters. One disadvantage of the all component system is that it demands more extensive use of wiring at the rear of the system. Even though such wiring connections may look formidable when you get finished, if you do it one step at a time, possibly testing as you go along by operating the connected component, you should have no problems.

System 22

In System 22 several modifications have been made; these are shown in Fig. 1–22. Instead of an am/fm/mpx tuner, the component is an fm/mpx type. Some hi-fi system users do not care for am, and aren't satisfied with the restricted audio bandwidth and the lack of stereo, so look on its inclusion as an extra cost feature they do not want.

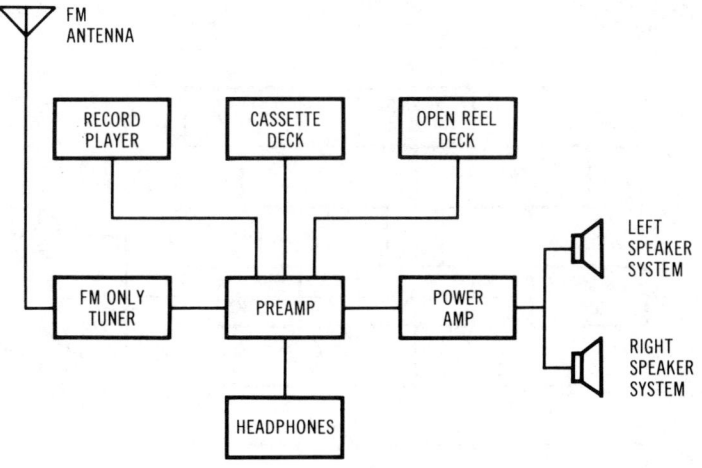

Fig. 1-22. System 22. This system uses an fm-only tuner.

System 22 also includes headphones and an open reel deck. However, another cassette deck could be substituted for the open reel unit.

System 23

On some preamplifiers you will find a provision for connecting not one but two record players. This is a useful feature if the hi-fi system is expected to play continuous dance music, without interruption. Thus, one record could be made to segue right into the next. With the addition of a microphone, as shown in System 23 in Fig. 1–23, the hi-fi system could be used for an in-home disco. The microphone not only can be used for making announcements but also for recording, but in mono only, since there is just one mike.

System 24

System 24, shown in Fig. 1–24, is a more complete arrangement since it uses two tape decks, has an equalizer and a pair of microphones, instead of just one. Stereo sound can be recorded via the microphones on either cassette or open reel tape.

System 25

System 25 (Fig. 1–25), is comparable to the directly preceding system, but note that a sound processor has been added. The system shows the use of two pairs of headphones, but the use of the headphones, and their number, is optional. One or more microphones could also be added.

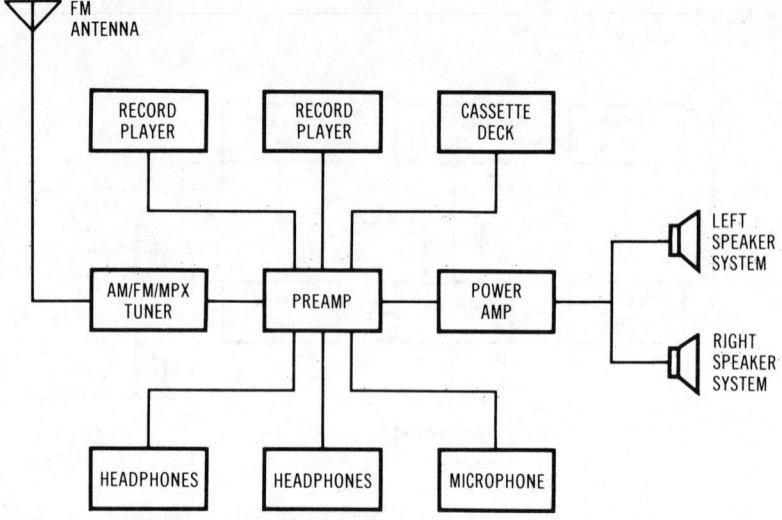

Fig. 1-23. System 23. Sometimes two record players are used to provide uninterrupted dance music.

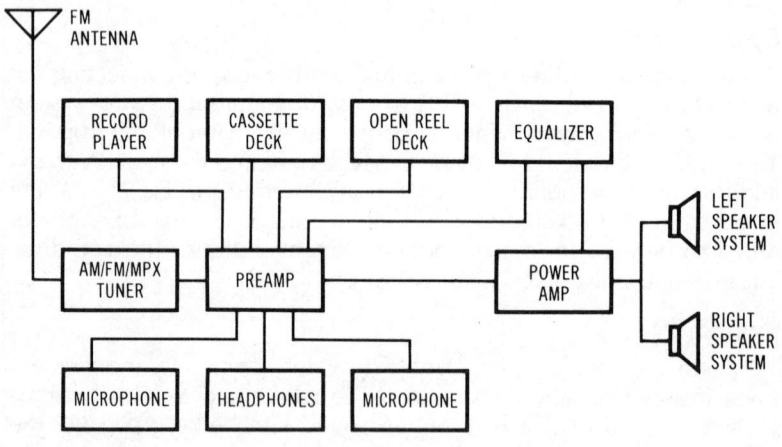

Fig. 1-24. System 24. An equalizer is included in the all-component system.

System 26

System 26, shown in Fig. 1-26, not only has a sound processor, but no less than three tape decks. Two of these are cassette; one is open reel. With an arrangement like this you can dub from one cas-

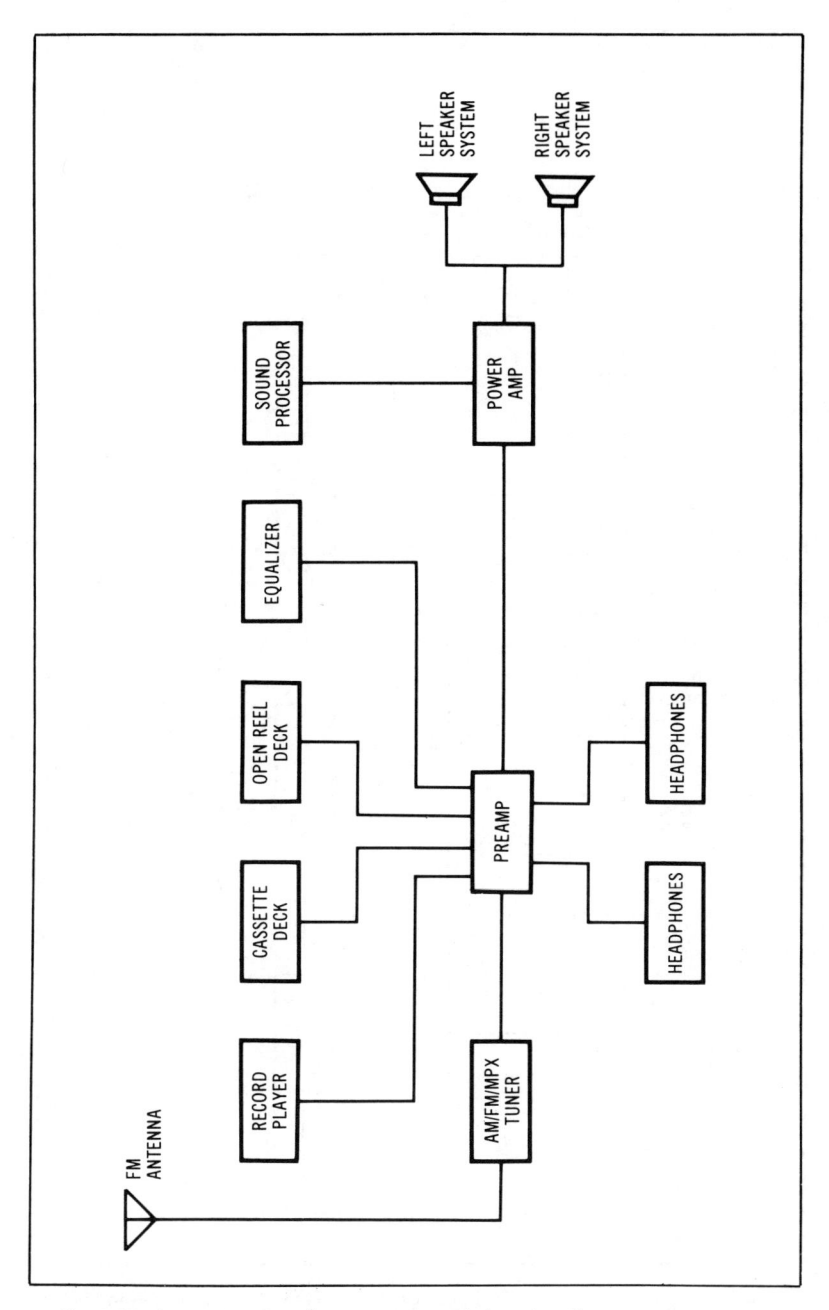

Fig. 1-25. System 25. Sound processor is added to the all-component system.

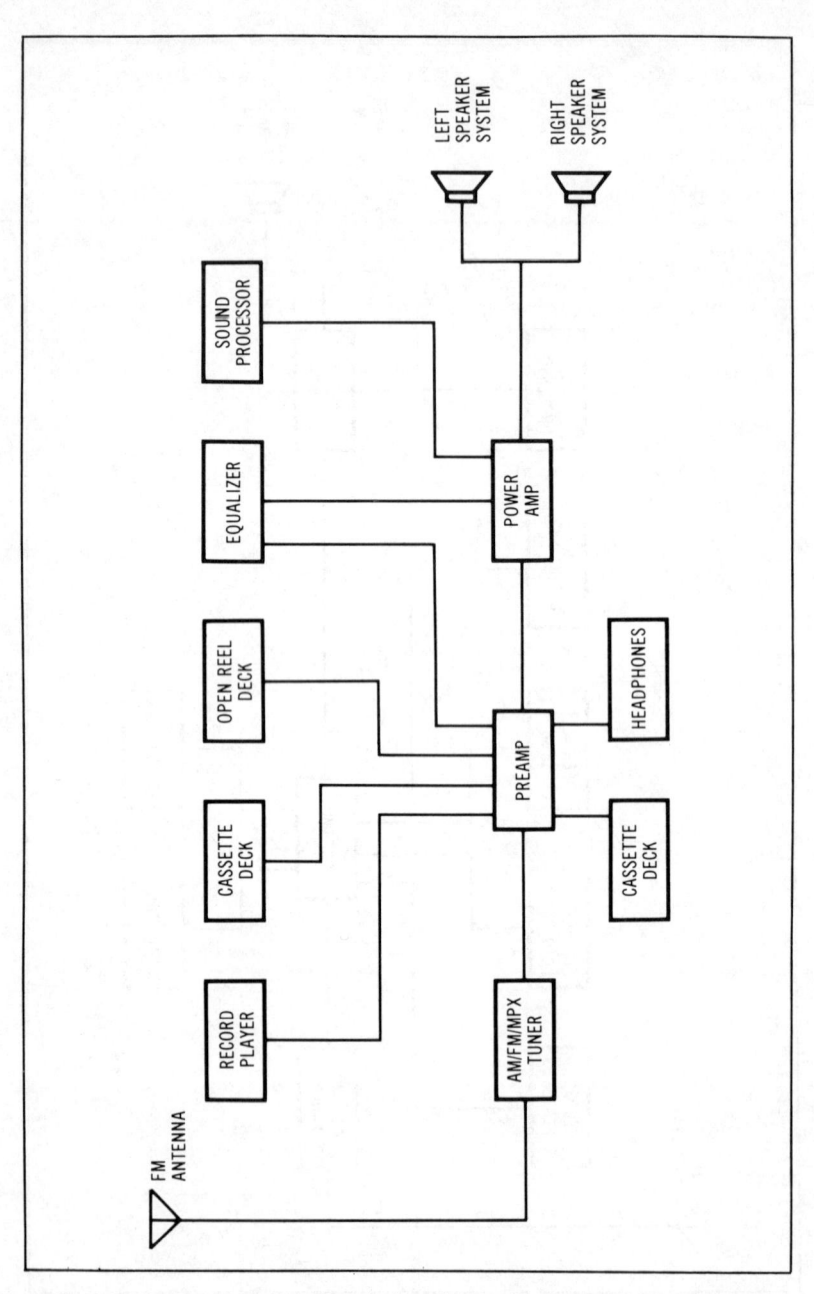

Fig. 1-26. System 26. Two cassette decks in this system permit dubbing from one to the other.

sette deck to the other, or from the open reel deck to cassette deck (or vice versa) with the dubbing action taking place while listening to an am or fm broadcast or to the record player.

System 27

System 27, in Fig. 1–27, sometimes called a multiamp system, is somewhat of a departure from any of the earlier systems. It is distinguished by the fact that it uses not one but two separate power amplifiers and an electronic crossover. When the audio signal from the preamplifier enters the electronic crossover, the audio band is divided by that component into two ranges, bass and midrange tones, and treble. The bass and midrange are fed into one power amplifier and from there to the speaker systems containing the woofer and midrange drivers.

The electronic crossover also delivers the treble tones to a separate power amplifier whose output is connected to a pair of left and right tweeter drivers.

Usually, the power amplifier for handling the treble tones is smaller and has a lower audio power output than the other power amp. Treble tones do not require as much electrical energy. Although the power amplifiers are shown in Fig. 1–27 as separate blocks, they can be mounted on one chassis and would appear externally as a single

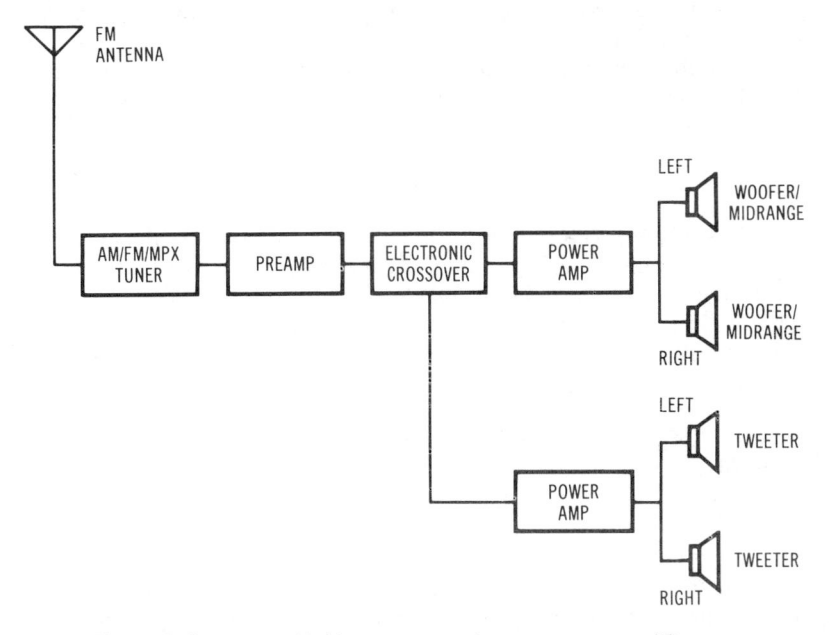

Fig. 1-27. System 27. Multiamp system using two power amplifiers.

unit. Electronic or active crossover and passive crossovers are described in greater detail in Chapter 5.

System 28

The multiamp system lends itself to the inclusion of a variety of hi-fi components as shown in Fig. 1–28, System 28. Note the large number of connections made to the preamplifier. This assumes that the rear of this component has the jacks that are necessary for accepting the connecting patch cords from all these components. One of the features of a preamp, then, would be on how accommodating that unit really is.

System 29

System 29 in Fig. 1–29 is a variation of that shown originally in Fig. 1–27. In this system the electronic crossover is only required to handle bass and midrange tones. Other components can be added to this system, as indicated earlier in Fig. 1–28.

System 30

System 30, illustrated in Fig. 1–30, uses three power amplifiers, one for bass tones, the other for midrange and still another for treble tones. No additional components are shown but you can add as many others as your system will accommodate, following the arrangement in Fig. 1–28.

Although three power amplifiers are indicated in System 30, these could all be mounted on the same chassis and so could appear to be a single unit.

IMAGING

Imaging refers to an audio system's ability to hold the image of an orchestra within a defined space in the listening room without the instruments in that orchestra "wandering" all over the listening area. That is, in a system that images well, a trumpet (or drum, or flute, or whatever) will have a specific place within the sound field and will remain in that *place* throughout the listening session.

In a system that images poorly, that same instrument may be difficult to pinpoint, or may begin a note at the right speaker and end it at the left.

Imaging also refers to the sense of depth a system portrays. Does the cymbal sound as though it's coming from the rear of the recording studio, or does it sound as if it is up front with the first string section? If it sounds like the latter, the system is imaging poorly.

Imaging is a factor in realism and is either part of or missing from the design of all audio components, including cartridges and speakers.

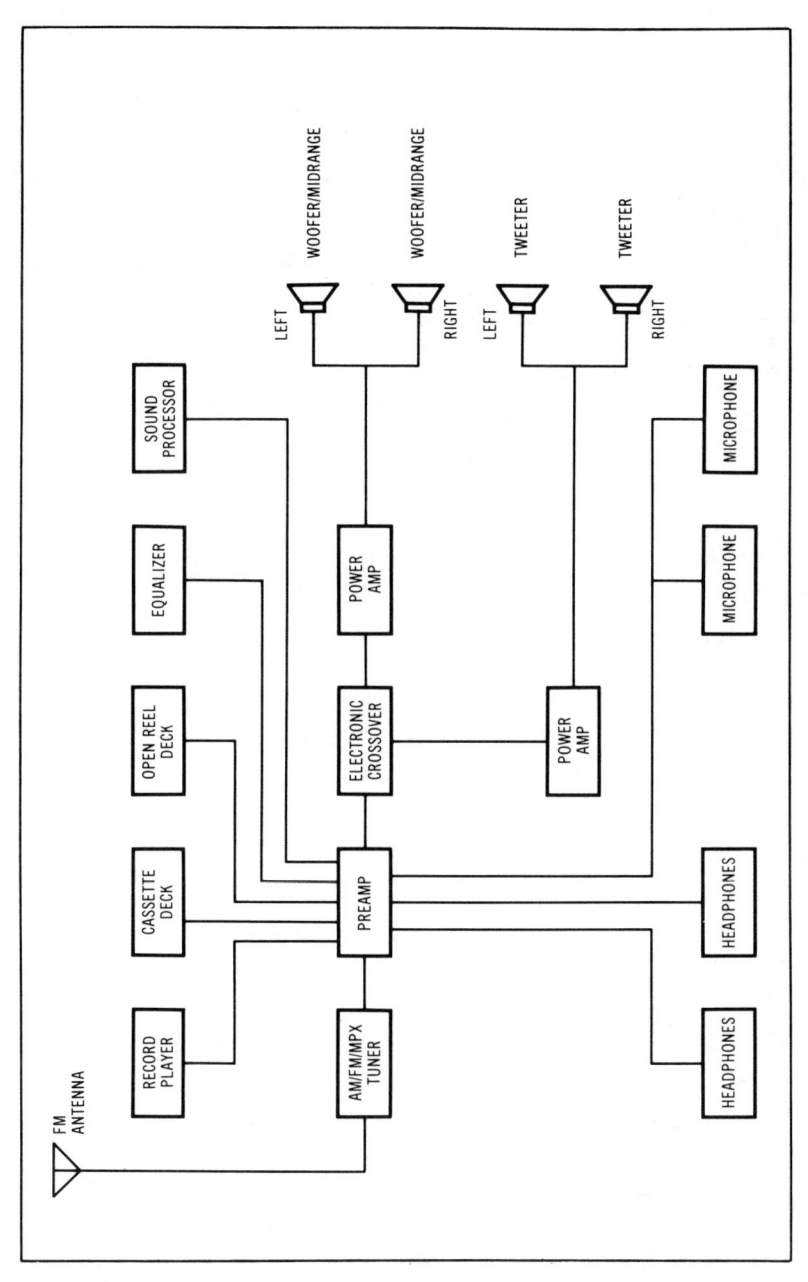

Fig. 1-28. System 28. The biamp system can include a number of hi-fi components.

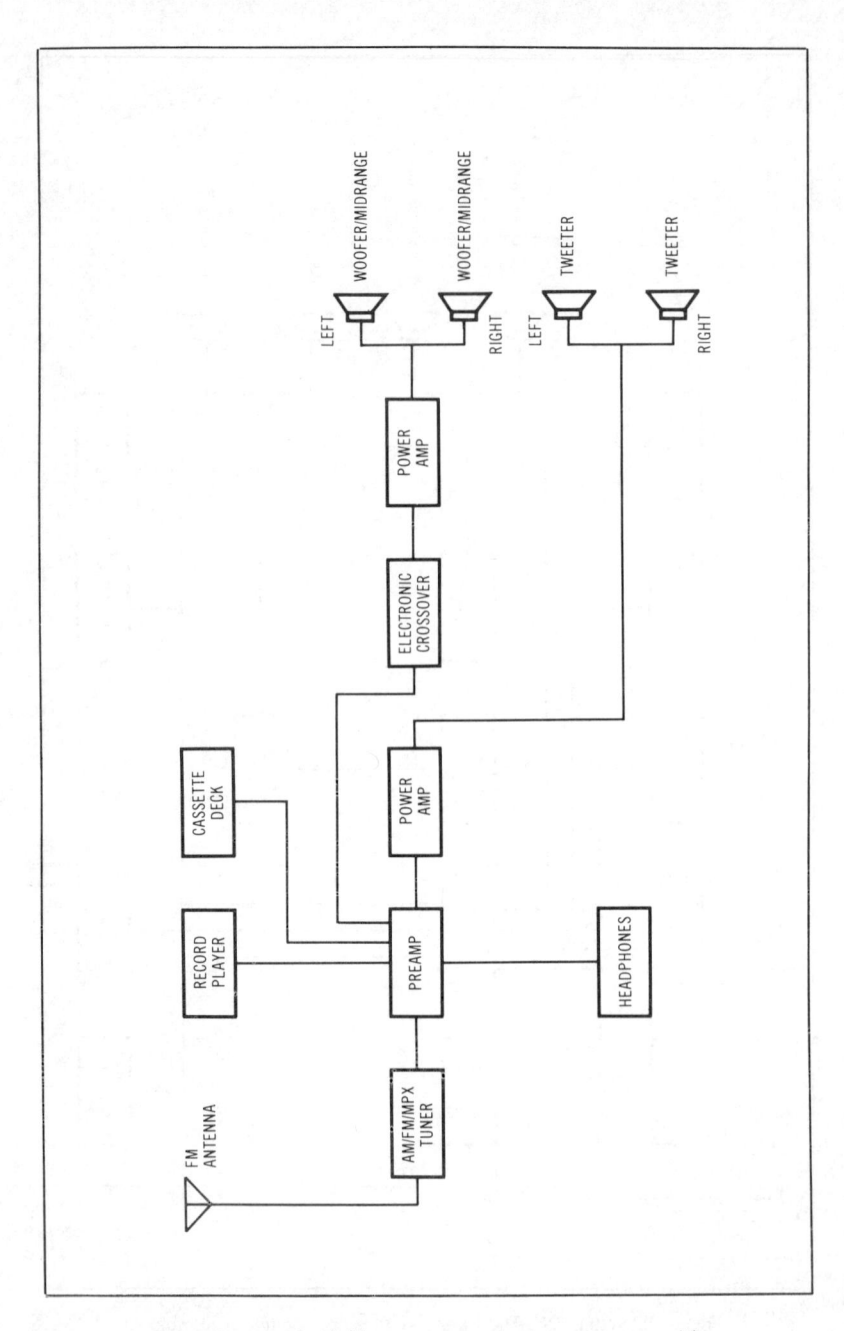

Fig. 1-29. System 29. Multiamp system using two power amplifiers.

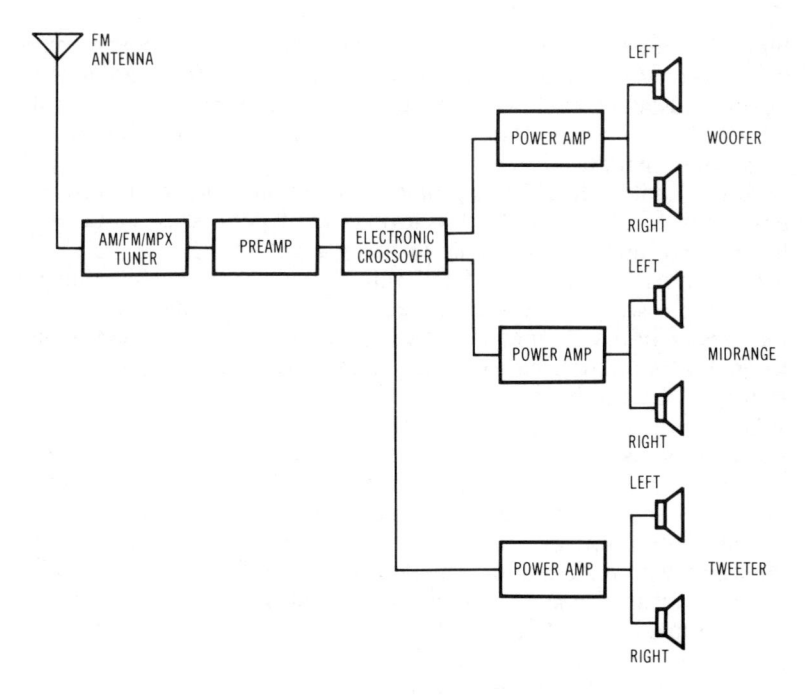

Fig. 1-30. System 30. Multiamp system with three power amplifiers. The arrangement is sometimes called a triamp.

INSTALLING YOUR HI-FI SYSTEM

One of the advantages of using separate components in a hi-fi system is that they lend themselves to flexibility of arrangement. Separate components can also be spaced for maximum freedom from electrical or mechanical interference between units. And, although it is convenient to have hi-fi components placed one above the other this is not an imperative. If you have shelf space that is deep enough the components can be mounted in a staggered, offset arrangement. If connecting patchcords aren't long enough you can always buy some that are longer or connectors that will permit you to join those you have. Patch cords for interconnecting components should be sufficiently long so you can turn components around to get at back panels without too much difficulty.

When possible, mount motorized components, such as turntables, cassette decks and reel-to-reel decks so they aren't affected by possible vibration from outside sources such as trucks and trains or inside sources such as people walking or dancing.

Since your record player will have a plastic dust cover, probably hinged on the back, allow enough room for the upward movement of this cover. Make sure that components whose controls require frequent adjustment, such as the tuner, receiver, or amplifier, are easily accessible. Illumination should be such that you should be able to read panel markings easily.

Your turntable should be absolutely horizontal, something you can easily check with a bubble glass level of the type used by carpenters. Inexpensive units are available in hardware stores.

Remember also that the total weight of the components can be quite impressive and when mounted on shelves located one above the other a hi-fi system can become quite top heavy. It's a good idea to anchor your shelving to a wall, preferably to one or more studs, instead of having it free standing.

You can get an etagere for a hi-fi system, mounting it on casters. For safety's sake anchor the etagere by using two or more large sized eye hooks at the rear. You can disconnect the hooks easily so you can swing the entire system around for making rear panel changes.

The best setup, of course, is a rack, mounted on casters and equipped with a glass door. The door may help in keeping children's busy, inquisitive fingers away from the controls of your system, and may be taken as notification by visitors that your hi-fi system is a personal item, as indeed it should be.

Some manufacturers supply complete hi-fi systems plus a rack, but you can also get racks and buy components from different manufacturers. It's just a matter of common sense to measure components to make sure they will fit in the rack. With a caster mounted rack and a wooden floor, swinging the entire hi-fi system around so as to make rear patch cord connections is quite easy.

Space is usually a problem. Some manufacturers sell microcomponents that can easily be mounted in a rack or on bookshelves. The trouble with using bookshelves is that getting around to the rear of the components and making the connections from one component to the other is a large scale nuisance.

OUTBOARDS AND ACCESSORIES

An outboard, also called an add-on, is any equipment that can be incorporated into an existing hi-fi system. An outboard is positioned on-line and becomes a permenant part of the system. Outboards are used to get certain operating features not included with the hi-fi system or to improve its performance. For this latter reason outboards are sometimes referred to as signal processors.

An outboard could include such different devices as graphic or parametric equalizers, electronic crossovers, subharmonic synthesizers, dynamic range expanders, time delay units, and noise reduction systems.

An accessory is any device essential to the hi-fi system or used for maintenance. Tape, whether cassette or open reel, is an accessory. Record cleaning systems, tape cleaners, bulk tape erasers, stylus timers, turntable mats, tonearm cartridge dampers, cables, are all examples of a few of the accessories that are available.

THE DO'S AND DON'TS OF COMPONENTS AND HI-FI SYSTEMS

A hi-fi system, including its collection of phono records and magnetic tapes, can be a considerable investment, third in the order of acquisitions, starting with a home, and then a car. So it makes economic and music enjoyment sense to follow certain precautions. These precautions are scattered throughout this book, but here are some that are generally applicable.

In-home hi-fi systems get all their operating power from an ac outlet. The amount of ac power your hi-fi system will demand from that outlet will depend on the number of components and the sum of all their power needs, but it can be substantial. Since most power outlets are twin types, that is, double receptacles, use either one or both exclusively for your hi-fi system and deny their use to any other kind of household appliance, especially to those that are electrical noise producers such as fluorescent lamps.

Nearly all hi-fi components are equipped with switched and unswitched outlets. If you make proper use of these you cannot only avoid a jumble of plugs from your hi-fi system but you can manage to get single switch operation or close to it. If you can possibly do so, avoid using a cube tap for connecting hi-fi power plugs.

At times you may need to disconnect the power plug from its outlet. Pull on the plug only, never on the connecting cord. Always make sure your hands are dry to avoid electric shock. If you plan to be away for a long time it is advisable to disconnect the power plug (or plugs) of hi-fi components from the outlet.

Some components have air vents across the top, back, or sides. If you drop any metal object inside do not touch the cabinet or box, do not touch the controls. Instead, disconnect the power plug from its ac outlet. Do not try to open any hi-fi component, or remove any panel, for to do so may well mean your warranty is nullified. Instead, contact your hi-fi dealer and ask for a unit inspection and test. Be sure to explain what has happened.

When you buy a hi-fi component staple your receipt to your warranty and keep both in a safe place along with your other important papers. Make a record of the serial number of the component, the date of purchase, the name of your dealer, and the amount you paid. This information will be necessary in the event of theft, whether or not you are covered by insurance.

Your hi-fi system will develop some heat so install it where there is adequate ventilation. Keep the system away from devices which are surrounded by magnetic fields of their own, including motors, transformers, and television sets.

From time to time dust your hi-fi system with a soft, lint free, dry cloth. Avoid using alcohol, thinners, or other volatile liquids for cleaning since these can erase markings on the faces of the components or disfigure them. Avoid using insecticide sprays near your units.

If you buy a rack for your hi-fi system, get one with a glass front that covers the entire system. It will help keep your hi-fi dust free.

CABLES/INTERCONNECTS BETWEEN COMPONENTS

The link between hi-fi components such as tuners and integrated amplifiers, or tuners, preamps and power amps, is made via cables, sometimes supplied by manufacturers along with their equipment. Cables can affect sound, sometimes to a minor degree or sometimes in ways that are audible even to those having "tin" rather than "golden" ears.

The length of the cable, its construction, its measured capacitance and resistance, the materials used in making the cables, even the type of solder used to connect the plugs at each end—all play a part in how that cable will affect the sound.

Connectors can deteriorate and become rusty at their connecting points. And, because the cables are out of sight, usually carefully hidden behind components, connector rusting isn't noticeable. However, it can lead to a gradual deterioration of sound, insidious because it is gradual. Some manufacturers specialize in cables, thus emphasizing how important these are to hi-fi systems. Gold is sometimes used, not as a conductor, but rather at the ends of the cables. Gold has excellent conductive properties and is noncorrosive.

SPECS

Not all hi-fi components are alike any more than all automobiles, or refrigerators, or air conditioners are alike. Manufacturers describe the technical capabilities of their hi-fi components by means of specifications, ordinarily abbreviated as specs.

The advantage of a spec sheet is that it enables you to evaluate the performance of a component before you buy it, thus giving you some idea of the capabilities of that component. A further advantage is that spec sheets give you the opportunity of comparing components made by different manufacturers.

Spec sheets are written in somewhat technical language but there just doesn't seem to be any help for it. However, as you begin to understand just what the specs mean, and they are described throughout this book, you will gradually appreciate their significance and importance. Understanding specs will enable you to separate the electronic wheat from the chaff as it were, differentiating between flamboyant claims and the hard facts of performance. It will make you a more discriminating buyer, and that's the best kind.

FEATURES

Not all hi-fi components have the same features. Thus, one receiver may have a pair of tone controls, one for bass tones, the other for treble. But still another receiver may have a third tone control made for the important midrange tones. Tone controls are features, but the inclusion of the midrange control is an extra added attraction.

The extent and variety of features is astonishing, and no component has them all. In some tuners and receivers, for example, you will find a digital readout of the frequency, supplementing or replacing the tuning dial; you may find soft touch controls that are clickless instead of the usual switches; you may find volume controls that have click stops instead of the usual continuously variable types. These are just a few examples, but as a buyer and user of hi-fi equipment you should be aware of them for they add to the pleasure, joy, and convenience of a hi-fi system. It is impossible to list them all for manufacturers know that features help sell equipment and they keep their research and development departments quite busy concocting new ones.

HERTZ

When sound energy is converted to its equivalent in electrical energy, the electrical output is an alternating current or alternating voltage waveform. A single, complete wave, known as a cycle (Fig. 1-31) consists of a pair of alternations, one of which is regarded as positive, the other as negative. Each alternation is half of a full cycle. The frequency is the total number of completed waveforms or cycles per second, including both the negative and positive half cycles.

Hertz (Hz) is a measurement of frequency and represents the cycle per second. The power line frequency used in many homes, offices, and factories is 60 Hz or 60 cycles per second or 60 cps. One hertz is one cycle per second, 400 Hz is 400 cycles per second.

There are some multiples that are also commonly used. The letter *k* (or kilo) means to multiply by one thousand. Thus, 5 kilohertz or

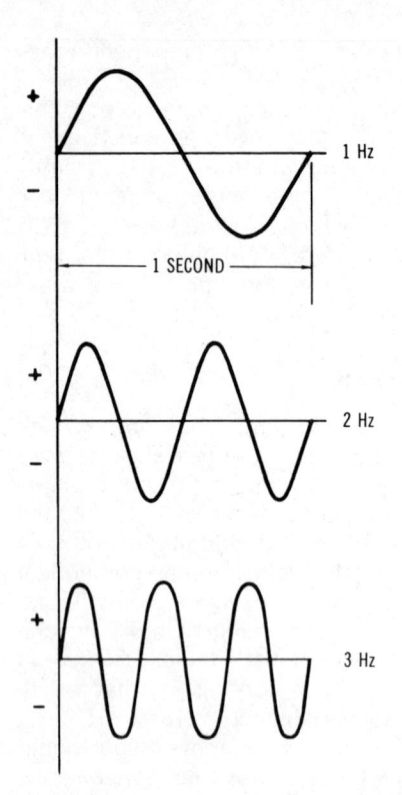

Fig. 1-31. As the frequency increases there are more waves per second. The top drawing is 1 Hz, the center is 2 Hz and the bottom is 3 Hz. The pitch of a sound increases with frequency. Bass tones are low frequency; treble tones much higher in frequency.

5 kHz means 5,000 Hz or 5,000 hertz or 5,000 cps. 100 kHz is 100,000 Hz or 100,000 cps. Another multiplier is the letter *m* and it indicates one million, or 1,000,000. The letter m is an abbreviation for mega and so 5 megahertz is 5,000,000 hertz, more conveniently written as 5 MHz.

SINE WAVES

The fundamental ac waveform is called a sine wave. The horizontal line drawn through the waveform separates the upper or positive (+) half from the lower or negative (−) half. With an increase in frequency there are more cycles per second. A pure tone, a tone having no harmonics or overtones, appears as a sine wave. Pure tones, except when used to achieve special effects, are boring.

HARMONICS

Harmonics, or overtones, are multiples of the fundamental frequency. The tone of a musical instrument having a fundamental frequency of 250 Hz could have a second harmonic or overtone at 500 Hz, a third harmonic at 750 Hz, a fourth harmonic at 1 kHz, and so on.

As we get beyond one of the upper-order harmonics, such as the fifth, the amplitude or strength of such overtones becomes quite small. The fundamental and its harmonics combine to produce complex waveforms (Fig. 1–32). It is these waveforms that give each tone its particular character. They enable us to distinguish between a tone produced on a piano and a tone of identical frequency made by a guitar or some other instrument.

The fundamental range of musical instruments is quite limited and extends from about 30 Hz to about 4 kHz. At the low frequency end there are very few musical instruments that have the capability of producing tones below 50 Hz. The human voice does not go much below 70 Hz while at the high-frequency end all musical instruments and the voice have a top limit of 4 kHz, while many are well below that. Although the fundamental range of instruments tops out at 4 kHz, their harmonics extend this upper limit considerably, and so the upper frequency limit of an instrument depends on the highest order overtone it produces at any moment.

While it is the fundamental frequency that determines the pitch of a tone it is the harmonics that add richness and quality. The number of harmonics produced depends on a variety of factors: whether the instrument is percussive or wind, for example, and also on playing technique. A violinist controls harmonic content by the movement of his fingers. In the case of the flute, playing it softly results in an almost

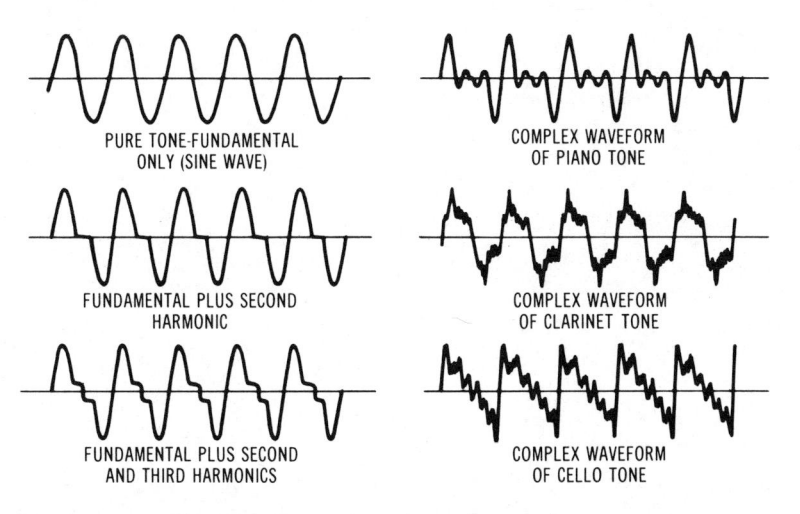

Fig. 1-32. Pure tones and complex waveforms.

pure tone. With louder tones, however, there are more harmonics. So our enjoyment of a particular tone depends not only on the number of harmonics produced but on the variation in the quantity of harmonics.

TIMBRE

That character of a sound which enables us to distinguish between different musical instruments, including the voice, is called timbre. Even if two instruments play the same tone—that is, each is playing notes having the same frequency and at the same loudness level, the notes have a different sound. Each musical instrument has its own particular pattern of overtones.

The various tones produced by musical instruments differ in two respects: in the total number of overtones they produce and whether these overtones are odd or even, or both. A harmonic that is twice the fundamental is an even overtone; one that is three times the fundamental is an odd overtone. An instrument such as the violin supplies a fundamental plus odd and even overtones. A trumpet produces a fundamental plus odd overtones. The total number of harmonics supplies the character of a tone. If a tone is accompanied by a large number of harmonics we hear it as bright or brilliant; if there are a few number of overtones it sounds restrained, muted, or mellow. Some even refer to it as dull.

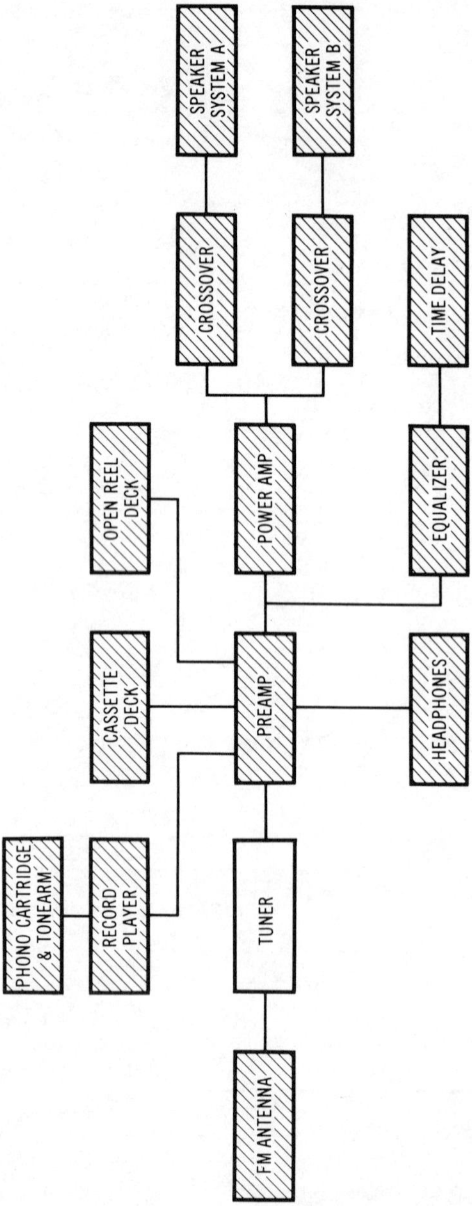
SPEAKER SYSTEM A
SPEAKER SYSTEM B
CROSSOVER
CROSSOVER
TIME DELAY
OPEN REEL DECK
POWER AMP
EQUALIZER
CASSETTE DECK
PREAMP
HEADPHONES
PHONO CARTRIDGE & TONEARM
RECORD PLAYER
TUNER
FM ANTENNA

2

Hi-Fi Tuners and Receivers

WHAT IS A RECEIVER?

Every hi-fi receiver (Fig. 2–1) is an integrated component made up of a tuner, preamplifier, and power amplifier. Buying a receiver is advantageous since you get three components mounted on a single chassis, completely interconnected, and with the assurance that they are designed to work together. Further, the receiver requires much less room than the individual components of which it is comprised, mounting easily into a rack or possibly onto a shelf. And the individual receiver may cost less than its tuner, preamps, and power amps purchased separately. So you can put up a rather good argument in favor of a receiver versus the separate component arrangement.

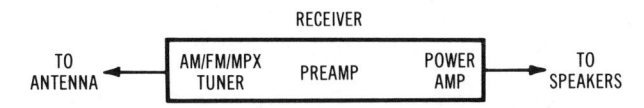

Fig. 2-1. A receiver consists of three components: tuner, preamp and power amp.

If you elect to buy a tuner, preamplifier, and power amplifier (Fig. 2–2) you will be faced with the responsibility of making sure that these components, if purchased from different manufacturers, will work well as an ensemble, will not impose an impossible space requirement, and will not overstretch your budget.

This sounds discouraging, and it is, but there is a payoff for going the completely separate component route. With separate units you may be able to get better specs per unit. If you choose a receiver you may be delighted with the specs for the tuner section, but less than enthusiastic about those for the preamp, and you may feel that the

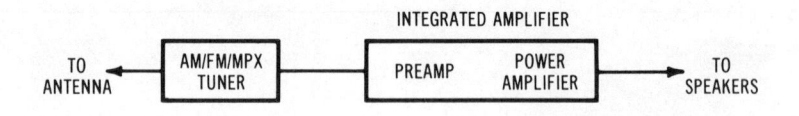

Fig. 2-2. The tuner can be followed by an integrated amplifier or by separate pre- and power-amplifiers.

audio power output of the power amplifier is not as much as you would like to have. This is no problem with separate components since you can mix and match, making sure that each has the performance characteristics you want.

You may, at some future date, want to modify your hi-fi system. Updating your receiver means disposing of three components. With the individual component hi-fi system you can upgrade the tuner, or the preamp, or the power amp, but not necessarily all three.

With a receiver you are locked into the features that are supplied with that receiver. With a separate component system you can select units that will give you more of the features you want for each of the three units. Hence, one of the most important decisions prior to getting a hi-fi system is to make a determination as to the kind of system you ultimately want to have.

WHAT IS A TUNER?

A tuner is not a stripped-down receiver. It is a high-fidelity component in its own right, specifically designed for a specific purpose (Fig. 2-3). The only difference between a tuner and a receiver is that a tuner isn't equipped with a preamplifier and a power amplifier. A receiver can drive a pair of speakers (or more) directly; a tuner cannot do so.

Because the tuner section of a receiver and the independent component tuner perform the same functions, the same specs are applicable to each, and each can have a variety of features.

COST DIFFERENCE

It is almost impossible to make a direct comparison between a receiver and the cost of such separate components as a tuner, preamp,

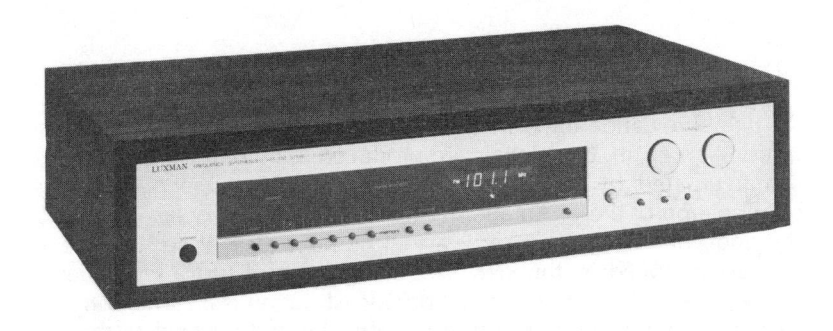

Fig. 2-3. A frequency synthesized fm stereo-am tuner. Special features include: quartz pll frequency synthesizer, 12-station electronic memory, memory storage indicator, signal strength indicator, digital frequency readout, stereo indicator, manual tuning buttons, automatic tuning button, mono/muting off switch, tuning keys, test tone switch for recording level, muting level control, am loop antenna. (Courtesy Luxman, Div. of Alpine Electronics of America)

and power amp. As a general rule, though, you can expect to pay more for the individual units. A receiver has a single power supply for each of its sections, but a tuner, preamp, and power amp come equipped with their own power supplies, so in buying separate components you are buying three power supplies versus just one.

The receiver, of course, has just a single cabinet, but individual cabinets are needed for the tuner, preamp, and power amp. The receiver uses just a single chassis; the separate components require three. So don't be too surprised to learn that the component route is the more expensive.

THE TUNER-AMPLIFIER

To emphasize the component aspect, some manufacturers refer to their receivers as tuner/amplifiers. The purpose here is to establish the fact that the components in such receivers can be regarded as comparable in specs and features to the separate units.

THE CASSEIVER

In some instances various components are included in the receiver, a method that helps eliminate some of the interconnections needed by a hi-fi system and also a device for lowering the overall cost. A receiver may be combined with an equalizer, or possibly with a cassette deck. The reduction in cost is partially due to the use of a common power supply. The joined receiver and cassette deck is sometimes called a casseiver.

THE AM SIGNAL

The abbreviation for amplitude modulation is am. In any broadcast station, am or fm, two signals are produced. One of these is an audio signal created by speech, singing, musical instruments, or some combination of these. An audio signal can't travel very far. Try shouting and your voice may carry a few hundred feet or so. Even loud sounds produced by thunder have a limitation of just a few miles.

The other wave generated in a broadcast station is a carrier, and, as its name implies, its function is to carry the audio signal to the antenna of your home receiver (Fig. 2–4). An audio signal can have a range of 20 hertz, or 20 complete cycles per second, to as many as 20 kHz, or 20,000 complete cycles per second. The carrier wave of an am broadcast station may be from 535 kHz–and that is 535,000 cycles per second–to as many as 1605 kHz and that is 1,605,000 cycles per second.

At the am station the audio wave is superimposed on the carrier wave and both are transmitted. In effect, the carrier functions as a truck while the audio signal is the merchandise being shipped. In your home, by adjusting the am tuning section of your tuner or receiver, you can select any frequency between 535 kHz and 1605 kHz, disregarding all the other signals that may be available. Your tuner or receiver accepts the signal you have selected, amplifies it, and at some point separates the carrier and audio signals. The carrier radio wave, having completed its transportation job, is discarded. The audio signal is strengthened further in the preamplifier and power amplifier and is finally delivered to the speakers.

THE AM TUNER

Fig. 2–5 is a block diagram of the front end, intermediate frequency (if) and demodulator section of an am/fm tuner. The selected signal is brought into an rf (radio frequency amplifier) and the signal

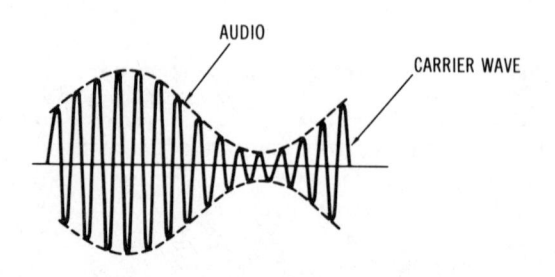

Fig. 2-4. Amplitude modulated (am) radio wave.

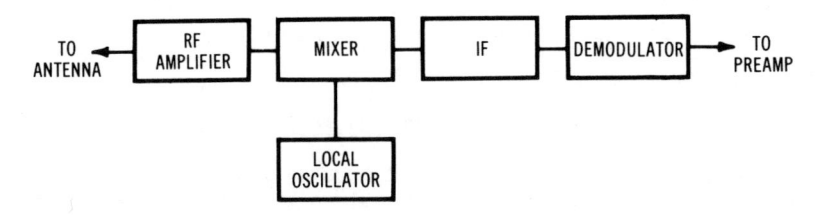

```
TO  ←  RF          →  MIXER  →  IF  →  DEMODULATOR  →  TO
ANTENNA   AMPLIFIER                                      PREAMP
                        ↑
                     LOCAL
                   OSCILLATOR
```

Fig. 2-5. The rf amplifier selects and amplifies the broadcast signal. It is then mixed with the signal generated by the local oscillator. The result of the mixing process is an intermediate frequency (if). The demodulator extracts the audio signal and discards the carrier.

having a strength of just a few microvolts (μV) is amplified. It is then fed into a mixer and that is exactly what this circuit does. It mixes the am signal with one that is generated right in the receiver itself by a circuit called a local oscillator. The mixing process produces a new, lower frequency signal called the intermediate frequency.

In effect, all we have done in the mixing process is to lower the frequency of the carrier, whatever that frequency may be, to 465 kHz or 465,000 cycles per second. The audio signal remains unchanged in this process. The local oscillator is a tuned circuit so that no matter what signal is brought in, the intermediate frequency (if) always remains the same. The intermediate frequency, plus the audio riding along with it, is amplified by a number of stages and is then brought into a demodulator or detector circuit.

In the demodulator, the intermediate frequency carrier is disposed of, and the audio signal is further strengthened by the preamplifier, and the power amplifier and is then finally sent along to the speakers.

An am broadcast signal has considerable traveling ability and it is no great difficulty to pick up stations that are several thousands of miles distant. The strength of local am signals is such that no outside antenna is needed. And, as you tune from one station to the next on the am band there is very little interstation noise.

However, am is single channel sound broadcasting. You can use two speakers with an am signal but both speakers will reproduce the same sound. Further, the band of audio signals is limited, so even if you listen to music on the am band, some of it will be missing. What you will get will be music having a range of 20 Hz to 5 kHz (5,000 cycles per second). This is quite limited since the useful audio range extends from 20 Hz to 20 kHz.

Electrical noise is another problem. Electrical noise is am, that is, it behaves like an am broadcast signal, and so your tuner or receiver, unable to distinguish between an am broadcast signal and electrical noise, will pick up both, amplify both, and deliver both to your speaker system. Electrical noise is that produced by fluorescent lamps, motors, storms, machinery, or automobile engines.

THE AM BAND

The am broadcast band covers the range of 535 kHz to 1605 kHz. Because of its limitations for the reproduction of high-fidelity sound, it is usually considered low-fi. However, it is useful for supplying news, sports programs, etc.

Some radio stations are am/fm, that is, they broadcast on both the am and fm bands. However, this does not mean the same program is broadcast on both. An am/fm station may simultaneously transmit a musical program on fm while broadcasting news on am.

THE AM ANTENNA

Because modern tuners and receivers are so sensitive and have so much amplification ability, outdoor antennas are seldom used for the am section. Instead, there is a built in antenna called a loopstick, attached to the rear. This is often mounted on a swivel of some kind to let you move the antenna away from the receiver. You should allow some few inches of space between the rear of the receiver and a wall, or other obstruction, to let you adjust this antenna for best am reception. It will have no effect on fm.

Generally, in urban areas this is all that will be needed for am reception. However, if the receiver isn't notably sensitive for am broadcasts and isn't located within 30 miles of the station (or stations) an outdoor antenna may be needed. This can consist of a length of antenna wire, possibly 20 to 50 feet in length, and connected by a single lead-in wire to the am antenna terminal of the receiver or tuner.

GROUND

At one time in the early days of radio when receivers did not have the sensitivity they have today a good ground connection was essential, with the ground a part of the antenna system. Today, a ground connection to a tuner or receiver is a hum reducing aid. If needed, run a wire from the ground terminal on the rear apron of the tuner or receiver to a radiator or cold water pipe. Good metal contact is needed. If necessary scrape paint away from the pipe so that the bare wire of the ground connection can make good contact.

FREQUENCY MODULATION

Frequency modulation or fm is just another method of loading the audio signal on a carrier wave (Fig. 2–6). The basic objective of fm is the same as that of am——and that is to overcome the inability of audio to travel more than just a short distance. And so, as it is also used in am, a carrier wave is used in fm to function as an "electronic truck" to deliver the audio to the tuner that is part of the receiver.

In am the method of loading the audio on the carrier, called modulation, is by changing the amplitude or strength of the carrier, hence the name amplitude modulation. In fm the strength of the carrier remains constant but its frequency is changed at an audio rate, hence the name frequency modulation.

In the receiver the am and fm sections are different. But after the audio signal has been recovered from the carrier it is sent through a common preamplifier and power amplifier. So combining an am tuner and an fm tuner into a receiver makes good economic sense, for they can both use the same audio amplifier section, and, of course, the same speakers.

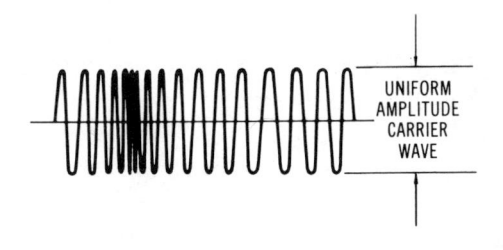

Fig. 2-6. In frequency modulation (fm) the carrier wave maintains constant strength but its frequency is changed in step with the audio signal.

SIGNAL SEPARATION

The process of signal separation starts in the rf amplifier of the tuner or receiver, whether that input is an am or an fm signal. Here tuned circuits begin the process of separating one broadcast signal from all the others picked up by the antenna. Some tuners have just a single tuned circuit; others have two, generally more. Obviously, the more tuned circuits the better the selectivity and the sensitivity. From this we get two important specs: sensitivity—the ability to pick up weak signals, and selectivity—the ability to choose a particular station, while at the same time suppressing adjacent signals and electrical noise.

THE FM ANTENNA

Manufacturers of tuners and receivers usually include a temporary antenna. Made of twin lead or transmission line (Fig. 2–7) of the type often used in connection with television antennas, it is supplied in the form of an antenna known as a folded dipole (Fig. 2–8). The two ends of the antenna are connected and then one wire from each lead at the center of the antenna is attached to the antenna input terminals of the tuner or receiver.

This antenna is flexible and can be pushed under a rug or attached to a wall baseboard. And, considering the sensitivity of the modern hi-fi tuner or receiver, it can bring in quite a few stations. The best that can be said for such an antenna, though, is that it is better than no antenna at all. In some locations there is no alternative and so this

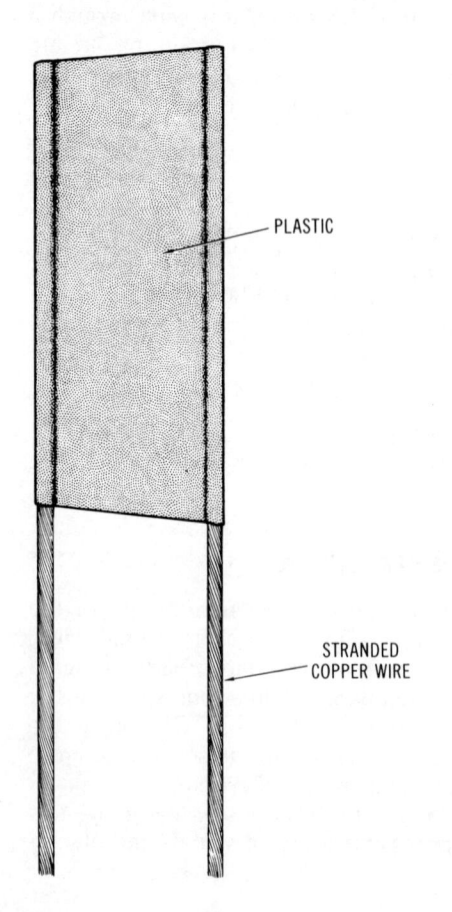

Fig. 2-7. Section of two-wire transmission line.

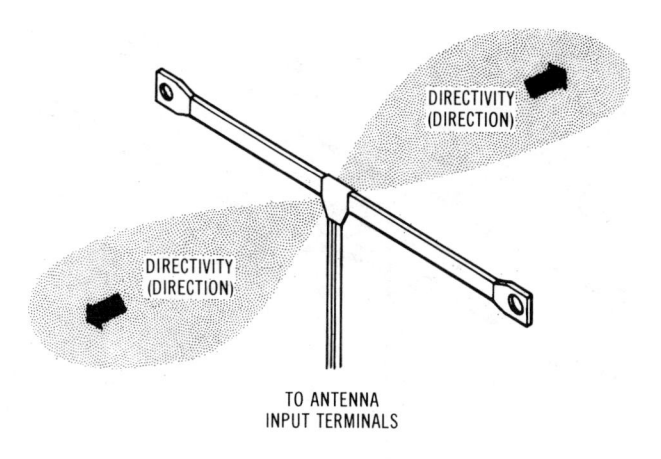

Fig. 2-8. The indoor dipole antenna is made of 300-ohm transmission line.

is the antenna that must be used. If at all possible, do not simply bunch the antenna and throw it on the floor behind the tuner or receiver. Instead, stretch the antenna to its full length and hold it in position along a baseboard by using pushpins or tacks. Do not mount it behind a radiator or any other large metal object. Because the antenna is quite flat you can also make its full length take up a position beneath a rug. If you can experiment with the location of this antenna, do so. You will get best reception when the antenna is broadside to the fm station. Quite often, though, a compromise in this "best" position will be necessary.

An outside antenna (Fig. 2-9), especially designed for fm reception, and not just an old tv antenna or some other kind pressed into service, will help the tuner or receiver achieve optimum performance. You will be able to pick up more stations, and you will be able to get stations as much as 75 to 100 miles distant. There will be a definite improvement in the signal-to-noise ratio, meaning that there will be less noise evident and stronger, cleaner sound.

Some antennas are motorized and can be rotated by using a control located in the home. This is another advantage that will help you select the fm stations you want, and, at the same time, discriminating against electrical noise pickup.

Fm Antenna Elements

An outside antenna for fm reception can consist of two metallic rods approximately two feet long with each connected to one end of the two wire transmission line going to the receiver's or tuner's antenna input terminals.

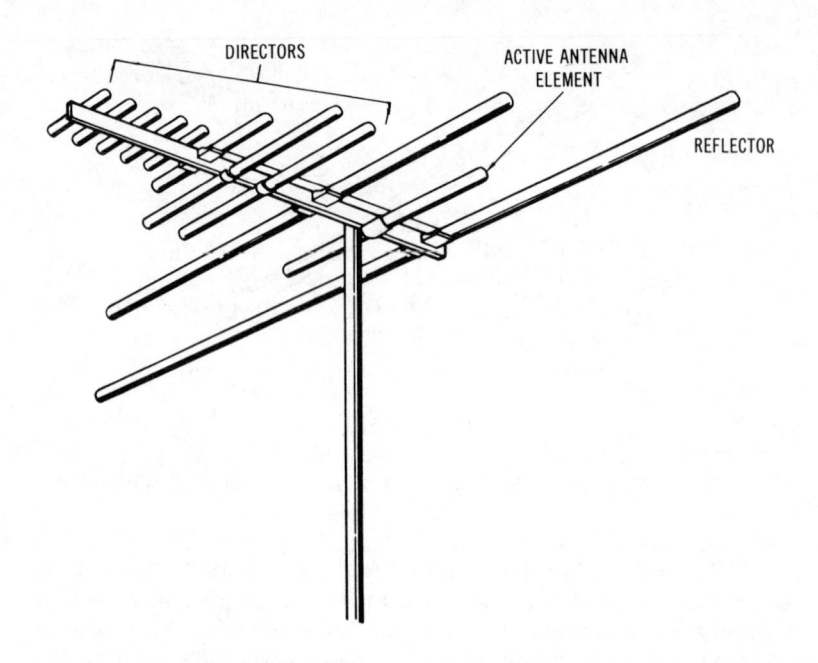

Fig. 2-9. Outside fm antenna consists of parasitic elements, such as reflectors and directors, and one active element, the antenna.

This is about the simplest kind of fm antenna, but a better one would be equipped with directors and a reflector. Directors and reflectors may be metallic tubing or rods, but have no physical or electrical connection to the antenna itself. Sometimes they are known as parasitic elements in contrast to the antenna, an active element.

When the antenna element receives a signal it produces a magnetic field which induces a voltage in the reflector. This produces a magnetic field around the reflector in phase with the energy radiated from the antenna and so helps reinforce the original signal. Reflectors are positioned parallel to and behind the antenna element and you can recognize them by the fact that the reflector is a bit longer than the antenna proper.

A director is also a metal rod (there are usually two or more) positioned parallel to and in front of the antenna element. Like the reflector, none of the directors are connected to the antenna in any way. A voltage is induced in the directors by the antenna element when it receives a signal. The resulting magnetic field is in phase with and aids the original signal. Thus, directors and reflectors increase antenna gain.

When a number of reflectors and directors are used the setup is known as a parasitic array and when four or more elements are used it is sometimes called a Yagi. With the addition of each parasitic element the power gain of the antenna increases approximately 1.4 times.

Connecting the Antenna

On the back of the receiver or tuner you will find terminals marked AM, GND, 75 Ω (75 ohms) and 300 Ω (300 ohms). These numbers do not refer to the kind or type of antenna used, but indicate the kind of connecting cable between the antenna proper and the terminals on the receiver. A 300-ohm line (also called two-wire transmission line) is the kind used most often on television sets. This type consists of a pair of wires separated by about ¼-inch and completely covered with light or dark plastic, or translucent plastic. The 75-ohm connecting cable consists of a solid wire running through the center of the cable. This wire is covered with a plastic material which, in turn, is housed in flexible wire braid. An outer covering of a dark plastic material is then put over the braid. In 75-ohm cable the two conductors are the center wire and the flexible shield braid.

No harm will be done if you accidentally connect a 75 ohm lead to the 300 ohm terminals, or vice versa, but you will get much better reception if you use the lead-in properly connected.

Two wire transmission line is more commonly used for connecting an outside fm antenna to a receiver or tuner. Two wire line is inexpensive, is easy to install, and the plastic coating over the wires makes them weatherproof.

The 75 ohm coaxial cable is more expensive, is somewhat more difficult to use, but the shield braid not only works as one of the signal conductors but acts as a shield around the "hot" central conductor and so it is less subject to noise pickup. It is advisable to use coaxial cable in installations where the fm antenna is near sources of electrical noise: machinery, neon or fluorescent lights, heavy traffic, elevators, and motors.

You can improve am reception by connecting a length of insulated wire to the am terminal. Strip the insulation from the end to be connected to the am antenna terminal, and if it is stranded wire twirl the ends so that the wire resembles a single, solid conductor. Wrap the wire in a clockwise direction around the screw of the am antenna terminal. This is the same direction you will use when tightening the screw. Make sure that no strands of wire reach out to touch any other terminals.

The drawing in Fig. 2-10 also shows the connections for the fm indoor antenna. The remaining terminal is a ground connection. If your hi-fi system produces some hum you may find it helpful to connect a wire between this terminal and ground, possibly a cold water pipe or radiator.

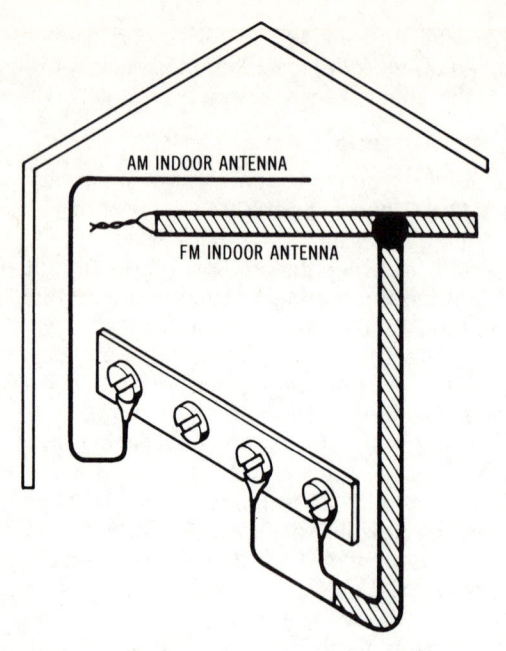

Fig. 2-10. Separate indoor antennas can be used for am and fm.

The connections for outside antennas are similar, and are illustrated in Fig. 2–11. The elements of the fm antenna should be broadside to the fm stations you prefer. This assumes they are all in the same general direction. If not, and if they form an arc or circle around your antenna, an omnidirectional antenna or a rotating type would be a better choice.

FM RECEPTION RANGE

Consistently effective fm reception range varies from about 20 to 30 miles but this can be extended by using a high gain antenna and a receiver having high input sensitivity. Reception range also depends on the terrain between the fm station (or stations) and the receiving antenna. Line of sight transmission and reception is the ideal arrangement, that is, there are no obstacles of any kind between the broadcasting and receiving antennas. Obstacles in the signal path may absorb some of the signal, reflect it, or do a little of both.

ANTENNA GAIN

The amount of signal an fm antenna can deliver to the antenna input terminals of a tuner or receiver compared to a standard antenna

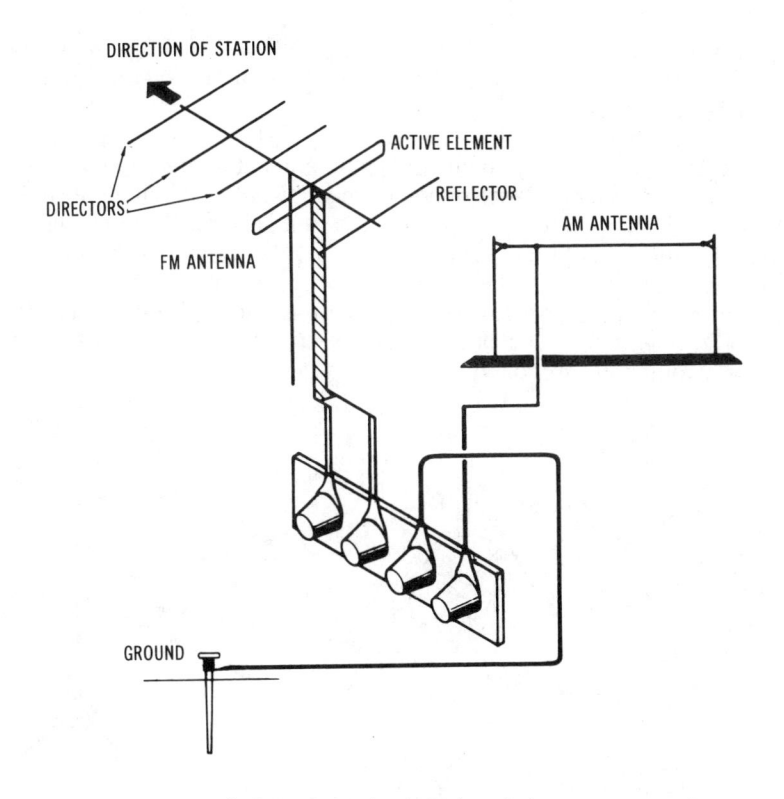

Fig. 2-11. Outside fm antenna should be broadside to station signal.

used as a reference is referred to as antenna gain. Some fm antennas have more "gain" than others. This doesn't mean they amplify the signal, but their design and construction are such that they can deliver a greater amount of signal to the receiver or tuner. A high gain antenna would be especially desirable in areas noted for having poor fm reception. It is also useful if you want to be able to reach out for weaker, more distant signals. And a high gain antenna also produces a better signal-to-noise ratio.

MULTIPATH RECEPTION

The ideal arrangement for the reception of fm signals is to have the receiving antenna along a clear line of sight with the fm transmitting antenna. This is an unlikely situation and generally there will be one or more obstructions between the two antennas: a building, a bridge, trees, a hill, etc. An fm signal striking these will bounce off and will be reflected and so the fm receiving antenna will pick up the same signal that is being radiated from the transmitting antenna from a number of different locations.

Since some of these signal reflections will take longer paths (Fig. 2–12) they will arrive at the receiving antenna at a later time. And so an antenna may get a series of two or more signals, all being transmitted from the same broadcasting antenna, but arriving at the receiving antenna at different times. These signals may reinforce each other, or they may cancel in whole or part. The result is distortion—changes in speech patterns. We can look on them as audio ghosts, similar to the ghosts you may see on your tv screen when the tv signal also takes a multipath approach.

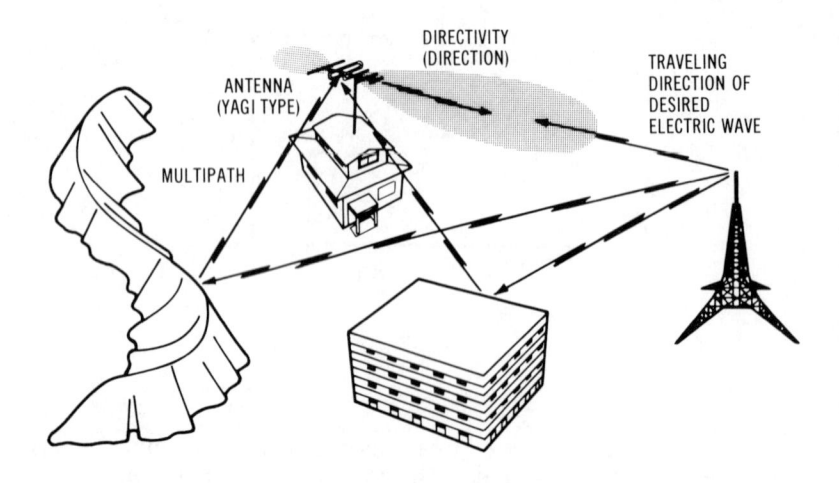

Fig. 2–12. Multipath signal competes with direct fm signal at receiving antenna—can produce fading and distortion.

THE SUPERHETERODYNE

The action of mixing the signal produced by the local oscillator in the am receiver with the incoming radio signal is called heterodyning and any tuner or receiver, whether am or fm, is called a superheterodyne–abbreviated as superhet. The superheterodyne (Fig. 2–13) has been with us for well over 50 years, is used by all radio receivers, ranging from the cheapest hand-held portables to the most expensive hi-fi tuners or receivers, so it isn't exactly new or limited in its application. Yet you will still find some advertising touting a hi-fi receiver as a genuine superheterodyne. As if it could be anything else!

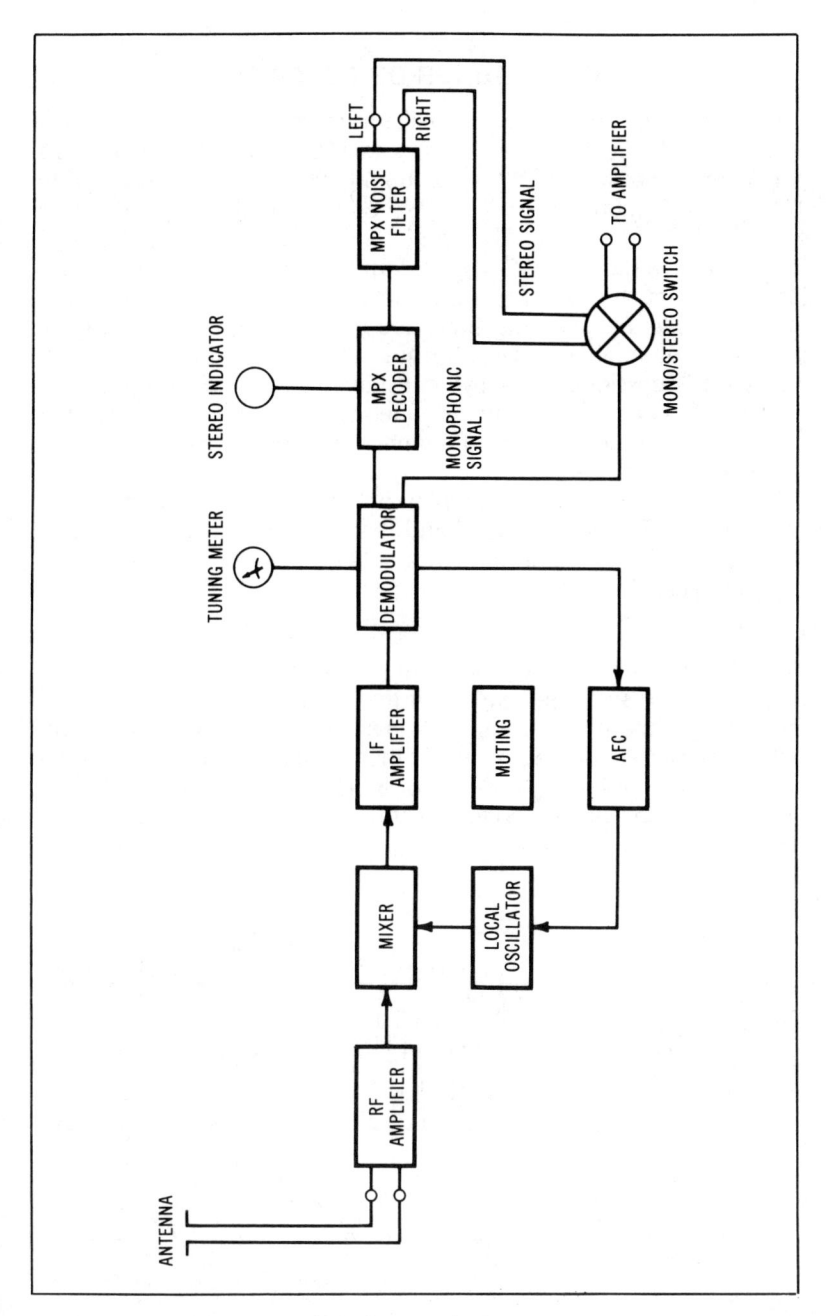

Fig. 2-13. Block diagram of a fm stereo tuner.

THE FM TUNER OR RECEIVER

In all superheterodynes, whether am or fm, the broadcast signal, 535 kHz (535,000 cycles per second) to 1605 kHz (1,605,000 cycles per second) for am, and 88 MHz (88,000,000 cycles per second) to 108 MHz (108,000,000 cycles per second) for fm is brought down to a single fixed frequency. This single fixed frequency is usually 465 kHz for am while it is 10.7 MHz for fm. For fm tuners and receivers, just as for am, this is done by a local oscillator and a mixer. The output, or intermediate frequency, abbreviated as if, is sent through a succession of amplifier stages. Because the if is tuned to a single frequency it is possible to achieve considerable signal selectivity.

One difference between tuners, whether am or fm, is in the number of if stages and their design. In a tuner that is both am and fm, separate if stages are used.

The purpose of the if is to amplify only the sharply defined section of the intermediate frequency band that contains the wanted sound information. The if is not only an amplifier, but also a filter, often a ceramic type.

Limiters

After the heterodyning process in which the incoming fm signal mixes with the signal generated by the local oscillator, the result of this action is the if signal which is then fed into a number of if amplifier stages. One of the features of the if in the fm section of a tuner or receiver is that it may have one or more limiters. These are circuits that clip the top and/or bottom of the signals going through the if stages, a kind of electronic haircut (Fig. 2–14). Since the audio in-

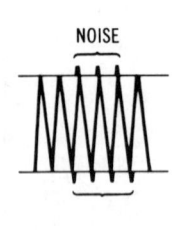

(A) Noise added to wave peaks.

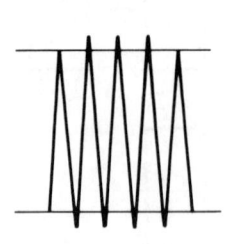

(B) Signal and noise after amplification in amplifier.

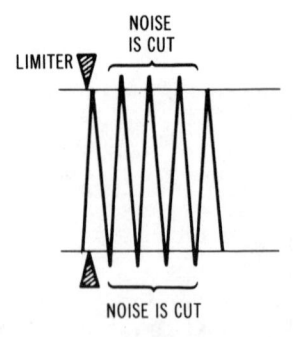

(C) Limiter removes noise peaks from top and bottom of signal.

Fig. 2-14. A fm signal.

formation is contained in the frequency changes only, not in the strength or amplitude of the signal, clipping the top and bottom of the intermediate frequency is, in effect, a way of removing any amplitude modulation (am). But that is exactly what electrical noise is—am—and so the effect is to remove any such noise that may have accompanied the fm signal. Limiters work when a signal of sufficient strength moves through the if and so noise limiting doesn't exist alone, but is related to the sensitivity of the tuner and the signal strength of the antenna.

Stereo

Stereophonic sound, abbreviated as stereo, is also known as two channel sound, or left and right sound, corresponding to the sound picked up by microphones positioned to the left and right of a sound source such as an orchestra. Quite often left channel sound is represented by the letter L and right channel sound by the letter R. The carrier that will transport the audio signal to the home hi-fi antenna will carry both left and right audio signals and arithmetically we can represent this as L + R. In an fm tuner not equipped for stereo all that will be heard will be L + R, the sum of the sounds of the left and right channels, and because these channels are combined into a single composite the sound out of the speakers will be heard monophonically. But this also supplies us with a clue on how to obtain stereo. Since the audio signal is transmitted as the sum of L + R, all we need do is to separate L + R in the receiver and supply those L + R audio signals to separate left (L) and right (R) speakers. But we will need more than just a circuit for separating L and R. We will need another circuit to warn or notify the fm receiver when such signals are being transmitted.

To accomplish this, when transmitting a stereophonic signal, that is, a carrier containing left (L) and right (R) channel sound, the fm station transmits a subcarrier wave (Fig. 2–15) separated from the main carrier wave by 38 kHz (38,000 cycles per second) together with a pilot frequency of 19 kHz (19,000 cycles per second).

Note that 19 kHz is within the upper limits of the audio range of 20 Hz to 20 kHz. The purpose of this pilot signal is to help recover the stereo broadcast in the multiplex circuit, but once this is accomplished the pilot signal must be eliminated; this is done by using a filter circuit. A filter that is poorly designed can remove some tonal harmonics in the upper treble register.

Phase Lock Loop

It is essential for the tuner or receiver to lock onto the pilot signal to enable the multiplex demodulator to separate the stereo information

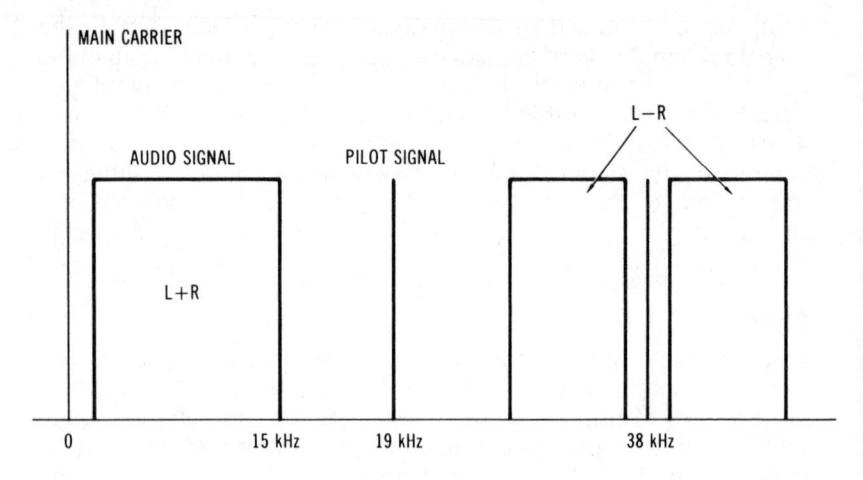

Fig. 2-15. A fm signal consists of radio frequency carrier, plus left (L) and right (R) audio signals. Also transmitted is a 19 kHz pilot signal and a 38 kHz subcarrier.

from the fm multiplex signal and to do so accurately. This is done by a phase lock loop (pll) circuit.

In the fm stereo tuner or receiver a detector circuit becomes activated when a stereo signal is transmitted. The receiver uses a multiplex decoder (mpx) to separate the left and right channels. A good tuner will supply 40 dB separation between 400 Hz and 1 kHz or 25 dB over the range from 20 Hz to 15 kHz. In some fm receivers the 19 kHz pilot signal is used to turn on a light or some other indicator to inform the user that a stereo signal is being received.

Not all receivers or tuners are equipped to handle stereo signals, and so such components often use the designation am/fm. This means they are monophonic units only, but can pick up both am and fm broadcasts. Most receivers and tuners are identified as am/fm/mpx, an indication that they not only pick up am and fm signals but are also capable of reproducing fm stereo. Stereo is important because all phono records are in stereo as are prerecorded cassette tapes. However, many fm stations still broadcast in mono only.

The MPX Decoder

The mpx decoder in the stereo tuner or receiver has delivered to it not one, but two audio signals. One of these is L + R; the other is L − R. These two signals are added and subtracted to obtain left and right channel sound, that is, stereo. If we add L + R and L − R, the two Rs will cancel because one is minus; the other is plus. So we are simply left with L or left channel sound. If we subtract L − R from L + R the two Ls cancel and we have 2R or right channel sound.

The two signals, L and R, both audio and both representing the output of the left and right microphones used in recording are then delivered to independent audio amplifiers connected to the output of the tuner. It is important to recognize at this point that left and right channel sound, once they have been recovered from the carrier wave, must be treated as separate entities. This means we must have separate amplifiers in the preamplifiers and power amplifiers, one each for left and right channel sound. This procedure isn't required for single channel or monophonic sound.

Logic Circuits

Logic, or decision making circuits, long used in computers, have been adapted to hi-fi components. Electronic circuits have a large variety of applications. We can use them as amplifiers, ac generators better known as oscillators, timing devices, circuits for delaying signals, wave shaping circuits, power supplies, etc. We can also have decision making circuits, an unfortunate term since it implies that such circuits can think. While electronic circuits can be designed to make decisions, they can do so because some design engineer has done the required thinking well in advance.

Decision making circuits, also known as logic circuits or gates, work because of existing conditions in that circuit. And, despite the fancy names they may have, they are just switches. An electronic switch may turn on when the voltage at its input reaches a predetermined amount, turning off automatically when the voltage becomes lower.

Decision making circuits can be electromechanical or purely electronic. The light switches in your home are mechanical, but the decision making function isn't part of the switch. Such switches are finger operated and you are the one who makes the decision to turn them on or off. The electromechanical type can include microswitches, microrelays, and micromotors. Here the action is initiated by a current flowing to these parts, causing them to operate, so like mechanical switches they aren't involved in the decision making process.

We reach the decision making process with the electronic switch or gate. An open gate is comparable to a closed switch; it is a gate that permits the passage of a pulse or current. A closed gate is the opposite; it means the signal cannot pass through. A closed gate is like an open switch.

Diodes

A diode is an electronic switch and is made of the same semiconductor materials as transistors. We can make the diode conduct, or not conduct, depending on the polarity and amount of an applied dc voltage, known as bias.

One of the simplest of all logic circuits consists of a single biased diode, an arrangement that permits the passage of current through the diode when an input signal voltage pulse exceeds the amount of bias applied to the diode (Fig. 2–16). If we put a negative bias of one volt on the diode, the input voltage pulse will need to be at least slightly higher than one volt positive. The positive voltage of the signal negates the negative bias voltage. All voltage pulses less than one volt will be blocked, that is, will not be strong enough to permit current flow through the diode. It is in this sense that the diode "decides" whether a pulse of current will pass through or not (Fig. 2–17).

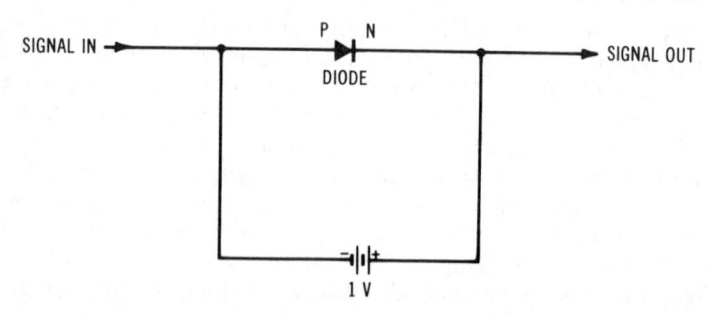

Fig. 2-16. Basic gating circuit. The bias or dc voltage on the gate is 1 volt, negative. The gate remains open until the incoming signal is 1 volt, or more, positive.

Fig. 2-17. A gate works as a switch, is either on or off.

And Gate

We can elaborate on this logic circuit by putting another diode in series with the first (Fig. 2–18). In this circuit both gates must be open for current to flow, a situation comparable to a safety deposit box in which two keys must be used at the same time to get the box open.

In this circuit using two diodes, since the first diode AND also the second must have the correct operating voltages, the arrangement is called an AND circuit. The first gate AND the second must be open.

Instead of using diodes we can have transistors to form an AND circuit. The advantage is that we manage to get a double function this way: the transistor works both as a gate and as an amplifier. Further, if a transistor is used as the gate, it not only controls the strength of the voltage pulses that pass through, but can also be used to invert these pulses, possibly changing them from positive to negative.

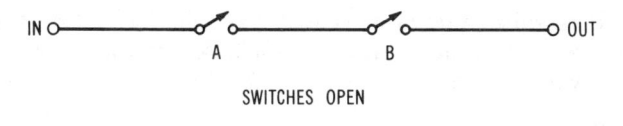

Fig. 2-18. Basic AND logic circuit. Switch A AND switch B must both be closed for a signal to pass from the input to the output. A closed switch is also known as an open gate.

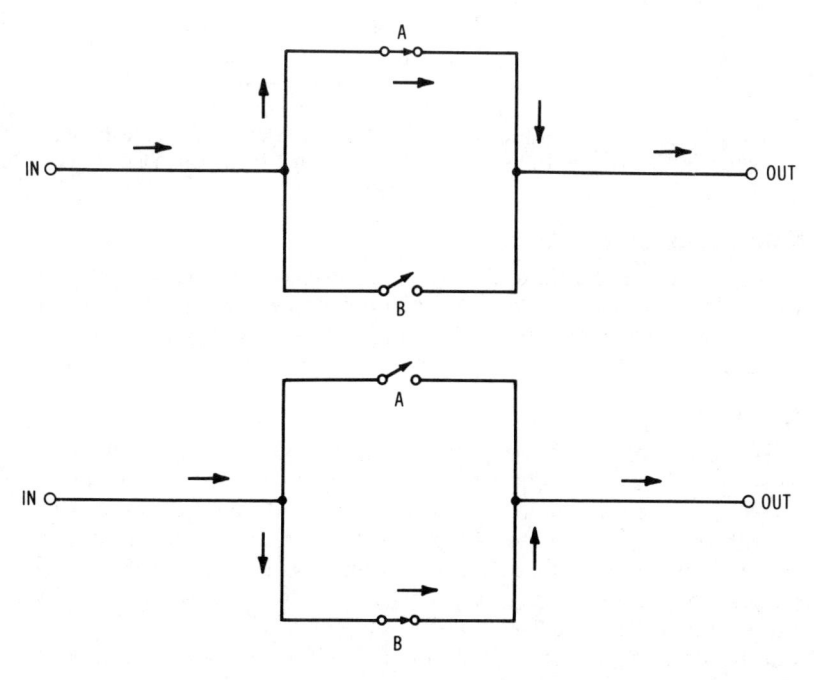

Fig. 2-19. Basic OR logic circuit. Either switch A OR switch B must be closed for a signal to move from point A to point B.

OR Gate

We can have diodes wired in parallel so that a current can pass through either one OR the other, or both at the same time; this circuit is called an OR circuit (Fig. 2-19). The idea is also used in home

73

wiring where either a downstairs switch OR one located upstairs will control a hall light. If these switches could be controlled by a photocell, they would be the mechanical equivalent of an OR circuit

AND/OR Gate

By combining an AND circuit with an OR we get an AND/OR (Fig. 2-20). In such a circuit the AND switch and one of the OR gates must be open for an electrical pulse to pass through. The gates can be diodes or transistors or some combination of these.

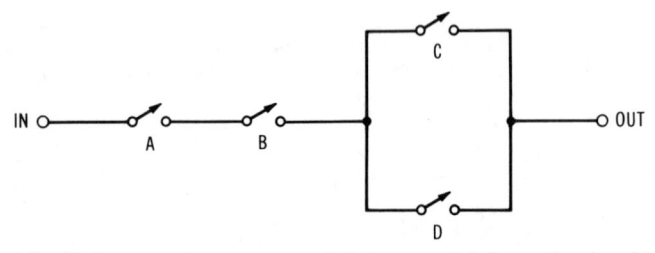

Fig. 2-20. Basic AND/OR logic circuit. Switch A AND switch B must be closed and either switch C OR switch D must be closed for a signal to move from the input to the output.

NAND Gate

Logic circuits can be operated by positive input pulses, but negative pulses can also be used. We can develop an AND circuit responsive to negative pulse input instead of positive, calling such a circuit a negative AND, or NAND.

NOR Gate

Similarly, we can have negative OR circuits, identifying them as NOR. And, just as we combined AND and OR circuits to produce an AND/OR, we can also combine NAND and NOR to give us NAND/NOR.

Both positive and negative types of gates are used since electronic pulses can fall into three classifications: positive pulses, negative pulses, and combined positive/negative. Positive pulses and negative pulses are dc. These pulses consist of the intermittent flow of current, but in one direction only. Combined positive/negative pulses are ac. With these the current flows first in one direction, then the other.

With the four basic types of gates, AND, OR, NAND, NOR, we have the means for automatic control of current flow. Further, we can arrange these gates in various combinations, using either diodes or transistors. Because they are electronic, not mechanical, they work with tremendous speed and since they have no moving parts they can have a long working lifetime.

Gating circuits can be combined with micro-type mechanical

components such as motors, relays, and switches. The gating circuits could operate these parts for specific amounts of time, turning them on or off as required. In this way we could have a gate operated relay or a gate operated motor. The gate would "decide" whether conditions were right for the relay to turn on or off, or for a motor to function. Generally, working conditions are such that the mechanical units are worked intermittently. In some instances, the motor or relay may be connected to a defeat switch on the front panel, giving the user the option of overriding the action of the gating circuit.

Gates can be single pulse input types or may have multiple input. With the multiple input gate, pulses from various circuits can trigger the gate, initiating some desired action.

TUNING THE TUNER OR RECEIVER

Tuning a tuner or receiver is the method for selecting a particular am or fm station. The word "tuning" is quite appropriate and if it reminds you of the way a violin is tuned then that is correct, for there is considerable similarity. A violin string is tuned or adjusted to produce a note of a particular frequency. A receiver or tuner is also "tuned" to put it in resonance with the frequency of a particular broadcast station.

Tuning Meters

Tuners or receivers may have either one or two meters or some substitute (Fig. 2–21). The meters are usually illuminated types with the lights turning on automatically when the power is switched on.

One of the meters is a signal strength type. It will help you tune in an fm signal properly if you adjust the tuning control so the meter pointer swings as far to the right as possible. The signal strength meter will also help you make some comparisons of the signal strengths of various broadcasting stations.

The tuning meter is a zero center type. Adjust the tuning control of the tuner or receiver until the meter pointer is at dead center. This will help avoid cutting fm sidebands, minimizing tuning distortion and supplying the best signal-to-noise ratio.

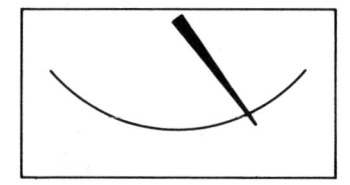

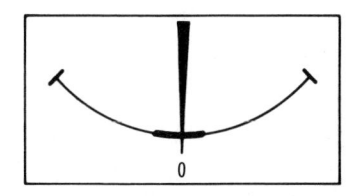

Fig. 2-21. Signal strength meter (left) and zero center tuning meter.

75

The use of the signal strength and zero center tuning meters (Fig. 2–22) makes several assumptions. One of these is that the local oscillator and mixer work to produce a precise value of intermediate frequency. But if the local oscillator is off frequency, even if by a small amount, the intermediate frequency will also be off, and so the signal will not be at its maximum strength if you try to tune precisely to zero on the center tuning meter. Theoretically, both meters should work together. The signal, as indicated by the signal strength meter, should be maximum at the same time that the tuning meter is at zero. Quite often, this is not the case.

The problem with meters is that they are ballistically slow and their inertia often does not permit making fine adjustments with the tuning knob. A simpler approach is the use of an illuminated center-tune indicator. This indicator lights only when the center frequency of an fm broadcasting station is tuned in. Other tuners and receivers may use a succession of horizontal lights. The objective is to tune so that the line of light is as long as possible.

The "line-of-light" approach is also used to replace the signal strength meter. This indicator operates for both fm and am broadcasts. If the indicator consists of a succession of horizontally arranged lights, maximum signal strength is shown when all of the lights are on.

Some tuners and receivers are now substituting LEDs (light-emitting diodes) for tuning and signal strength meters. While LEDs for tuning and signal strength may look more attractive they do not appear to offer electronic benefits. A possible argument in favor of the meters is they are sometimes made multifunctional, that is, they can also be made to indicate multipath distortion. While frequency synthesized tuners have no need for tuning meters, they may come equipped with signal strength and multipath meters. The advantage of the signal strength meter is that it supplies an indication of the presence of a weak station that the frequency synthesized tuner or receiver may ignore.

Both types of meters, the mechanical using a moving pointer and those that are light operated, are subject to error. In the moving pointer meter there are errors related to the scale and the reading indicated by the meter. Thus, the movement, or the scale itself, may not be accurate, or, the deflection of the pointer may not be strictly proportional to the voltage it receives.

Similar errors can occur in light operated systems if the lights do not step correctly corresponding to steps in the received voltage because of tolerances in the circuit. The result is then identical to errors of scale with mechanical meters. It isn't unusual to find errors of 2 dB or more, even in apparently high quality meters with 1 dB or less between markings. Mechanical meters also suffer from the inability to stop instantly, a characteristic called overshoot.

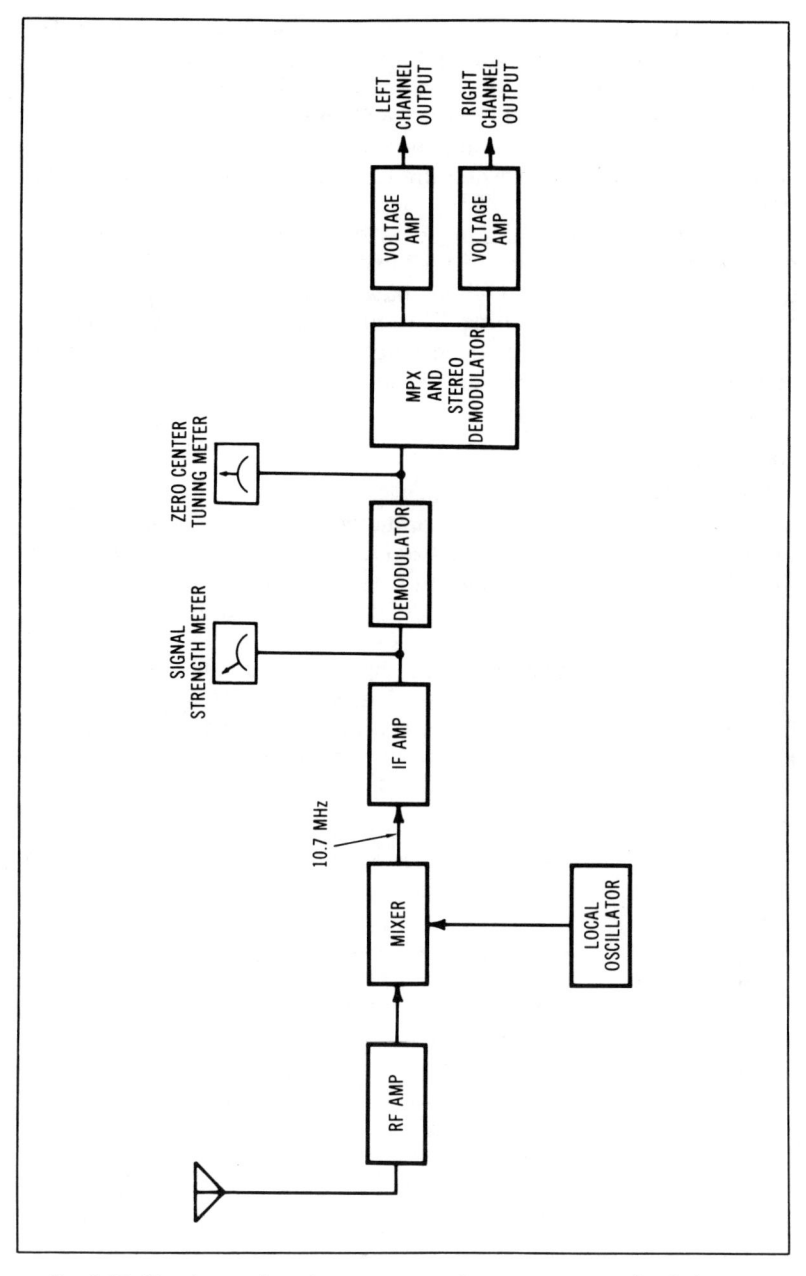

Fig. 2-22. Signal strength and zero center tuning meters are used to help tune signal in precisely to avoid tuning distortion.

Various other methods are used to get around the limitations of meters. One component uses an illuminated tuning pointer instead of the usual center tune meter. A pair of yellow-colored arrows appear on either side of the pointer which is illuminated in red. The yellow arrow is a guide as to the direction in which to tune. When both yellow arrows are extinguished the tuner is right at the center of the station's frequency.

And that is the nub of the matter. We want to be able to tune right to the center of the station's carrier frequency. This is very difficult to do manually, so various methods have been suggested to overcome the limitations of such tuning. One tuner uses a servo lock mechanism. Once the station has been tuned in reasonably well, the servo lock mechanism takes over and *locks* the receiver onto the station's frequency.

Digitized Tuner

The problem with a tuning dial, even those with large numbers, is that it can be difficult to read, particularly since the tuner is seldom positioned at eye level. The digitized tuner overcomes this by using a digital frequency display, as indicated in Fig. 2–23. The display shows the frequency of the broadcast station being received, am or fm.

This frequency display does not improve the tuning accuracy, nor does it change any of the tuner's or receiver's specs. It does make it easier to select am or fm stations whose operating frequency you know or which you can obtain from published station listings.

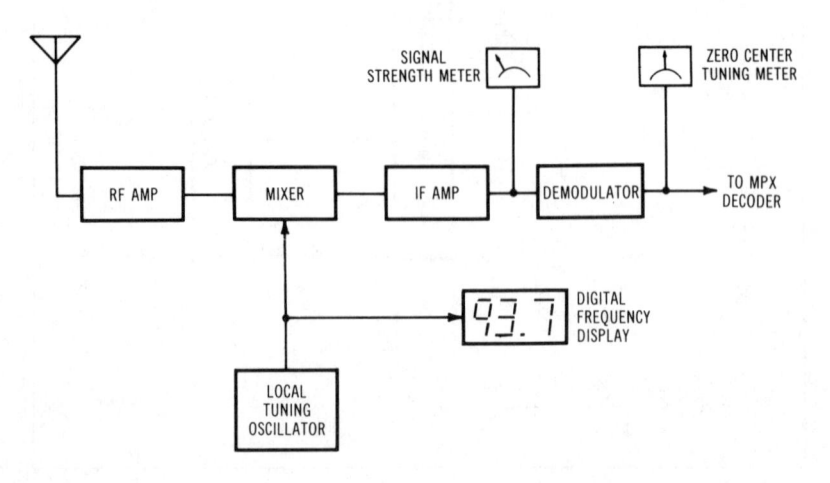

Fig. 2-23. Use of digital display does not improve performance of this fm tuner or receiver.

A number of tuners use four digit type readouts. These glow in bright red or green and indicate the tuned-in frequency. Digital display is used by tuners and receivers that are frequency synthesized types and also by those that do not have this feature. The digital display consists of a number of light-emitting diodes (LEDs).

In the United States the tuning range of the fm band is 88.1 MHz through 107.9 MHz with a 200 kHz bandwidth. This means that the low order digit of the readout is always an odd number; thus, 102.7, 103.1, etc. In tuners and receivers that have digital display, but which are not frequency synthesized, the tuners must still have signal strength and zero tuning meters. You must fine tune the station just as you would with a tuner or receiver equipped with a tuning dial. It is also possible to get a tuner that displays the call letters of the tuned station in addition to the frequency.

Frequency Synthesis

Like the digitized tuner or receiver, a component using frequency synthesis also displays the carrier frequency of the received broadcast signal, either am or fm. However, unlike a digitized tuner or receiver components using frequency synthesis do show considerable improvement in performance.

In frequency synthesis, as indicated in Fig. 2–24, the usual free running local oscillator is replaced by a frequency synthesizer whose operation is governed by a crystal controlled oscillator. The crystal oscillator itself works on a single fixed frequency and this frequency can be any value selected by the manufacturer. The great advantage of quartz crystal control is that the frequency of the crystal can be precise and is much more stable than that of ordinary oscillator circuits.

The output of the crystal oscillator is brought into a frequency divider circuit and is then used to control a frequency synthesizer circuit whose output is always 10.7 MHz above or below the rf signals brought into the mixer from the preceding rf amplifier stage.

There are a number of advantages in digital frequency synthesizer tuning. Tuning distortion is either eliminated or substantially minimized since the accuracy is far better than manual tuning using variable capacitors. The usual zero tuning meter or indicator is eliminated.

However, there can be disadvantages as well. One of the problems with some synthesizer tuners and receivers is that the crystal oscillator and frequency divider can produce radio frequency (rf) signals within the tuner or receiver. These frequencies tend to leak into other parts of the component, causing distortion, interference, and a loss in signal-to-noise ratio.

In a way, frequency synthesizer circuitry is similar to the way in

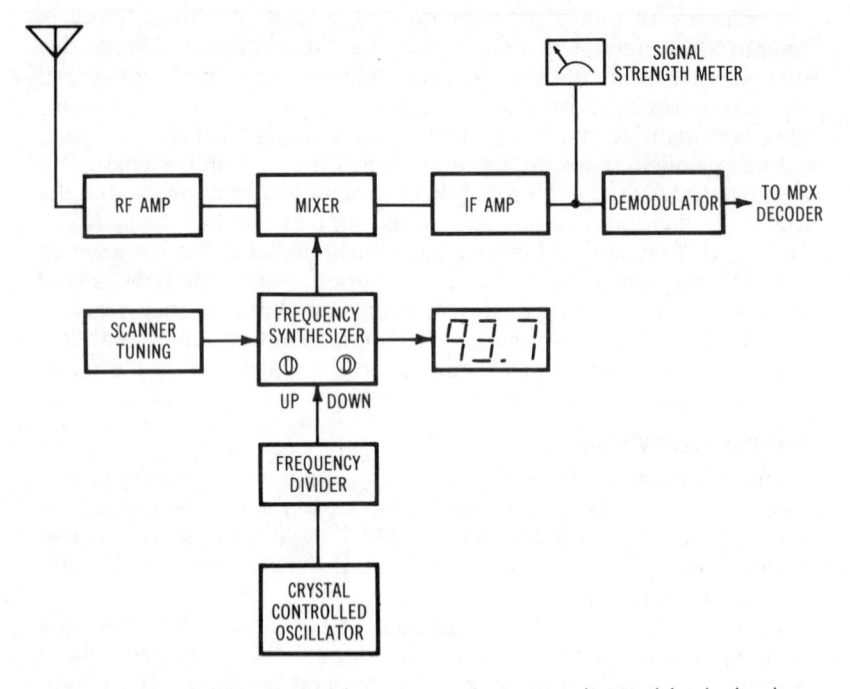

Fig. 2-24. Crystal controlled frequency synthesizer is substituted for the local oscillator.

which broadcasting stations manage to remain on their assigned frequencies. This is through the use of a crystal oscillator. This oscillator is an ac generator and is used to produce the frequency or some submultiple of the frequency of the carrier broadcast by the radio station. The crystal is a bit of mineral, such as quartz, which has the wonderful property of vibrating physically when a voltage is applied to it. These vibrations result in an ac voltage whose frequency remains remarkably constant.

In another method the received signal is changed to digital pulses. These are counted as they are sent through a logic gating circuit. The tuning accuracy is 0.00003% or 30/1,000,000; that is, 30 parts per million.

Scanning

Tuning is the process of selecting one station from among all those whose signals are being broadcast. Scanning is the process of going up and down the tuning dial until you find the station you want so you can tune it in.

For most receivers and tuners scanning is done manually. However, automatic scanning is used in some frequency synthesized digital tuners and receivers. By depressing an auto-scan button, the tuner or receiver scans the full range of the fm or am bands and then automatically reverses. The tuner remains locked into each station momentarily, giving the user a chance to defeat the scanning operation and to hold a wanted station. With this method, though, a minimum signal level, a threshold level, is set. Stations below this level are passed over in the scanning process. Absence of "lock in" is also true of manually operated frequency synthesized digital tuners, that is, the station must have a minimum signal strength.

In some tuners there is no tuning dial or conventional tuning knob. Instead there is a combination of touch buttons and a tap on the forward or reverse button will change the frequency by 0.2 MHz. Using a crystal chip the tuned frequency is compared to a reference crystal. This is manual scanning, comparable to using a tuning knob.

A somewhat less expensive version of the frequency synthesized digital tuner is one that could be called semiautomatic. Tuning is done manually with the help of a zero center tuning meter and then a frequency locking circuit holds the signal in place.

Memory

Tuners and receivers are now making use of memory, also called preset tuning. In one receiver, twelve stations—six fm and six am—can be preselected and locked into memory for instant recall via a pushbutton. The advantage of memory is that favorite stations are available for instant recall simply by touching or pressing a control.

Programmable Tuners

A programmable tuner or receiver is one that can turn itself on or off and can also do the same with a tape deck. If you want to record a particular program while you are away the programmable tuner or receiver can handle this assignment.

Dual If Bandwidth

One of the advances in fm tuner and receiver design has been the incorporation of user-selectable dual if bandwidth. At one time a figure of 1% for fm stereo distortion was considered to be good. With dual if bandwidth tuners have been able to lower this distortion percentage by more than a factor of 10.

All tuners or receivers, digital or conventional, convert an incoming fm signal, regardless of its original frequency (and this can be from 88 MHz to 108 MHz) to a frequency of 10.7 MHz, the intermediate frequency. This if signal is still frequency modulated fm so that, depending on the loudness of the music or speech being broadcast, it can swing back and forth around the 10.7 MHz center frequency by as much as plus/minus 75 kHz. The selectivity and the distortion of the tuner as a whole are affected by the bandpass characteristics of the if amplifier stages. The bandpass is 150 kHz total (plus/minus 75 kHz).

Ideally, a bandpass amplifier, such as the if section of a tuner or receiver, should completely reject all frequencies outside its passband and should amplify all frequencies inside its passband without distortion or phase shift. In shape its characteristic would resemble that shown in Fig. 2–25. However, a bandpass with perfectly steep sides and a perfectly flat top is impossible to achieve and conventional if bandpass sections have response characteristics more nearly resembling Fig. 2–25B or Fig. 2–25C. Fig. 2–25B has admirable selectivity because of its very steep sides, but the area around 10.7 MHz is very small, so a frequency swing of plus/minus 75 kHz would be subjected to very considerable variations plus the loss of some audio. Fig. 2–25C has a reasonably wide top, and so could produce rela-

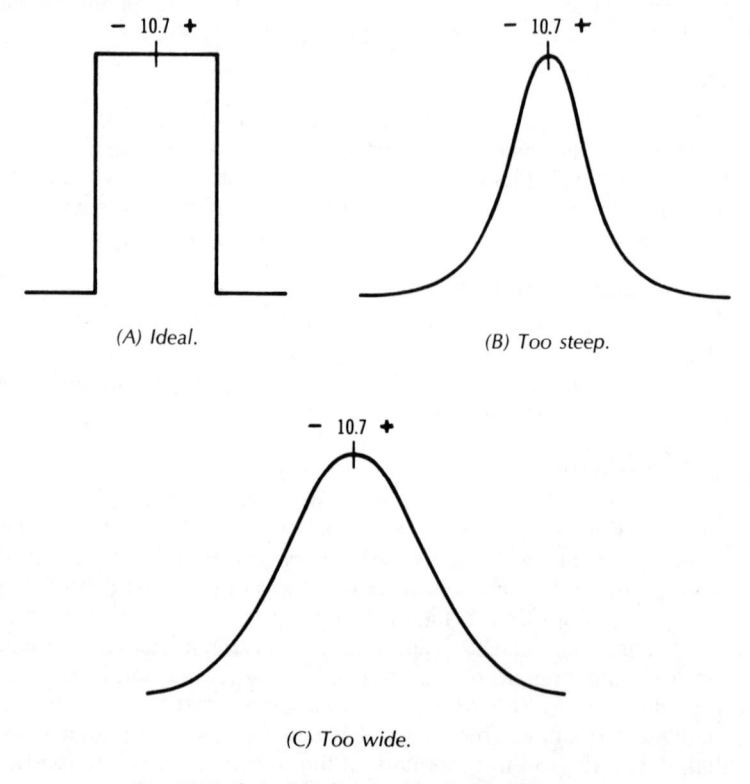

(A) Ideal.

(B) Too steep.

(C) Too wide.

Fig. 2-25. If amplifer bandwidth characteristics.

tively low distortion, but its slope outside the desired passband is so gentle that stations on adjacent and alternate channels could interfere with the desired signal.

Recent studies of filter characteristics have helped achieve steeper sides and flatter tops. A wide bandwidth produces less distortion, and better stereo separation, while a narrower bandwidth produces better selectivity and marginally higher sensitivity.

When only one if bandpass characteristic is incorporated, the user is dependent on the designer's best idea of a compromise, which may or may not be best for a given listening condition. Some tuners and receivers include two different bandwidths, allowing the user to choose either very low distortion by using the wideband mode or better selectivity by selecting the narrowband switch position.

In some components the choice of if bandwidth is automatic with the tuner or receiver making the decision as to whether bandwidth should be wide or narrow. A switch may be included so that the automatic feature can be defeated, permitting manual adjustment.

In one tuner, the if bandwidth control is continuously variable. A horizontal row of fine green lights indicates narrow to wide bandwidth. The longer the row of lights, the wider the bandwidth.

Anti-Birdie

With the inclusion of an anti-birdie filter, so-called "birdie" interference noises triggered by adjacent fm stations can be completely removed when the if bandwidth is switched to its "narrow" position.

Preemphasis and Deemphasis

Electrical noise exists for the most part in the treble tone range. However, tones in the treble range do not have as much acoustical energy as the lower frequency bass and midrange tones. Because of these two conditions, low audio energy in the treble and a relatively high noise level, the signal-to-noise ratio is poor.

At the fm transmitter, this signal-to-noise ratio can be improved by increasing the strength of treble tones, a technique that is known as preemphasis. However, this upsets the relative strength of treble to bass and midrange tones, a condition that is corrected in the receiver by a deemphasis network. In the tuner or receiver this network or circuit, quite a simple one, attenuates treble tones so we have a correct ratio of treble to bass and midrange once again. The advantage is that as the treble tone strength is reduced so is much of the noise hiding in that region.

Switches

Feather touch switches are popular. These are turned on or off simply by touching them with a finger tip. The pops and clicks of mechanical switches are eliminated. In some units a light above the switch indicates that the switch has been turned on.

CONTROLS

A control is a method by which you can change the operating conditions of your hi-fi system. A volume control is so-called since it lets you govern the sound output from your speaker system. But not all controls are immediately identifiable as such. The on/off power switch is a control and so is the function selector. All controls are features but not all features are controls. Thus, a receiver may have input jacks for two sets of headphones. This is a feature since most receivers have just one such jack. Further, a headphone jack does not give you any control over the functioning of your hi-fi system.

During the early days of high fidelity, the preamplifier, used as a separate component, was the control center of a high-fidelity system. This doesn't mean that the other components had no controls at all—they did. In some instances a control must be associated with a particular component for there is just no other way. Thus, the tuning control of a receiver or tuner must be an integral part of that component. The on/off switch on a tuner is not only a control for that component but may also be used to control the line power input to a following preamplifier and power amplifier.

Why So Many Controls?

A hi-fi system will have numerous controls, possibly as many as 50. Some of these are one-time adjustments; others need frequent manipulation. But we need them for a number of reasons. One of the first is that we have so many different sound sources: am/fm/mpx broadcasting, phono records, cassette tapes, open reel tapes, microphones, etc. Each of these sound sources requires different treatment and they get it through controls. We need a control to let the hi-fi system know which of these sound sources we are planning to use.

We need tone controls to compensate for the acoustic characteristics of our listening rooms. We need a control to compensate for the fact that our ears have differing sensitivities at different frequencies to soft and loud sounds. We need a control for adjusting the sound output from left and right speakers.

Controls are not put on a hi-fi system to aggravate the listener. They are necessary because listening to music is a completely subjective experience. No two people hear the same way. No two people have identical musical tastes.

The trouble with controls is that you cannot ignore them. If you do then you might just as well have no high-fidelity system at all. Further, some controls must be adjusted fairly often. And, if anything, the trend in high fidelity components seems to be toward more controls. You will find controls on tuners, on receivers (Fig. 2-26) on preamplifiers, on power amplifiers, on record players, and you will also find them on some speakers.

Power Switch

This is the basic control in any hi-fi system. You will find it on most audio components, but not all. Speakers aren't equipped with such a control and remain permanently connected to the audio power output terminals of the power amplifier. A power amplifier may or may not have an on/off power switch.

Outlets

At the rear of your receiver or tuner you will find a number of convenience power outlets. Since most homes have relatively few line power outlets, these are supplied on audio components to eliminate or minimize the number of power line cords that would need to be inserted in outlets.

There are two types of outlets on the rear of components such as tuners and receivers. One of these outlets is known as unswitched. This means you can use it as a power outlet but it also indicates that the component using this convenience outlet must have its own on/off power switch. Thus, you might connect the line cord of a record player to an unswitched outlet on the rear of the tuner, but that record player would need to have its own on/off power switch.

The other type of outlet is called switched and can be regarded as a control. Thus, when you turn on the power to your tuner or receiver you are also, at the same time, turning on power to the component whose line cord is plugged into the switched outlet. For example, it would make sense for a preamp and a power amp to make use of the switched outlets on the rear of a tuner. In this way, all three components could be turned on and off at the same time simply by using the tuner switch alone. Alternatively, the preamps and power amps, equipped with their own on/off switches, could use the unswitched outlets on the rear of the tuner. This means, though, that these components would need to be turned on every time you decided to use your tuner, or record player, or tape player.

Some tuners and receivers are equipped with glow lamps, generally red, as a reminder that power has been turned on. Others do not have such a lamp but rely on the fact that there will be a glow from the tuning dial or some other part of the receiver or tuner to act as a warning.

Tone Controls

Every listening room is an integral part of a hi-fi system. The sound that leaves your speakers will be partially absorbed, partially

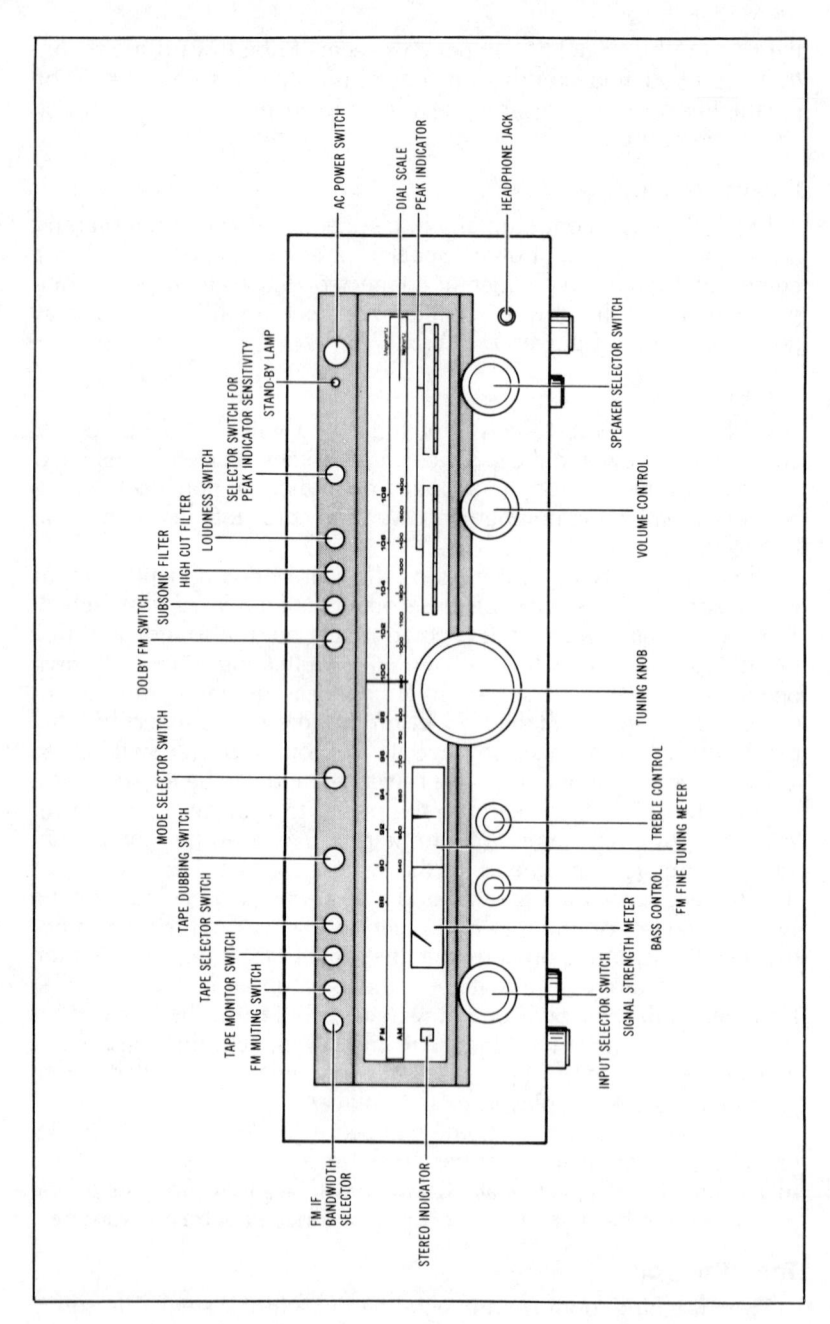

Fig. 2-26. Operating controls and features of a receiver.

reflected, by the floor, walls, ceiling, furniture, persons, and objects in that room. The sound absorption and reflection is frequency selective and so the room may make bass tones sound muddy, can make some treble tones weaken, and can be a general all around nuisance.

The purpose of tone controls is to give you a chance to fight back against this usurpation. Most receivers are equipped with bass and treble tone controls while a few have a third control for midrange tones.

A tone control may be equipped with a "flat" position and in this setting it is as though the tone control was completely out of the unit. The purpose of a tone control is to emphasize bass, midrange, and treble tones as required by the type of music played and acoustic conditions in the listening room. Some audio listeners mistakenly set these controls to their flat position with the idea that it is unwise to tamper with musical selections.

That isn't the purpose of the flat position at all. Its function is to let you make an easy comparison between the effect of the settings you have chosen for the tone controls and the complete absence of tone controls. In that way you can gauge their effectiveness.

There are two types of tone controls: continuously variable and detent. Continuously variable means you can go through the complete motion of the tone control, smoothly, from start to finish. The detent is a click-stop type and, if it is accompanied by suitable markings around the control, lets you get back to a previously determined setting quickly.

You may also come across a receiver with four tone controls: two for left channel sound and two for right channel sound. However, such an arrangement is the exception.

Filter Switches

There are two types of filter switches, low and high. Some receivers have just one of these; others have both. The low filter removes low frequency noise, such as hum and rumble, or else weakens it considerably. The high frequency filter disposes of high pitched nuisances such as tape hiss and record surface noise. You pay a price when you use these switches, however, for they remove part of the signal along with the noise. Quite often, however, these switches will permit listening to sound that would otherwise be impossible because of a high noise level.

Not all receivers have filter switches but when they are included they appear under various names such as mpx noise filter or hi-blend.

Left-Right Balance Controls

These govern the amount of sound out of the left and right speakers and are needed for a number of reasons. You do not have equal hearing sensitivity in both ears and so you hear better on one side than the other. The amount of sound you can get out of a speaker depends on its efficiency and, even with identical speakers, one may be a bit more efficient than the other. Finally, your listening position may be nearer to one speaker than the other.

Adjust the left-right balance control to give you the best stereo separation. Stereo effect tends to be lost when the sound output of one speaker overwhelms and dominates the other.

Main Balance Control

If your hi-fi system uses four speakers, with two positioned in front and two in the rear, you will need some method for controlling the balance of sound between the front and rear speakers. Through manipulation of the left-right balance control and the main balance control you can adjust the sound output of the speaker system to supply you with the effect you find most pleasing.

Volume Control

The purpose of the volume control is to set the overall amount of sound, regardless of the number of speakers. Advancing the volume control will increase the output from all the speakers, regardless of their location, whether left or right, front or rear. The volume control operates independently of the left-right balance control or the main balance control.

Loudness Control

This is an on/off type and is not generally adjustable. It is often thought of as a volume control, not only because loudness and volume are synonymous, but because the loudness control does affect volume. Physically and operationally the two, volume and loudness, are distinct and different. The volume control is continuously variable; the loudness is either on or off. The loudness control should only be turned on when listening at low volume levels. At such levels human hearing discriminates against bass and treble tones and so the loudness switch helps bring them up.

Tape Monitor Switch

The purpose of this switch is to let you monitor—that is, to listen to—a tape recording in the process of being made. Keep this switch turned off unless you are recording. Otherwise you will get no sound, for example, if you are trying to listen to a broadcast or phono record.

Selector Switch

This control helps you select the sound source you want to listen to. It will have a phono position. Set the switch to this position when you want to listen to phono records. The fm auto position means the receiver can pick up either mono or stereo broadcasts. When set in fm auto, the receiver may have a readout that becomes illuminated automatically when a stereo broadcast is received.

One of the advantages of setting this control to the fm auto position is that it cuts down on interstation noise when you tune from station to station. Unfortunately, it also cuts down on signal strength and you may go right by the station you want. In that case, set the control to fm mono and tune in your station. You will hear the station in mono, but you can then switch to fm auto. If the station then becomes too noisy, switch back to fm mono. Better fm mono than no reception at all.

Muting Control

This control cuts down on signal strength by predetermined amounts. It may be a continuously variable control but a better arrangement is one that is stepped, that is, that has click stops. This lets you get back to a predetermined setting easily. A muting control will let you enjoy more quiet tuning on the fm band without blasting as you go from station to station. You can turn the muting control off once you have selected the station you want.

In some receivers there is no variable muting control but a switching arrangement. With the switch in its on-position you can have a predetermined amount of muting, such as 20 dB for example. You can also use the volume control to cut down on volume level if you wish. But a muting control that is clearly graduated, or that is a switch type, will let you cut down on sound level easily and quickly; something you might want to do if your listening is interrupted by a phone call.

Output Level Control

On some tuners you will find a pair of such controls, one for am, the other for fm. These regulate the tuner's output voltage so as not to overload a following component, such as a preamplifier. It is also used to match the sound volume of the tuner relative to that of other sound sources, such as tapes or phono records.

Tuning Control

This is usually one of the larger controls on the front of a tuner or receiver. In better components it is mounted on the same shaft with a rather heavy, precision-machined flywheel. The use of the flywheel permits easier and smoother tuning. The tuner control governs the movement of a pointer along the am/fm tuning scale or dial.

Most fm dials are now linear, meaning that tuning is no longer crowded at one end of the dial. A linear dial is one that has equal distances between all am and fm frequency markings. A linear dial is easier to read and permits more accurate and quicker tuning.

Auto Dx/Local

Most tuners and receivers have fixed amounts of selectivity for all signals. An auto dx/local switch permits the selection of local stations or distant ones, optimizing the reception of strong or weak stations.

Auto Blend

A weak, noisy stereo broadcast can be improved by using auto blend circuitry that automatically monitors the signal. When the signal being received is extremely weak, the auto blend circuitry reduces high frequency hiss that is often characteristic of such signals, but at the same time helps maintain adequate stereo separation. When the auto-blend control is in its off position, however, maximum stereo separation is obtained.

Switchable FM Muting

This control permits the automatic rejection of stations that are weak or excessively noisy. It also eliminates the distracting hiss that occurs when tuning between fm stations. The muting control should be turned off if you want to tune in weak or distant fm stations.

Stereo Indicators

Not all fm signals are stereo. Actually, quite a few are mono. To avoid the problem of trying to decide whether reception on the fm band is mono or stereo some tuners and receivers are equipped with a stereo indicator. This glows with the word "stereo" when the signal is actually two channels. If you know you are listening to a stereo signal there are adjustments you may wish to make in speaker positioning to let you have maximum stereo effect.

Test Tone Switch

While this is user adjustable it isn't a control. The test tone switch lets you adjust the recording level of an associated tape deck. When this button is depressed a recording calibrator is operated. This calibrator produces a test tone of the level equivalent to 440 Hz, 50% modulation at the output terminal. Using this test tone you can adjust the recording level on your tape deck so that you get a 0 reading on the vu (volume unit) meter of that deck. Using too high a sound level as input to your tape deck when recording will produce distortion.

However, in the case of actual broadcasting, the modulation ratio may possibly be higher, and that, incidentally, may create an excessive recording level.

Headphone Jacks

Although generally not regarded as such, a headphone jack can function as a control when insertion of the jack automatically disconnects the speaker system. Some tuners and receivers have provision for just one pair of headphones; others for two pairs. And some tuners have no headphone jacks at all.

You can use headphones for monitoring your tape recordings or for private listening to any sound source. On some receivers one of the headphone jacks is designed to let you listen to the front speakers and another to let you listen to speakers located at the rear.

FEATURES

These are not all of the controls you may find on a tuner or receiver. Some have more, others fewer, but those that have been discussed are regarded as the most important.

There is often no sharp dividing line between controls and features. A tuning meter, for example, is a feature but it is used to control the functioning of a tuner or receiver. A control, though, is usually thought of as something you must manipulate manually, although there are some controls that are under the supervision of the tuner or receiver.

TUNER VERSUS RECEIVER CONNECTIONS

Because the receiver is actually three components in one, you will find more connections on the rear of a receiver than on a tuner. See Figs. 2–27 and 2–28.

Speaker Terminals

Speaker terminals are supplied in pairs, one pair for each speaker. On some receivers you will find just one pair of terminals; on others two, and on a rare few as many as three. If there is more than one pair they will often be identified by some letter of the alphabet, A, B, or C. If there are two pairs of terminals they will be under the control of a front panel switch, permitting you to select pair A, pair B, or A and B together. If there are three pairs of speakers the procedure is the same, but with a greater possible number of combinations.

Speaker terminals are polarized and may be marked "plus" and "minus," or (+) and (−), may be color coded red for plus and black for minus, or may simply have the plus terminal color coded red. These should be connected to correspondingly identified terminals on the speakers.

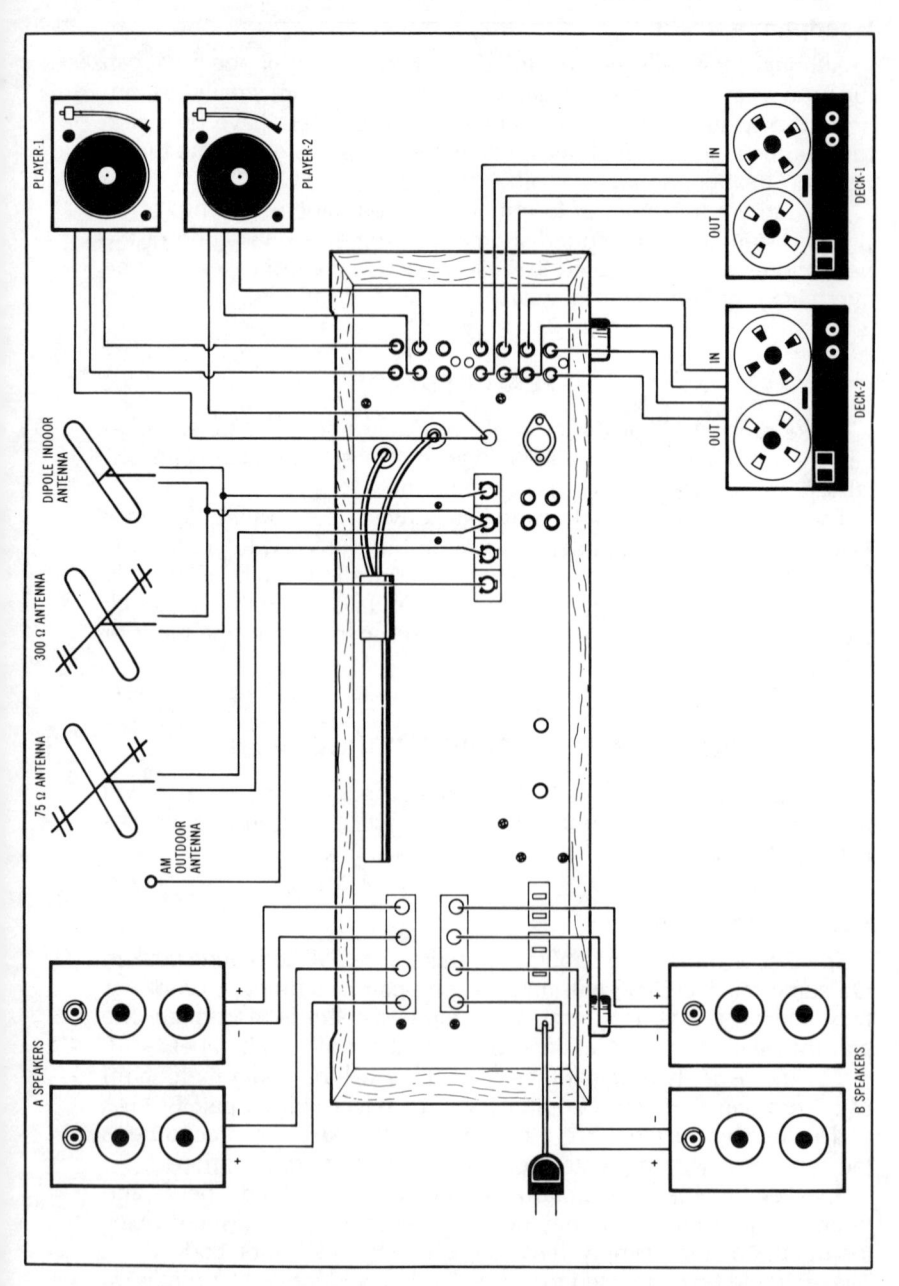

Fig. 2-27. Connections to the rear of a receiver. Cassette deck can be substituted for one of the open reel decks.

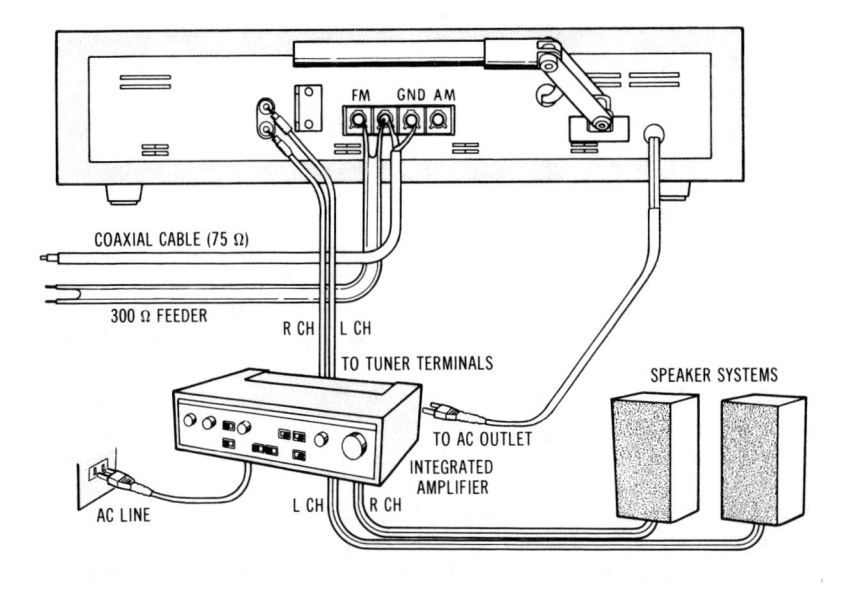

Fig. 2-28. Connections to the rear of a tuner.

Speaker Wire

It isn't necessary to use shielded wire to connect the speaker terminals of the receiver to the speakers. You can buy speaker wire made especially for this purpose but you may also use lamp cord. You can use No. 18 or No. 20 lamp cord. This is stranded wire. Make sure, when cutting the wire, that you do not cut any strands away accidentally. When connecting the wire to the speakers make sure that there is no possibility for the strands to touch an adjacent speaker terminal.

Power Cord

The power cord carries line power to the receiver or tuner and to the switched and unswitched outlets on the rear. If, after installing your hi-fi system you think you hear some hum, try transposing the plug of the line cord in the ac power outlet.

Speaker Protection Fuse

An excessively large, possibly momentary audio current from the output of the audio power amplifier in the receiver could burn out the voice coils of the speakers. This could result from the coincidence of several factors: the peak of a strong bass tone, a frequency at which the input impedance of the speakers will be very low, and the volume control left turned to an advanced setting. To protect the speakers, you will find one or more fuses on the rear apron. If these "blow" or "open" replace them with identical units. Do not use fuses of higher rating since you will then lose the protection these supply. If the fuses open repeatedly or if they need to be replaced often, it would be best to consult the service department of the dealer from whom you purchased the receiver.

Phono Terminals

These are for connection to a record player. In some receivers there are a pair of such terminals so that two record players can be connected. This is an advantage if you are using your hi-fi system for supplying dance music and you want continuous music without waiting.

The phono terminals are connected to the input of the preamplifier section in the receiver. The preamplifier requires a certain level of signal so it may operate properly, something not all phono cartridges can supply. Usually moving magnet and moving iron types of phono cartridges have enough signal output to keep the input of the preamp quite satisfied. Phono cartridges of the moving coil type are characterized by very low signal output, and so for these you may need to use a special transformer, or a pre-preamplifier, also known as a head amplifier.

Aux Terminals

The aux (auxiliary) terminals are the input terminals for signals to be delivered by a tuner, the line output of a tape recorder, or the audio output of a tv receiver.

Recording Output Terminals

These terminals supply an output signal for use with any tape recorder, whether cassette or open reel. There are two requirements, though. The audio amplifier in the receiver must be getting a signal from any one of the several sound sources. This could be an am or fm broadcast, a phono record, or the sound voltages supplied by the microphone. The input selector switch must be set to its corresponding position. This switch chooses the sound source. Thus, if you plan to record from a phono record, set the input selector to its phono position.

Dubbing

Dubbing is the act of transferring a signal from one tape to another. Some receivers are equipped with a tape dubbing switch. This should be correctly set so that the signal can be recorded on the blank tape in the tape recorder.

Monitor Terminals

These are used for connection to the output of a tape deck. For playback you will need to depress a tape monitor control on the front panel of the receiver.

Antenna Attenuator

There is such a thing as excessive signal and it can result in distortion. Excessive signal is possible on the am band if the receiver has a very sensitive front end, possibly by using transistors known as MOSFETs (metal-oxide-semiconductor field effect transistors). If the receiver is also in an area that is close to the transmitter antenna this, plus the use of MOSFETs, can easily produce an overload condition.

WHAT DECIBELS ARE ALL ABOUT

Most of the specifications (specs) of hi-fi components are supplied in terms of decibels, abbreviated as *dB*. A decibel is always a ratio, a comparison of the relative strengths of two voltages, or currents, or powers.

There are various reasons for using decibels. One is that decibels are related to the way we hear. If we start with a very soft sound it doesn't take much additional audio power to make us notice the increase. But, as the sound becomes louder, it takes more and more audio power for us to notice any change. How much of a power increase also depends on what we are listening to. If we are listening to a flute being played softly, then an increase of 1 dB in acoustic power will be noticeable; but if what we are hearing is a group of instruments, then about a 3 dB increase will be needed. But as the sound becomes still louder it takes more and more of an increase in acoustic power for us to note the difference. Our ears, then, aren't linear devices.

We can use decibels to compare not only currents, voltages, and power, but sound as well. Because decibels are always ratios, that is, a comparison of one value with another, we must have a starting point for sound. That starting point is known as the threshold of audibility, the point at which we just about begin to hear. We arbitrarily set this point as our reference and it is this reference we use for comparison purposes. If we say that a sound has a value of 40 dB, this means it is 40 dB compared to the threshold of audibility, or 0 dB.

Table 2–1 is a listing of decibels and their power ratios. Note that each ratio is obtained by multiplying the preceding ratio by 1.259. Thus, 3 dB is the same as a ratio of 1.995. To obtain 4 dB we multiply this ratio, 1.995, by 1.259.

$$1.995 \times 1.259 = 2.5117$$

This has been rounded off to 2.512 in Table 2–1 and corresponds to 4 dB.

It would seem that decibels are complicated but they help us avoid some tedious arithmetic. If we use ratios we must multiply by an odd number such as 1.259. With decibels we just add. It is easier to go from 3 dB to 4 dB than to multiply 1.995 by 1.259.

Another advantage of using decibels instead of ratios is that in some instances ratios can involve very large numbers. Table 2–2 supplies an idea of the number sizes. In hi-fi it is possible to become involved in power ratios of a million to 1 or ten million to one. It is much easier to handle numbers such as 60 dB or 70 dB.

To show how practical decibels are, let us suppose you have a 30 watt amplifier and decide to replace it with a 40 watt unit because you think it will give you louder sound. The ratio of 40 to 30 is 40/30, or 4/3, or 1.333, or between 2 and 3 dB. The difference in sound would be barely perceptible. Of course there are other factors that would make a 40 watt amplifier superior to one having an output of

Table 2–1. Decibels and Their Power Ratios

Decibels	Power Ratio
0	1.000
1	1.259
2	1.585
3	1.995
4	2.512
5	3.162
6	3.981
7	5.012
8	6.310
9	7.943
10	10.000

Table 2–2. Decibels and Their Corresponding Power Ratios

Decibels	Power Ratio
0	1
10	10
20	100
30	1000
40	10,000
50	100,000
60	1,000,000
70	10,000,000
80	100,000,000
90	1,000,000,000
100	10,000,000,000

30 watts, but an increase in apparent sound loudness isn't one of them.

We can also use decibels for making a comparison between two voltages as indicated in Table 2-3. Each ratio is obtained by multiplying the preceding ratio by 1.122.

Table 2-3. Decibels and Voltage Ratios

Decibels	Voltage Ratio
0	1.00
1	1.122
2	1.259
3	1.413
4	1.585
5	1.778
6	1.995
7	2.238
8	2.512
9	2.818
10	3.162

SPECIFICATIONS

Specifications, more usually referred to as specs (Fig. 2-29), detail the electronic performance characteristics of a high fidelity component. Unlike features which may vary from one tuner or receiver to another, specs are guidelines to all components. This doesn't mean that all spec sheets are alike or that they all contain the same detailed information. Some specs may be omitted because the manufacturer may not regard them as important or because they may reflect unfavorably on the product or simply because they may have been overlooked.

Since specs are supplied by the manufacturer they may be suspect since they are self-serving. Generally, though, specs are fairly accurate. In any event they are often verified by independent testing agencies and there have been some instances in which manufacturers understated the performance specs of equipment.

In either a tuner or receiver separate specs are supplied for the am and fm sections. The specs described here are those for tuners and for the tuner portions of receivers. Specs for preamplifiers and power amplifiers used in receivers will be covered separately in the chapters set aside for those components.

Root-Mean-Square (rms)

The abbreviation for root-mean-square is rms; and rms is one way of measuring a varying voltage, current, or power. In this technique, a number of instantaneous values are taken over a complete cycle of

[FM SECTION]
* 50dB Quieting Sensitivity: 75usec. 14.2dBf (2.8µV)   50usec. 15.3dBf (3.2µV)
* IHF Usable Sensitivity: 10.3dBf (1.8µV)
* Signal to Noise Ratio at 65dBf: 75dB
* Frequency Response: 30Hz — 15kHz (±1dB)
* Total Harmonic Distortion: [mono]   [stereo]
    100Hz 0.06% (wide)   0.15% (wide)
    1kHz   0.06% (wide)   0.1% (wide)
    6kHz   0.12% (wide)   0.2% (wide)
    1kHz   0.2% (narrow)   0.5% (narrow)
* Capture Ratio: 0.9dB (wide)   1.9dB (narrow)
* Image Response Ratio: 85dB
* IF Response Ratio: 90dB
* AM Suppression Ratio: 62dB
* Stereo Separation: 45dB (wide, 100Hz),   48dB (wide, 1kHz)
  40dB (wide, 10kHz),   30dB (narrow, 1kHz)
* Spurious Response Ratio: 100dB
* Adjacent Channel Selectivity: 12dB (narrow, ±200kHz)
* Alternate Channel Selectivity: 80dB (narrow, ±400kHz)   60dB (narrow, ±300kHz)
  48dB (wide, ±400kHz)
* Subcarrier Product Ratio: 65dB
* SCA Rejection Ratio: 60dB
* Output Voltage: 1V
* Muting Threshold: 10µV

[AM Section]
* IHF Usable Sensitivity: 250µV/m
* Image Ratio at 1MHz: 50dB
* IF Rejection Ratio at 1MHz: 40dB
* Signal to Noise Ratio: 50dB

[General]
* Power Consumption: 2.8VA (CSA rated)
* Dimensions: 490(W) x 405(D) x 180(H)mm
  (19-5/16'' x 15-15/16'' x 7-3/32'')
  (including Legs, Rear Protrusions & Knobs)
* Weight: Net   14.9kgs (32.8 lbs.)
  Gross 16.5kgs (36.3 lbs.)
* Additional Features: Speaker Selector Switch, Headphone Jack,
  Tape Dubbing Circuit, Tape Monitor Circuit,
  Peak Indicator Sensitivity Selector, FM
  IF Bandwidth Selector, FM Muting Off Switch,
  Mode Switch, AM Loop Stick Antenna, Protection
  Circuit, etc.

Specifications and appearance design subject to change without notice.

Fig. 2-29. Receiver specifications.

the wave. These values are then squared, that is, multiplied by themselves. The results are added and then an average figure is calculated. The final answer is the square root of this average; rms is described in more detail in Chapter 4.

Weighted Root-Mean-Square (wrms)

The human ear has a varying sensitivity to sounds of different frequencies. To take this characteristic into consideration when making test measurements the results are weighted, or changed accordingly. Thus, a wrms measurement is a subjective result, not the actual result as produced by test equipment.

Sensitivity

Sensitivity is the ability of the receiver or tuner to reach out and pick up weak, possibly distant stations. It is the smallest amount of broadcast signal which can be converted into a satisfactory amount of sound out of the speakers. So what we are concerned with here is signal strength. You will also find sensitivity indicated as IHF sensitivity indicating that the measurement was made in accordance with testing guidelines recommended by the Institute of High Fidelity.

An fm receiver or tuner is noisy in the absence of a signal. When a signal is received it will suppress noise and the amount of this suppression is supplied in dB. This decibel number indicates how much louder the signal is than the noise. Figures of 30 dB and 50 dB are often used and so the spec is sometimes called quieting sensitivity instead of simply sensitivity. Fifty dB of quieting of the noise is far preferable to 30 dB. A spec that doesn't include the amount of quieting isn't very meaningful.

For the fm section look for 3.0 microvolts (3 μV) for 50 dB quieting sensitivity. This spec means that if the antenna input of the tuner or receiver is presented with a signal that has a value of three microvolts or three one-millionths of a volt it will produce a useful, listenable sound from the speakers and that the noise will be suppressed by 50 dB.

Figures of 3 μV are quite good but you will find some tuners and receivers indicating specs down to as low as 1.6 μV. Lower figures are better, provided they also produce 50 dB of noise quieting. From a practical viewpoint you may not notice the difference between 3.0 μV and 1.6 μV. A tuner or receiver with a good outside fm antenna and having a quieting sensitivity of 3.0 μV will supply better performance than a tuner or receiver with a sensitivity of 1.6 μV but using an antenna tacked along a baseboard.

Table 2-4. Tuner or Receiver Input Sensitivity—μV vs dBf

μV	dBf	μV	dBf	μV	dBf
1.5	8.71	2.6	13.478	3.7	16.548
1.6	9.28	2.7	13.804	3.8	16.776
1.7	9.8	2.8	14.134	3.9	17.012
1.8	10.3	2.9	14.436	4.0	17.2
1.9	10.77	3.0	14.74	5.0	19.17
2.0	11.198	3.1	15.01	10.0	25.19
2.1	11.6	3.2	15.3	30.0	34.74
2.2	12.04	3.3	15.7	40.0	37.23
2.3	12.424	3.4	15.8	50.0	39.17
2.4	12.8	3.5	16.07	100.0	45.19
2.5	13.15	3.6	16.3	1000.0	65.19

Decibels are used for the measurements of either voltage or power ratios. Sensitivity can be measured in either. If a sensitivity figure is specified in dB it is understood that the reference is to a voltage measurement. To indicate that the measurement has been made in terms of power, the abbreviation dBf is used instead. A spec may indicate sensitivity numbers in either dB or dBf or both. Table 2–4 is a comparison of dB and dBf figures. The amount of quieting, however, is always supplied in dB only. Thus, a spec for a quality receiver's sensitivity may be: 50 dB quieting sensitivity: 3.0 μV (14.8 dBf).

Sensitivity figures are sometimes supplied separately for the am section of the tuner or receiver. No quieting figure is included. The spec simply indicates the amount of signal voltage needed for useful speaker output. Look for a minimum figure of 300 microvolts. Lower figures are better.

Signal-to-Noise Ratio

This spec is a useful comparison between noise voltage and signal voltage and is measured in dB. Technically, it is a comparison between a 400 Hz fully modulated signal and the noise component. Signal-to-noise ratios are poorer for fm stereo than for mono. Consequently, some manufacturers supply only one spec without indicating whether it is for mono or stereo reception. However, if you see just one spec you can be quite sure it is for mono.

For signal-to-noise ratios, higher numbers are always better. Look for a minimum of 60 dB mono, and a bit less than that for stereo. Anything above these numbers is definitely acceptable. Sixty dB is a ratio of 1,000 to 1. Refer to Table 2–5.

Capture Ratio

It is possible for two fm stations, one weaker and somewhat distant, and one stronger and local, to be operating on identical or nearly identical frequencies. Capture ratio is the ability of the turner or receiver to suppress the weaker signal while responding only to the stronger.

Table 2–5. Signal-to-Noise Ratios vs Decibels

S/N	Decibels
3.2	10
10.0	20
32.0	30
100.0	40
316.0	50
1000.0	60
3162.0	70
10000.0	80

Capture ratio is also indicative of performance characteristics. Thus, in the case of multipath reception, a receiver or tuner with good capture ratio will be able to suppress the multipath signal and favor the direct transmission. In so doing it will reduce or eliminate multipath distortion. And, as a further benefit, a good value of capture ratio is related to the component's suppression of random noise.

Capture ratios are measured in dB and the lower the figure, the better. Look for a value of 3 dB, but some units are even better and have values of 2 dB or lower.

Frequency Response

Frequency response is a measure of the range of audio frequencies that will be selected and amplified by the tuner or receiver. An fm broadcast station transmits a range of audio signals extending from 30 Hz to 15 kHz. What the receiver will do with these frequencies is another matter. A quality tuner or receiver will be able to handle this range of audio signals with a minimum amount of deviation. The deviation is expressed in dB and the smaller the deviation the better. A quality frequency response spec would read: 30 Hz to 15 kHz within plus/ minus 1 dB. Any other spec whose frequency response is narrower, possibly 50 Hz to 13 kHz and whose deviation is greater, possibly plus/minus 2 dB or 3 dB, is definitely not as good.

Frequency response figures for am broadcast signals are generally not supplied.

Selectivity

This spec is a measure of the ability of the tuner or receiver to choose a desired station, and that station only, from among all the signal voltages from various stations that are present on the antenna. It is also a measure of the ability of this component to reject any signals whose frequencies are close to that of the station being tuned in. For this reason the spec is sometimes described as alternate channel selectivity. Expressed in dB, higher values are better. Look for a minimum of 50 dB, although a quality tuner or receiver will have a spec as much as 65 dB, and you will find some with higher figures.

Whether alternate channel selectivity is important depends on your location. If you are in an area crowded with fm stations then alternate channel selectivity numbers are something to look for when choosing a receiver or tuner. If you are in a fringe area, or a region that has relatively few fm stations, then the spec is less significant.

Adjacent channel selectivity is a related spec. Look for values of about 5 dB, but higher figures are better.

There are a number of specs relating to the ability of the tuner's or receiver's rejection of unwanted signals. These include image rejection, spurious response or rejection, and am suppression.

Image Rejection

Every superheterodyne receiver contains an ac generator called a local oscillator. The signal produced by the local oscillator mixes, beats with or heterodynes the incoming broadcast signal, producing an intermediate frequency or if, as described earlier. But it is also possible for an unwanted broadcast signal to mix with the local oscillator signal resulting in an intermediate frequency that will be accepted by the if stages. However, just one of the broadcast signals is the one that is wanted. Image rejection is the ability of the tuner or receiver to reject the unwanted signal. Image rejection is supplied in dB—larger numbers are better. A quality tuner or receiver will specify 75 dB or higher.

Spurious Response

The local oscillator in the component is a miniature transmitter and so its signal can sometimes escape through the front end. This self-generated signal can cause interference to neighboring radios or tv sets and, equally bad, be picked up by your own antenna. It will then be treated and handled just as though it was a legitimately broadcast fm signal. Spurious response is the name given to the tuner or receiver's reaction to this signal. Rejection of the signal is measured in dB and the larger this number the better. A ratio of 80 dB is good, but 90 dB is much better.

AM Suppression

Am or amplitude modulation in this case does not refer to am broadcasting but rather to noise signals. Noise signals behave as though they were radiated by an am station. Notorious offenders are neon lights, electric advertising signs, motors, generators, vacuum cleaners, fluorescent lights, and automobile motors. Am suppression is measured in dB and higher values are always better. Look for a minimum of 40 dB, but 50 dB or 60 dB are much better.

Stereo Separation

Stereo separation (also called channel separation) is a measure of the ability of the multiplex decoder in the fm tuner or receiver to separate the left and right channels of sound in stereo broadcasts. The FCC requires that stereo separation must be 30 dB between left and right channels throughout the audio range. Some tuners and receivers do better than that, though. Channel separation in the middle audio frequency range from 400 Hz to 1,000 Hz is important so here you should look for about 40 dB. A quality tuner or receiver would show 38 dB at 100 Hz, and 45 dB at 1 kHz and 33 dB at 10 kHz. Higher figures are better.

Output Voltage

The amount of output audio signal available from a tuner is important because it must be enough to satisfy the signal voltage input requirements of the following preamplifier stage. For fm the output voltage should be a minimum of 500 mV (500 millivolts or a half a volt). For am look for a minimum of 200 mV, but 250 mV (quarter of a volt) is better.

Integrated Circuits

In the early days of transistors these units were individually mounted on printed circuit boards. To save space and increase operating speed transistors and diodes were subsequently etched on a tiny bit of silicon, known as a chip, with the arrangement called an integrated circuit or IC. In the IC manufacturing process chips are made several hundred at a time on a single silicon wafer about two inches in diameter. The individual chips are then cut apart, but are tested first.

Subsequently, chips underwent a microminiaturization process. The number of transistors and diodes on an IC chip increased from about 15 or 20 to as many as 500. These were called LSIs (large scale integrated circuits). A VLSI or very large scale integrated circuit has about 5,000 transistors on a single chip.

MODERN RECEIVERS

The latest trend in receivers is pictured in Fig. 2–30. This 90-watt per channel, digital-synthesized receiver also has a wireless remote control.

Fig. 2-30. The Luxman Model RX-103 Receiver and remote control. *(Courtesy Luxman, Div. of Alpine Electronics of America)*

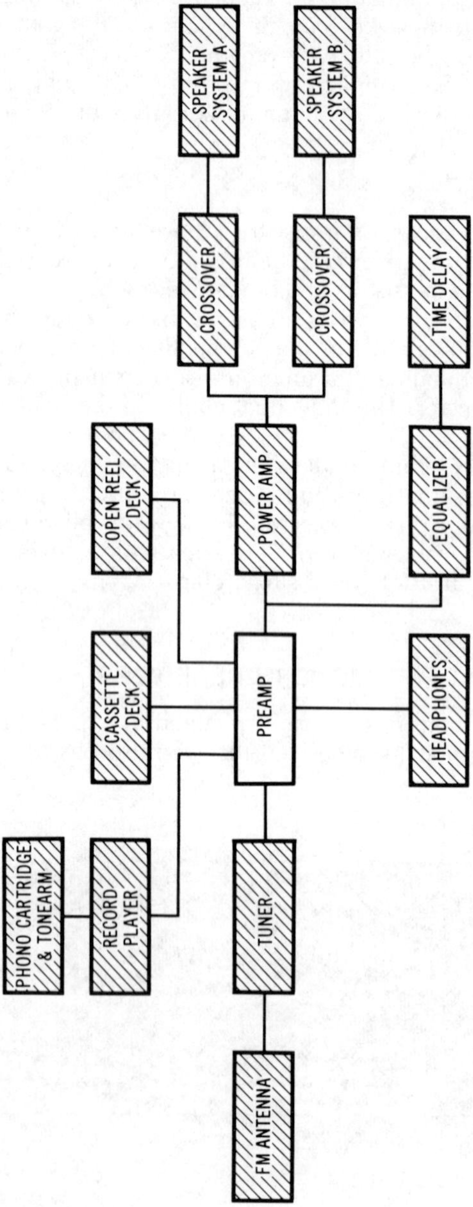
SPEAKER SYSTEM A
SPEAKER SYSTEM B
CROSSOVER
CROSSOVER
TIME DELAY
OPEN REEL DECK
POWER AMP
EQUALIZER
CASSETTE DECK
PREAMP
HEADPHONES
PHONO CARTRIDGE & TONEARM
RECORD PLAYER
TUNER
FM ANTENNA

3

Preamplifiers

Every hi-fi system can take advantage of some or all of a number of different sound sources. These could include signals from tuners, phono records, tape, whether cassette or open reel, and microphones. In all instances the signals supplied by these sources are audio voltages, but before these voltages can be applied to speakers, they must be amplified.

The preamplifier strengthens all these signals, but it does more than just that; for the preamplifier selects the sound source to be played and determines the mode of operation—mono or stereo. The preamplifier also alters the ratio of bass to treble or treble to bass via its tone controls; it is used to control overall sound volume and, via its balance control, the amount of sound from each speaker (Fig. 3–1).

PREAMPLIFIER INPUTS

The preamplifier is the input point for all audio signals, whether those signals are from the output of a tuner, a phono cartridge, or a tape deck. The number of available input connections on the rear of the preamplifier will vary, with more connections available for higher priced components. Thus, inputs for different sound sources for a preamp can start with a minimum of 1 phono, 1 tape, and 1 tuner. Average inputs are often 2 phono, 2 tape, tuner, and aux (auxiliary), but some preamps are better equipped in this respect, offering 3 phono, 3 tape, tuner, 2 mic (microphones), and 2 aux.

INPUT SIGNAL STRENGTH

Each sound source input to the preamplifier is an audio frequency and each can have a different level, that is, a different strength. A

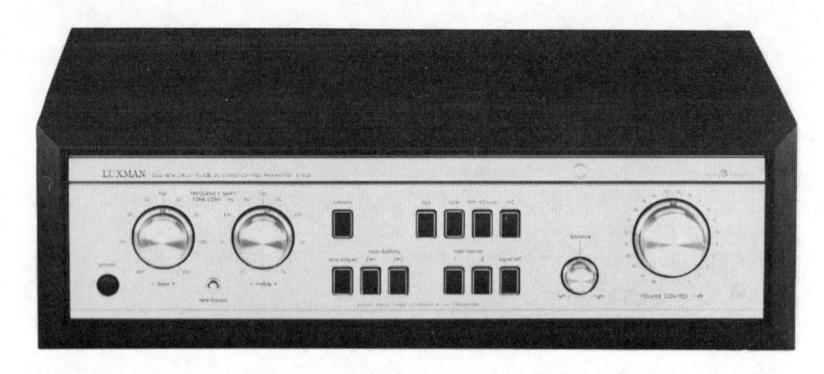

Fig. 3-1. A stereo preamplifier/control center. Special features include: direct coupled design, moving coil cartridge input, frequency-shift type controls, tone control defeat switch, dual tape monitoring, tape dubbing, phono impedance selector switch, subsonic filter, and balance control. *(Courtesy Luxman, Div. of Alpine Electronics of America)*

moving magnet (sometimes abbreviated as mm) phono cartridge will supply one or more millivolts (thousandths of a volt). A moving coil cartridge (sometimes abbreviated as mc) will supply a signal that is generally less than a millivolt, although there are some exceptions. A tuner or tape deck will deliver a half volt to a volt or more. But the 1 volt of audio from the tuner is a thousand times the strength of the 1 millivolt from a phono cartridge, so the preamp input must be capable of handling a wide range of input signal strengths.

The initial stage of the preamp brings all signals up to approximately the same level before the signal is moved ahead for further amplification. Ideally, the preamp will have a flat frequency response over the entire audio range and will amplify all signals at all frequency points equally. For a quality preamp this is so, except for the phono input which requires equalization.

One input signal source, that from a moving coil cartridge, supplies such a small amount of signal strength that the signal must first be fed into a special amplifier. Known as a pre-preamplifier or head amplifier this additional component strengthens the very weak signal supplied by the moving coil phono cartridge to the point where it is acceptable to the moving coil input circuit of the preamplifier. The head amplifier can be a separate component but it can also be supplied as an integral part of the preamplifier. In some instances a special step-up transformer is used instead of the pre-preamplifier.

PHONO EQUALIZATION

The amount of movement of the cutting stylus used in making a phono record depends on the frequency of the audio tone at any moment. For bass tones the excursion of the stylus is wide, and, while this movement could be accommodated by the disc, it results in fewer record grooves. To prevent this, the recording engineer deliberately reduces the strength of bass tones. Bass frequencies aren't changed; simply the amplitude of these notes.

Cutting a phono record produces surface noise, the result of the movement of the cutting stylus. To overcome this annoyance, treble tones are deliberately strengthened. Again, the frequency of the treble range isn't changed, but the strength or amplitude of tones in this range is made much stronger. Treble boost is exactly the same as the treble boost used by fm broadcast stations.

As a result, no phono record is cut "flat." for what we have is deemphasis of the bass and preemphasis of the treble region. The amount of bass cut or attenuation at the lowest audio frequency, 20 Hz, is about 15 dB. The cut gradually weakens until 1 kHz, and from that point on we get gradually increasing treble boost until it is about 15 dB at 20 kHz.

Because of the way in which phono records are cut some provision must be made in the preamplifier to neutralize or overcome this prior preemphasis and deemphasis. An equalizer amplifier is used in the preamplifier just for this purpose. The equalizer follows a procedure that is the opposite of the one used in cutting the record. It emphasizes bass tones by about 15 dB at 20 Hz and attenuates the treble tones by the same amount at 20 kHz. A graph of the curve used when cutting the record is an approximate mirror image of that used in playback. The equalization curves are known as RIAA or Recording Industry Association of America curves. And so it is the function of the equalizer amplifier in the preamp to produce a flat response from phono records. This is shown in Fig. 3-2.

EQUALIZATION ACCURACY

Not all phono amplifiers in preamps follow the RIAA equalization curve as accurately as they should. The equalization spec should indicate the amount of deviation, and this is supplied in decibels. A deviation of only plus/minus 0.2 dB is considered good. The frequency range, generally 20 Hz to 20 kHz, should also be specified.

NONLINEAR VERSUS LINEAR INPUTS

The input signal from a record player is known as nonlinear since the signal must be corrected. Of course the signal is linear after the equalizer has had a chance to do its work. The audio output of the equalizer is fed into the audio voltage amplifier section of the preamplifier where it is handled as a linear signal.

(A) Phono record's recording curve.

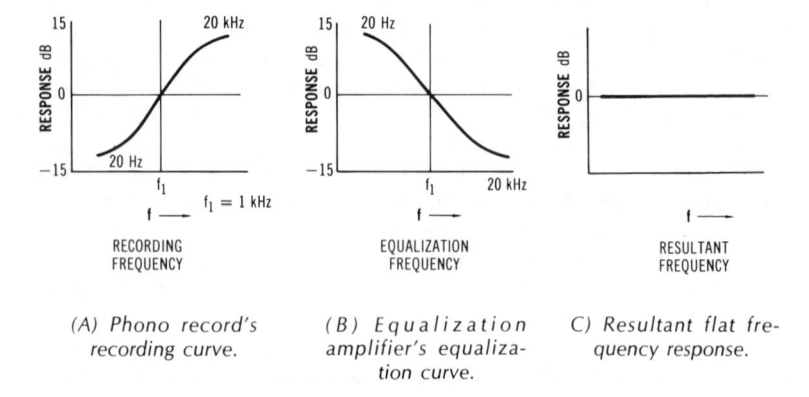

(B) Equalization amplifier's equalization curve.

C) Resultant flat frequency response.

Fig. 3-2. Neutralizing preemphasis and deemphasis. Symbol f_1 in the graphs is 1 kHz.

The input signals from tape decks, from the tuner, or microphone(s) do not require equalization. There is no prior processing of these signals and so, when they are brought into the preamplifier they are made to bypass the equalizer amplifier.

PREAMP FEATURES

Preamp features vary from one manufacturer to the next but here are some that are offered: two-way tape dubbing (some have only one-way tape dubbing); cuing facilities; selectable time control turnovers; separate tone controls for left and right channels; headphone jacks (some have provision for two pairs of headphones); rack mountable; remote control; stepped attenuator volume control; a head amplifier for moving coil cartridges; LED level display; feather touch controls; illuminated flow chart on the front panel showing the movement of the signal from input to output. You may also find a preamp having a built-in graphic equalizer.

DUBBING

Dubbing is the process of transferring the signal from a previously recorded tape to a blank tape. You can do tape dubbing if you own two tape decks. Some preamplifiers are able to accept two tape decks and with some you will not only be able to dub but to listen to a completely different sound source, such as fm or a phono record while waiting for the dubbing to be completed.

TAPE MONITORING

The front panel of the preamplifier will have a tape monitor switch with two positions. Sometimes this switch is simply marked *tape monitor* or it may be marked *source* and *play*. Ordinarily, when listening to all other sound sources this control must be in the *source* position. For tape playback, put the switch in its *play* mode.

CONTROLS

At one time a preamplifier was the complete control center for a hi-fi system, but now you will find many more controls on the other components as well. As a result, in some instances, especially when components of different manufacturers are used, there may be some duplication of controls.

Volume Control

This control sets the overall volume level of all the speakers, whether you use front speakers only, front and rear speakers, or have another set of speakers in an adjacent room. Some speakers come equipped with sound level controls so you can set limits on volume, especially important if you plan to use speakers in locations away from the preamplifier.

Balance Control

You may find a pair of these, sometimes individually, sometimes in concentric form. The advantage of the concentric form, with one control nesting in the other, is that it demands less front panel space and makes that panel look less cluttered. The purpose of the balance control is to adjust left-channel speaker sound and right-channel speaker sound independently. Once you have made these adjustments you can then set the volume control to give you the overall sound level you want.

There are several reasons why a balance control is needed. One of the speakers, either left or right, may be positioned so that it cannot take advantage of the sound reflecting properties of walls and floor. Further, you cannot assume you hear equally well with your left and right ears, a factor that becomes more evident as you get older. Still another reason is that you may not be seated or you may not be able to be seated equally distant from each of the speakers.

Fader Control

If you have rear speakers in addition to those in the front, a fader control will let you adjust the sound balance between the front and rear speakers.

Tone Controls

On most preamps you will find a pair of tone controls, one for bass, the other for treble. Some are continuously variable; others are click stop detent types. Nearly all have a zero or flat position, a point at which the tone controls are inoperative. On some preamps you will find a pair of tone controls for left-channel sound and another pair for right channel sound. These are usually in the form of concentric double knobs.

The bass and treble tone controls may be separate for the left and right sound channels with one set for the left speaker and another for right channel sound. This means there will be a total of four such controls, but to keep the front panel of the preamp from becoming too cluttered, the tone controls may be nested.

In most preamps there will be only two tone controls: a bass and treble. The bass control is for left and right speaker sound, and so is the treble tone control.

Some tone controls are slider types that can be operated individually or in tandem. And while preamps are generally tone control equipped, it is possible to get a preamp that omits them entirely.

Stereo Image Control

Preamplifiers, like other components in the hi-fi system, do not all have the same number or type of controls. One that is unusual is a control used for adjusting the stereo image. The control governs the sound so that it ranges from mono to stereo and is continuously adjustable. The effectiveness of this control depends on the phono record being played. With some it can help produce a wide stereo image; less so with others.

Left/Right Reverse

With this switch you can permit left/right channel sound to come out of left/right located speakers as you wish. Ordinarily, left channel sound is supplied from speakers at the left in your listening room; right channel sound from speakers at the right. But with reverse you can switch this usual listening order.

Loudness Control

This is a commonly misunderstood control and is often mistaken for the volume control. At low volume levels your ears aren't as sensitive to extremely low and high frequencies. If you're in the mood for some quiet listening, turn the loudness control on (Fig. 3–3). In this way you will be sure to hear the bass and treble tones. For ordinary listening with the volume control turned up, keep the loudness control turned off, otherwise the sound will be boomy down at the bass end and shrill in the treble region.

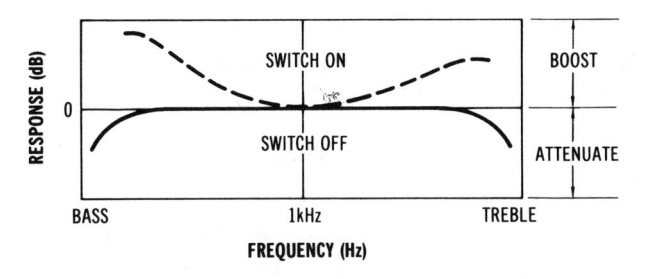

Fig. 3-3. The dashed-line curve indicates frequency response with loudness control switch on.

The loudness control lets the preamp supply extra bass plus a slight boost in the treble when you use your high fidelity system at low sound outputs. However, you can also compensate by using your bass and tone controls while keeping the loudness control turned off.

Contour Control

The problem with the loudness control is that it is an on/off type and so you either have the loudness control active or else it is off. A contour control also supplies a loudness function, but does so in graduated steps. The procedure is to set the volume control to the listening level you wish and then to adjust the contour control to bring up bass and treble accordingly. The contour control is a step-type control so that you can have a number of different positions. For each position the amount of listening equalization is adjusted automatically in the preamplifier. The contour control is least effective at frequencies between 500 Hz and 2 kHz and becomes more effective as the limits of the audio spectrum are reached, that is, at 20 Hz and 20 kHz.

On/Off Power Switch

The on/off power switch means that the amplifier has control of ac line power independently of any other component. The line plug can be connected to an ac outlet or to an unswitched outlet if one is available on the tuner. While the plug can be connected to a switched outlet on the rear of these components, it wouldn't be practical to do so. If, for example, the plug was inserted in the switched outlet of the tuner, then the tuner's power switch would need to be turned on, even though you might only be planning to use the record player or the tape deck.

Selector Switch

Since the preamplifier works with signals from a number of different sound sources it must be able to choose from among them. The selector switch gives you the choice of listening to one or possibly two phono inputs, a tuner, or an auxiliary signal possibly supplied by a cassette deck or an open reel deck.

FILTERS

The preamp can be equipped with switches for cutting filters in and out of the preamp circuitry. Filters are often characterized by the amount of attenuation or rolloff they supply, usually either 6 dB or 12 dB per octave. Less often a preamp will have a filter of 18 dB/octave in the bass region. Sometimes the attenuation rate, also called a slope, is made variable and is under the control of the user. The higher the slope rate, that is, the greater the attenuation in terms of dB/octave, the more effective the filter. A filter with a slope of 20 dB or 25 dB would be more precise than a 12 dB/octave filter.

The point at which low and high filters take action varies from one preamp to another. A high filter might begin at about 10 kHz, a low filter around 100 Hz or lower. The starting point depends on the attenuation rate per octave. Those with high attenuation rates, possibly 12 dB instead of 6 dB per octave, can begin working at lower and higher frequency points because of the steepness of their attenuation slope (Fig. 3–4). Such filters are advantageous since the frequency range of the audio signal that is unaffected by filter operation is wider.

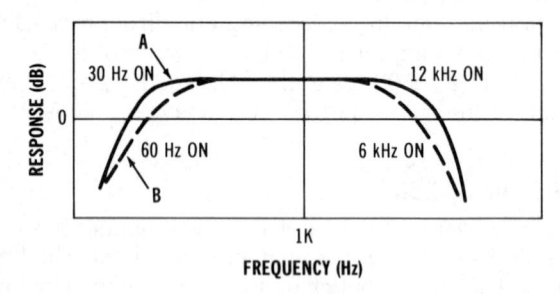

Fig. 3-4. The rolloff in curve A is sharper than curve B and permits a broader frequency response. Curve A could be 12 dB/octave, curve B only 6 dB.

Low Filter

If this control appears on the front panel of a preamplifier it may have some other name such as bass, low cut, or rumble filter. It is intended to attenuate audio signals below some selected low frequency such as 50 Hz or 100 Hz or lower. Its purpose is to minimize hum or rumble as well as subsonic signals, possibly produced by a record player.

The problem with the low filter is that it removes wanted audio frequencies at the same time it lowers hum or rumble. A better approach is to find the cause of hum or rumble and eliminate it at the source. By its action the low filter cuts down on the working audio frequency range of the hi-fi system. Keep this switch turned off if at all possible.

High Filter

The high filter, also called scratch or high cut, comes under the same adverse criticism as the low filter. The high filter attenuates treble frequencies above some selected point such as 8 kHz or 10 kHz. Its main purpose is to weaken treble noise such as tape hiss or phono record surface noise. Again, it is better to attack the problem at its source since the high filter removes wanted treble tones at the same time that it weakens unwanted signals. Keep this switch turned off if possible. However, some high filters have little audible effect on the music; that is, they do their job quite well.

OCTAVES

An octave is a doubling or a halving of frequency beginning with some selected frequency point. If we start at 16 Hz, then from 16 Hz to 32 Hz, with the starting frequency doubled, would be one octave. Similarly, from 32 Hz to 64 Hz would be still another octave, and from 64 Hz to 128 Hz the next octave. Or, we can start at some other frequency, possibly 300 Hz and move lower in frequency. We would have one octave at 150 Hz, another octave by going to 75 Hz.

HEAD AMPLIFIERS

Some phono cartridges, such as the moving coil (mc) types, have low signal outputs and so the audio voltages they produce aren't enough to meet the minimum phono sensitivity requirements of most preamps. As a result some manufacturers are including a pre-preamplifier, also known as a head amplifier, into preamplifiers and integrated amps. The purpose of the head amplifier is to strengthen the low output of the moving coil cartridge to the point where it can swing the input of the preamplifier. Some audiophiles claim that the moving coil phono cartridge produces better sound than moving magnet phono cartridges, currently the more widely used type.

PREAMP SPECS

Most manufacturers issue a set of specs for their preamplifiers. If the preamp is part of a receiver or an integrated amp, these specs are often supplied as a separate listing. Fig. 3-5 shows a representative spec sheet.

Phono Input Sensitivity

With an fm tuner or receiver manufacturers try to make the input sensitivity as good as possible because the input signal is not only very small, in the order of microvolts or millionths of a volt, but we have no control over it. We must accept the broadcast signal as it is and the tuner or receiver must work with what it gets.

With a preamplifier we have a different state of affairs. We deliver to that preamplifier a signal from the tuner, or a tape deck, or a record player, or a microphone. These are all signal producers and we do have control over the amount of signal they supply. So we now have the responsibility of making sure that these signal sources de-

Output Voltage:	Pre. Out;	typical 1V, max. 18V
	Rec. Out;	typical 130mV, max. 18V
Total Harmonic Distortion:	phono MM;	no more than 0.007% (Rec. Out. 3V. 20 — 20 kHz)
	aux, tuner;	no more than 0.007% (Pre. Out. 3V. 20 — 20kHz)
Frequency Response:	phono;	20Hz — 20,000Hz ±0.3 dB (Rec. Out)
	aux, tuner;	10Hz — 80,000Hz —0.5 dB
Signal-to-Noise Ratio:	phono MC;	74 dB (IHF-A weighted, input short-circuited)
	phono MM;	88 dB (IHF-A weighted, 10mV input)
	aux, tuner;	100dB (IHF-A weighted, input short-circuited)
Input Sensitivity: (Pre. Out; 1V)	phono MC;	200 µV
	phono MM;	2.0mV
	aux, tuner;	130mV
Input Impedance:	phono MC;	390 ohms
	phono MM;	50K ohms
	aux, tuner;	60K ohms
	monitor;	60k ohms
Output Impedance:	Pre. Out;	100 ohms
	Rec. Out;	100 ohms
Phono Overload Voltage:		300 mV
Tone Control:		Bass Turnover Frequencies; 150Hz, 300Hz,& 600Hz
		Treble Turnover Frequencies; 1.5kHz, 3kHz & 6kHz
Linear Equalizer:	Up-tilt;	2 points
	Down-tilt;	2 points
Low Boost:		70Hz & 150Hz (6dB max.)
Subsonic Filter:		15Hz & 30Hz 6dB/oct.
High Cut Filter:		9kHz & 15kHz 6dB/oct.
Additional Features:		Dual Tape Monitoring Circuit, Tape Dubbing Circuit, Built-in Step Up Transformer for MC cartridge, Tone Bypass Switch
Power Consumption:		17W

Fig. 3-5. Representative preamplifier specification sheet.

liver the right amount of signal input to the preamp; not too much, not too little.

The phono input sensitivity of a preamp is the minimum amount of signal input that it requires from a phono cartridge. Phono input sensitivity is measured in millivolts and ranges from a little less than 1 mV to about 4 mV. A preamp may have inputs for two phonos, rather than just one but we cannot assume that their input sensitivities will be identical.

You can, of course, deliver less than the minimum amount of audio signal to the input of the preamp. No damage will result, but the audio power amplifier following the preamp will simply not deliver its amount of rated audio power output.

Some preamps are equipped with phono inputs having adjustable sensitivity which can be operated by a front panel control. An input of 10 millivolts is conventionally used in running comparison checks between preamps.

For phono input sensitivity, the lower the figure the better. However, an extremely sensitive phono input also means an increased possibility of picking up more electrical noise, such as the surface noise of a phono record. A preamp having a highly sensitive phono input also demands more care in the shielding of the phono cartridge against stray voltages that could result in sound output. The advantage of a highly sensitive phono input is the unusually good responsiveness to delicate musical nuances and very low voltage levels that may be recorded on a phono record.

Phono Overload

The word minimum implies the word maximum and the input of the preamp is no exception. The maximum input audio signal from the phono cartridge is known as phono overload voltage. This voltage, also in millivolts (mV), is the largest amount of signal we should deliver to the phono input terminals of the preamp. To do otherwise will overdrive the amplifier and the result is distortion.

Phono overload figures run from as little as 100 mV to as much as 1,000 or even higher. But the problem with input sensitivity and overload is that the input signal doesn't have a constant amplitude but can vary considerably from almost zero to some maximum or peak. Reference is not always made in specs as to the frequency at which sensitivity and overload figures are produced. However, the input signal specs for a preamp do dictate the choice of phono cartridge. That cartridge must be able to meet the minimum input phono signal requirements of the preamp.

The spread between phono input sensitivity and phono overload should be as wide as possible. Phono input sensitivity should be as low as you can get; phono overload as high as you can get. The ratio between the two should be 10 to 1 as a minimum, but a higher ratio is even better. Ratios of 100 to 200 to one or better are available in quality preamps.

Total Harmonic Distortion

Musical tones have harmonics and it is these harmonics that give a tone its timbre, the distinguishing characteristic that enables us to recognize the tone. Harmonics, also called overtones or eigentones, are multiples of the fundamental frequency and can be even or odd. The even order harmonic of 1 kHz is 2 kHz and is the second harmonic. The fundamental is sometimes called the first harmonic. The third harmonic of 1 kHz is 3 kHz or three times the fundamental.

An amplifier, such as a preamp, can distort the fundamental, thus producing a succession of harmonics which are also distorted, or can distort the harmonics of the fundamentals. This doesn't mean that all the harmonics are distorted, but that is also a possibility. Since the harmonics are no longer related to the fundamental, they sound unpleasant.

Total harmonic distortion or the sum of all the harmonic distortion products in preamps, abbreviated as thd, is specified in percentage and ranges from a low of 0.0008% to a high of about 0.5%. The lower the percentage figure, the better. The figures supplied here could be considered extremes, since a thd figure of 0.01% is much more common.

What you can hear in terms of distortion depends on whether you are testing your ears with pure tones or with music. With pure tones your hearing mechanism would need to be very sensitive to be able to detect total harmonic distortion below 0.15%. For music, the bottom tolerance level is a bit higher, with about 0.2% as the point at which such distortion becomes aurally perceptible. If one amplifier is offered with a thd of 0.02% and another of 0.05%, the difference in specs can be meaningless since it is possible your ears won't be able to recognize the difference.

Intermodulation Distortion

Intermodulation distortion, abbreviated as im, measured at rated output, is caused by the interaction of bass and treble tones, producing new tones that aren't harmonically related to the original music. It is easier for the ear to detect intermodulation rather than harmonic distortion.

Harmonic Versus Intermodulation Distortion

Distortion, like immorality, is relative. You don't equate a driver who goes through a red light with a murderer. And, so, in a comparable way, not all distortion has the same nuisance value.

Amplifiers are supposed to be linear devices. This means they amplify all tones applied to the input without modifying them (other than amplification) in any way. But if an amplifier is nonlinear it will produce harmonics of the tones that are presented to its input. The harmonics of a tone having a frequency of 1,000 Hz will be 2,000 Hz, 3,000 Hz and so on. Some of these harmonics are cancelled by the circuitry of the amplifier, others are not. An instrument may produce a fundamental of 1,000 Hz, but the tone may be a complex one having harmonics of 2,000 Hz, 3,000 Hz, etc. But if the nonlinear amplifier also produces harmonics of the 1,000 Hz tone, how are you to know the difference? While the distortion will exist, quite possibly you may not even notice it.

The trouble is that the spuriously generated harmonics may not be satisfied to let well enough alone, and so they may modulate the original tones producing a condition called intermodulation distortion (im). This results in sum and difference signals; the harmonic added to and subtracted from the original. These sum and difference signals are *not* harmonically related to the original sound, and so you cannot only hear them, but can become quite irritated by them. To get an idea of what intermodulation distortion may sound like, whistle a single tone. Now deliberately whistle above and below that one, making sure the new tones are "flat." The result is horrible.

And so, for an amplifier you are going to look for intermodulation distortion (im) figures. Look for these at rated power and also at 1 watt. You want to know the performance of your amplifier at two extremes. The im at rated power should not be above 0.5%, and there are many amplifiers that are far below this figure. The im at 1 watt should also be far below the full power percentage figure. Some amplifiers have an im of 0.1% at 1 watt, but you can find some amplifiers that are a fraction of this amount——0.05% is rather good.

Frequency Response

The hearing range of an average pair of ears is seldom much below 60 Hz with a top limit of about 15 kHz. However, many quality preamps have a frequency response extending from a few hertz to as much as 100 kHz. Extremely wide frequency response seems to be the rule rather than the exception in quality preamps.

The frequency deviation from flat can be 1 or 2 dB, plus or minus, but is often just a fraction of a decibel. The frequency response can be checked from any of the inputs of the preamplifier but is usually measured from the auxiliary input. The RIAA equalization curve extends from 30 Hz to 15 kHz and while it would be interesting to see what the frequency response would be for phono, this information isn't supplied by preamp manufacturers.

The frequency response of a preamp should also indicate the deviation in decibels. Often the deviation will be plus/minus 2 dB, but some are much better in this respect. A frequency response number without an accompanying deviation figure isn't meaningful. Look for deviation figures that are as small as possible.

Slew Rate Limiting

A cause of high frequency distortion, in both preamps and power amps, is known as slew rate limiting. Usually we think of limiting as signal clipping but slew rate limiting refers to speed. As an example, your car might perform well at 60 mph, but when pushed faster may begin to vibrate or show other symptoms of distress. We get a comparable situation in amplifiers.

Slew rate is a measure of the rate of change of voltage per microsecond. If the music input requires a voltage change faster than the slew rate capability of the amplifier then the output is no longer the same as the input.

If you throw a ball into the air, its rate of change, its ascent, is greatest when traveling vertically. Before descending, the ball hangs in the air for a brief fraction of a second, neither moving up or down. At that precise moment its rate of change is zero. We can apply the same analysis to a sound wave. At its peak it is neither increasing nor decreasing so at that moment its rate of change is zero. Its maximum rate of change is at the time when the waveform crosses its horizontal axis.

Slew Rate Distortion

If the slew rate of the amplifier cannot accommodate the rate of change of the electrical equivalent of the musical waveform we get harmonic distortion.

If you run a slew rate test with a sine wave input, the output of the amplifier should also be a sine wave. If the output has a triangular shape, as one example, it can be analyzed and this will show it to consist of a sine wave plus some harmonics. But since these harmonics did not appear in the input we say the amplifier has harmonic distortion, or to be specific, slew rate distortion. These harmonics can easily be out of the audible range, but they mix or modulate audible signal voltages and as a result we get distortion that can be heard. The idea is comparable to the modulation used by broadcasting stations in which a radio frequency (inaudible carrier) is modulated by an audio signal.

Transient Intermodulation Distortion

Another form of distortion in amplifiers is transient intermodulation distortion, abbreviated as tim, caused by excessive negative feedback and a slew rate that is too low.

Slew rate is affected by negative feedback when it is used excessively, which is why slew rate and negative feedback both figure in the production of transient intermodulation distortion.

Both total harmonic distortion and transient intermodulation distortion will have an effect on what you hear. Audibly, tim adds harshness to sound and can result in listener fatigue. If you have two amplifiers with identical amounts of thd, the one having lower tim will sound better.

Noise

There are several major sources of noise in connection with preamplifiers. One of these is radio frequency interference, abbreviated as rfi, and can be due to noise signal pickup by the connecting wires between the turntable and the preamplifier. Another cause of noise is related to pickup of ac line hash from power cords and transformers positioned close to signal inputs on the preamp.

This doesn't mean that noise pickup is inevitable. It isn't. But if you have rfi, try using shorter connections, if you can, between the preamp and its signal sources.

SIGNAL-TO-NOISE RATIO

Signal-to-noise (s/n) ratio is measured in the phono position with signal-to-noise ratio figures sometimes supplied for auxiliary input. Phono signal-to-noise ratio should be 70 dB or more: auxiliary should be 80 dB or more. Higher figures are better.

Impedance Matching

The output impedance of one high fidelity component should match the input impedance of the component to which it is connected, although there are some exceptions to this rather broad statement. Still, the output impedance of the phono cartridge should be the same as the input impedance of the phono input of the preamp. This input impedance is generally indicated as 47 K (47,000) ohms and so this should be the output impedance of the phono cartridge. A typical value is 47,000 ohms (kΩ), but it can also be 100 kΩ or sometimes an intermediate value. Some preamps have switchable input impedances of 47 kΩ and 100 kΩ, thus allowing greater leeway in the selection of a phono cartridge. You can also get a preamp having an adjustable input impedance range from 30 kΩ to 100 kΩ.

NEGATIVE FEEDBACK

Negative feedback (nf) is a process in which part of the output signal of an audio amplifier circuit is fed back to the input circuit.

Feedback can take place over a single stage or over a number of stages. In all cases, the fed back signal must be out of phase with the input signal. The effect of negative feedback is to decrease the amplification of the stage, or stages, and to reduce distortion. Negative feedback also widens the frequency response.

Negative feedback is sometimes used by circuit designers as a corrective measure. Thus, an amplifier may suffer from a tendency to oscillate at some particular frequency, possibly due to in-phase signal leakage from the output to the input. Even if the circuit does not oscillate, positive feedback (or regeneration) distorts the frequency response, narrows the frequency response, and increases gain. It also tends to make the amplifier unstable, and can add a noticeable sharpness to the sound.

Negative feedback is not without its problems, despite its obvious advantages. While it reduces total harmonic distortion it can introduce transient intermodulation distortion. And, as mentioned earlier, it also affects slew rate. Since transient intermodulation distortion is much more apparent it is much more disturbing to listeners than total harmonic distortion (thd).

Too much feedback can also destabilize an amplifier, inducing oscillation which, in turn, can result in damaged speakers. Moving in the other direction and using too little feedback can also reduce an amplifier's damping factor (described in more detail in Chapter 4) which increases speaker induced distortion. Too little feedback also increases thd and makes noise and rumble more obvious.

The problem for amplifier designers, then, is to use the correct amount of negative feedback. Too little is bad, but so is too much.

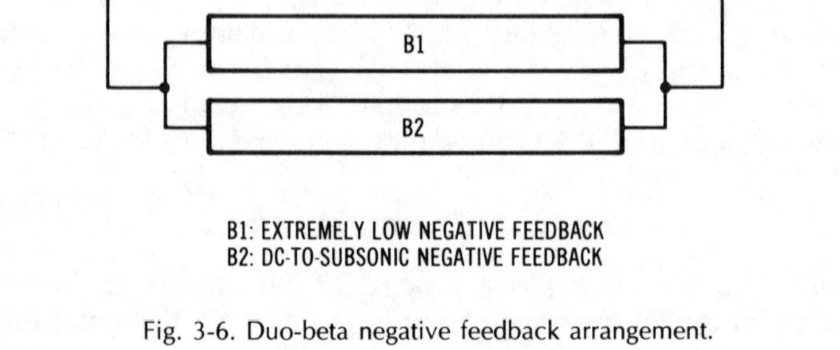

DUO-BETA CIRCUIT: PRE & POWER AMP

B1: EXTREMELY LOW NEGATIVE FEEDBACK
B2: DC-TO-SUBSONIC NEGATIVE FEEDBACK

Fig. 3-6. Duo-beta negative feedback arrangement.

DUO-BETA NEGATIVE FEEDBACK

Duo-beta feedback, an arrangement shown in Fig. 3-6, uses a low order of negative feedback which just about eliminates tim and increases the frequency response. It is indicated as B1 in the diagram. Beta 2, shown as B2 in Fig. 3-6, is a higher order feedback loop that operates only at subsonic frequencies, about 5 Hz and below. As a result damping factor is improved, low frequency noise is filtered without loss of gain, and dc drift is minimized.

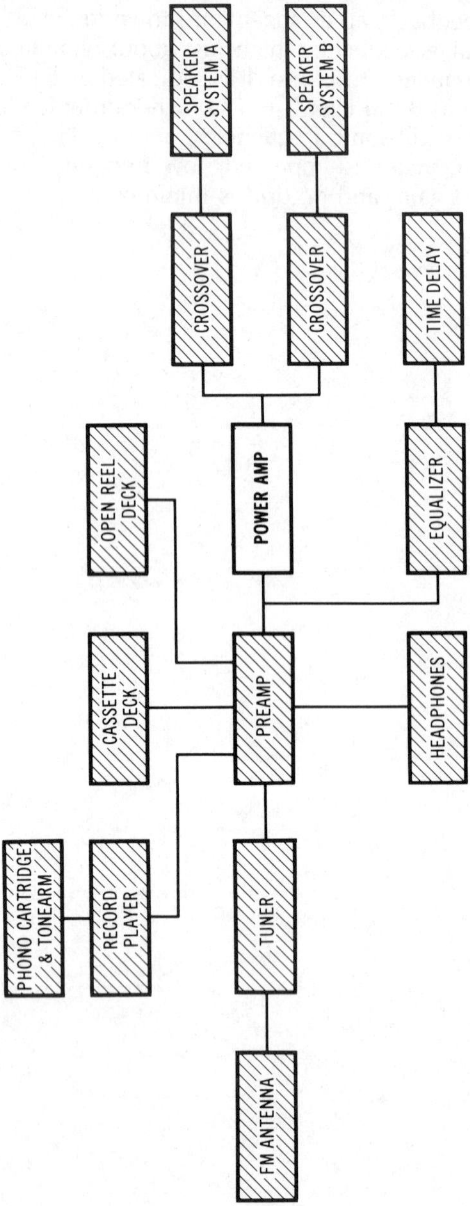

SPEAKER SYSTEM A
SPEAKER SYSTEM B
CROSSOVER
CROSSOVER
TIME DELAY
OPEN REEL DECK
POWER AMP
EQUALIZER
CASSETTE DECK
PREAMP
HEADPHONES
PHONO CARTRIDGE & TONEARM
RECORD PLAYER
TUNER
FM ANTENNA

4

Integrated Amplifiers and Power Amplifiers

In their travels from the antenna, the am and fm broadcast signals have passed through various hi-fi components and in the process have undergone some changes. At the antenna, the signal, actually two signals comprising the carrier and audio, has a strength of only a few millionths of a volt.

The separation of the radio frequency carrier and its accompanying audio signal takes place in the tuner or receiver and from that point on there is only one objective and that is to strengthen the audio. This is partially achieved in the tuner and the process continues in the preamplifier. The audio signal, amplified a million or more times, is now ready for entry into the power amplifier (Fig. 4-1).

Of course we have other audio signals as well: those supplied by a record player or a tape deck, but these have much higher initial strengths and are fed directly into the preamplifier. A function control on the front panel of the preamplifier selects the signal that would be handled by that component. And so, going into the input of the power amplifier from the output of the preamplifier we have a number of audio voltages: tuner, record player, tape deck, or possibly that supplied by a microphone (Fig. 4-2). The routing of these into the power amplifier is one at a time with the choice being made by the preamplifier. The power amplifier has no say in the matter and accepts what it is given.

The amplifier in the receiver is really a two-part unit, consisting of a preamplifier (or preamp) and a power amplifier (or power amp). The preamp builds the audio signal until it is of quite respectable magnitude, delivering it full grown to the power amp. The power amp adds the electrical energy needed by the speakers. In some re-

Fig. 4-1. A zero-feedback stereo tube power amplifier. Features include: triode output and driver tubes, zero feedback and 16 dB feedback operation via selector switch, 30 watts per channel rms into 8- or 4-ohms with 16 dB feedback, 25 watts per channel with zero feedback, input level controls for each channel, signal-to-noise ratio better than 95 dB. *(Courtesy Luxman, Div. of Alpine Electronics of America)*

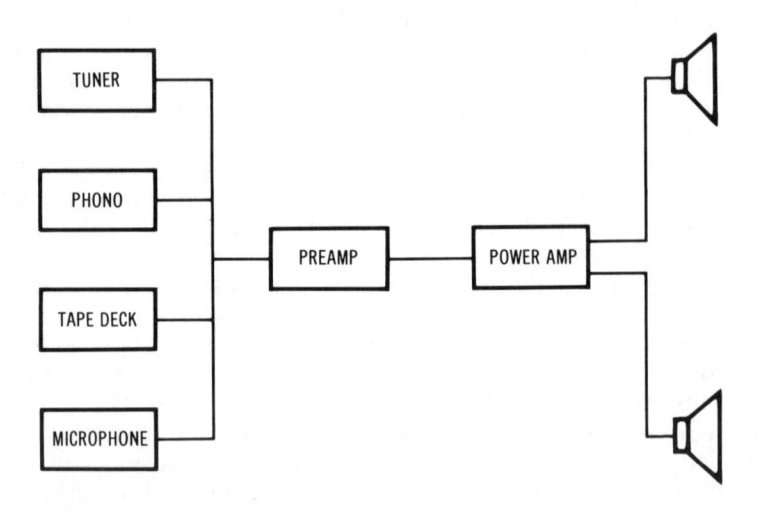

Fig. 4-2. In the hi-fi chain, the separate pre- and power amps or the integrated amp are positioned between the sound sources and the speakers. A front panel switch permits selection of one of the sound sources.

ceivers, the preamp and power amp are electronically separate units and so you can take the signal voltage from the preamp and use it to operate a separate, independent stereo amplifier. This is the first step toward a more elaborate sound reproduction process called multi-amplification, in which the audio spectrum is divided into segments, with each having its own stereo amplifier.

VOLTAGE VERSUS POWER

Sometimes the preamplifier is called a voltage amplifier to distinguish it from its following component, the power amp. A voltage can exist without current flow, but power must have both current and voltage. It is possible to have voltage without the expenditure of energy. Thus, you can carry a flashlight, but as long as its operating switch is turned off there is no current flow. But you still have voltage and that voltage can be measured across the terminals of the flashlight battery. No energy is being delivered, and there is no light.

When you move the switch of the flashlight to its on position, current flows from the battery through the lamp. Electrical energy is supplied to the lamp, and, like the transducer it is, converts that electrical energy to light energy.

The power amplifier is comparable to a flashlight. It delivers an audio current to the voice coils of the speakers to which it is connected. There is also a voltage across the output terminals of the power amplifier. Since the speakers are transducers, they convert the audio power supplied by the power amplifier and give us sound energy.

So here we have the basic difference between a preamp and a power amp. The preamp supplies voltage to the input of the power amp. The power amp supplies both voltage and current to the voice coils of the speakers.

The audio signal supplied to the power amp by the preamp is simply a control signal. The purpose of all the prior amplification is to make that signal strong enough to control the rather heavy currents flowing through the output transistors (or tubes) in the power amp. And, because the control signal varies at an audio rate, the currents flowing through the output transistors also vary at the same rate.

Every stereo power amplifier is not one, but two amplifiers, one for left channel sound, the other for right channel sound. Crosstalk is possible in some stereo amps with sound signal leaking from one channel to the other. One way of preventing or minimizing crosstalk is to use separate power supplies for each of the stereo amps. The power supplies can be completely separate and independent of each other—the ideal and most expensive arrangement—or, the two power supplies may use a common power transformer followed by independent rectifiers and filters.

AC LINE POWER

A power amplifier does not manufacture power. All it does, and all it can do is to take the electrical energy available at an ac (alternating current) outlet and convert that ac to dc (direct current) for use by the amplifier. That dc is then converted into audio energy by the power amplifier. So that is all a power amplifier is, a converter of line energy to audio energy.

No energy converting device is 100% efficient and power amplifiers are no exception. You should not assume that an amplifier rated at 50 watts requires only 50 watts of electrical power from the ac line. Some of the line's ac power is inevitably lost in the form of heat. Further, the energy demands of the power amplifier fluctuate and are minimum when the power amplifier is idling, that is, when it is not handling a signal, to maximum at rated power when it is operating wide open.

If an amplifier is designated as capable of delivering 100 watts rms livering 100 watts rms continuous power output per channel, a fair assumption would be that the line power is 75% of this amount with the amplifier idling, and in this example would be 75 watts. Multiply the audio power output by a factor of six to get some idea of the approximate amount of line power needed with the amplifier being fully driven. In this case the audio output is 100 watts per channel, and six times this amount is 600 watts. That is the demand being made on the power line.

Obviously, you are going to operate your power amplifier somewhere between these two extremes.

This means you should be somewhat selective about the power outlet you are going to use for your power amplifier. If you plan to use either a switched or an unswitched outlet on the back of some of your other hi-fi components, make sure that the outlet has a high enough power rating. If not, then it is a safer practice to use a separate baseboard power outlet to obtain the power for the power amplifier.

IMPORTANCE OF THE POWER SUPPLY

The amplifier power supply is a sort of reservoir of electrical energy between the baseboard ac power and the demands of the audio signal. The control signal, the audio signal from the preamp, may be small, but at other times may be quite large. As a result we have a varying demand for energy from that power supply. If the power supply cannot meet that demand called for by the amplifier, the result will be distortion.

A high amplitude bass tone, for example, delivered to the input side of power transistors, means that those transistors must receive a correspondingly high amount of current from the power supply. The power supply should supply this demand for electrical energy and do so with the least delay.

IMPORTANCE OF THE LOAD

Every time you turn on a light or an appliance in your home you are making a demand for electrical power. Fortunately, the electric generators that supply energy to your home are huge and are quite capable of giving you all the electrical energy you want—when you want it.

In hi-fi, a speaker is a load on the power amplifier. Like the lamp or appliance connected to your ac outlets, the speaker looks on the power amp as its source of energy. But the power amplifier in your hi-fi system, unlike the generators at your local power utility, does not have such a seemingly inexhaustible amount of available electrical energy. This means that the load on your power amplifier, the speakers, must take into consideration the amount of audio power that the amplifier can supply. If the speakers at one particular moment want 30 watts of audio power and the most your amplifier can supply is 20 watts, the amplifier will make every effort to supply it, but will distort the signal in the process. If the demand for audio power is sufficiently high, the power output transistors in the power amp will literally die—they will burn out. They may not do so if the power amp is equipped with suitable protector circuits to keep this from happening.

ADDING MORE SPEAKERS

Since speakers are a load on the power amp, adding more speakers to the original pair you have in your hi-fi system means increasing the load on the power amp. And the more speakers you add, the "heavier" the load, that is, the greater the demand for audio power from the amplifier. If your amplifier is designed to handle this greater demand, well and good. If not, the result will be distortion, and, in the absence of protector circuitry, transistor burnout.

SHORT CIRCUITS

On some speakers the connecting terminals are quite close together, an unhappy situation since it means that the strands of connecting wires from the power amplifier may touch accidentally. This is the heaviest possible load and the result could be an output transistor burnout in the power amp.

PROTECTOR CIRCUITS

Most speakers are rated at 8 ohms impedance, but there are some that have an impedance of 4 ohms. The lower the impedance, the greater the load. Impedance, though, isn't constant and varies with frequency. Thus, at certain audio frequencies, the impedance of a 4-ohm speaker could possibly drop to as low as 2 ohms, and this is getting dangerously close to a short circuit condition. If the impedance does drop momentarily to 2 ohms and at the same time the power amp is triggered by a strong bass tone—a demand for a large amount of energy—the greatly increased flow of current could cause burnout of the power transistors.

To protect the expensive output transistors of the power amp from destruction, power amps are equipped with protector circuits. These act to disconnect the power amp at the first indication of trouble. The protector can be nothing more than a simple fuse housed on the back apron of the amplifier. Restoring the power amp to a working condition simply means replacing the fuse. If the fuse keeps opening as fast as you replace it then it is quite likely there is a short circuit condition.

There are various other methods of circuit overload protection. The power amp may contain a voltage-current limiter, also known as a VI limiter (V=voltage; I=current), which senses load impedances below two ohms without being affected by reactive loads, such as electrostatic speakers, which are rated at 4 ohms or greater.

Sometimes the power amplifier relies on an ac power line fuse for protection. The problem with this method of protection is that such fuses can have a rather wide tolerance and so may withstand a much heavier overload current for a longer time than other protective methods.

COOLING

Because power output transistors develop large amounts of heat they are mounted on heat sinks. The purpose of a heat sink is to radiate as much heat as possible and the greater the area of the heat sink the more heat that can be dissipated.

Heat sinks can be flat plates fastened to the outer case of the output power transistors, or, more generally, have a wavy surface. In some power amps a thermal cutout device shuts down the amplifier if the heat sink reaches an unusually high temperature. This is something the power transistors will do if they are forced to carry heavy currents will do if they are forced to carry heavy currents due to an overload.

In some instances, particularly for high-power amplifiers such as one rated for 150 watts or 200 watts per channel, supplementary cooling other than heat sinks may be used. This supplementary cooling can be obtained from a forced air two-speed fan. The fan should be as quiet as possible and develop minimum electrical and mechanical noise.

CLIPPING

There are some combinations that can make momentary heavy demands on the power amplifier, possibly a bass tone having a high strength or amplitude. If the power amp is incapable of meeting this demand for power, it will clip the audio waveform. This means the power amp will handle as much of this demand for power as it can but will level off at some point. At that point the power amp will no longer follow the shape of the input signal but will flatten. The amount of clipping will depend on the power amplifier. A unit having a high power rating will either not clip at all, or clip much less than a lower powered amplifier. Clipping, sometimes called overload, results in distorted sound out of the speakers.

There is an alternative. If, for example, you listen to a phono record characterized by heavy bass, turn down the volume control until the apparent distortion no longer exists. The penalty you pay is reduced dynamic range. So you have a choice between distortion and lower dynamic range. Another solution is to use a power amplifier having substantially higher power output capabilities.

OUTPUT CAPACITOR-LESS CIRCUITS

The output power transistors of the power amp receive their dc operating voltage from the power supply and, at the same time, deliver audio voltages and currents to the speakers. These audio voltages and currents are ac.

To keep these two different voltages away from each other and to keep the dc, intended only for the power transistors, away from the speakers, some form of blocking device is used. Some power amplifiers have a capacitor for this purpose, a radio part that permits the passage of ac but blocks dc.

Capacitors used to connect the output of the power amp to the speakers would offhand seem to be ideal for that purpose. However, capacitors are frequency selective and while they do indeed block the passage of dc they have a greater opposition to low frequencies than to high. This characteristic, known as capacitive reactance, varies inversely with frequency, and so tones of increasing pitch have less difficulty in making the passage from the power amp to the speakers. Unfortunately, this tends to nullify the objective of high fidelity whose purpose is to amplify and pass on all tones equally.

Changes in circuit design finally permitted the elimination of the coupling capacitor and so some audio power amplifiers are known as ocl, or output capacitor-less circuits. Because ocl circuits do not show frequency favoritism they are preferred in better audio power amplifiers.

AMPLIFIER EFFICIENCY

In power amplifiers the output transistors are called upon to handle fairly large amounts of audio power. The power is handled in a better manner if two transistors (sometimes more are used) handle the work load. There are various ways in which the power transistors can do this and the method they use is known as class of operation.

We ordinarily do not think of a hi-fi system as an energy guzzler and, as a matter of fact, most hi-fi systems are fairly efficient in converting the power they get from the ac power line into useful audio power which we hear from speakers. We are more concerned with energy in connection with power amps than we are with other hi-fi components for it is in the power amp that we finally get the development of audio power.

The efficiency of a power amplifier stage is the ratio of audio watts out versus dc power input. If, for a given amount of signal input at a specified frequency the audio output of an amplifier is 50 watts while the same amplifier requires 100 dc watts from its power supply, the efficiency is 50/100, or 0.5. If we multiply this answer by 100 we will have it in terms of percentage: 0.50 × 100 = 50, or 50%. So this amplifier output stage is 50% efficient.

The maximum efficiency is 100%, unattainable since the movement of current, no matter how small, always develops some heat. Unfortunately, the opposite is true for we can have 0% efficiency, a condition in which all of the dc energy delivered to a power output stage is wasted as heat, something that happens in the absence of an input signal.

The average efficiency of an audio power amp is about 20%. Some amplifiers have a better rating than this; other do much more poorly. A 20% energy efficiency rating means that of every 100 watts of electrical power delivered to the power amp, only 20 watts is delivered to the voice coils of the speakers. The remaining 80 watts is given off in the form of heat. Not only is this a waste, but we are also put to the trouble of dissipating those 80 watts. That heat, doing no useful work, is radiated into the space around the power amp by the heat fins attached to the power transistors.

AMPLIFIER CLASSES

If we have a power amplifier using just a single pair of output transistors we can have various working arrangements for them. They can both work all the time, together, or one can work while the other rests. Then the resting transistor goes to work and the other transistor takes time off.

What we have here are two extremes, either both working, or alternate sharing of the work by the two transistors. There are an interesting variety of working conditions that come between these extremes.

Class A

Class A operation of the output power transistors is one of the extreme conditions mentioned above. This type of operation has much to recommend it and it is used in some power amps. The transistors operate linearly, that is, they introduce very little distortion. There is also very little crossover distortion, a condition that can exist in transistors that work alternately.

The problem with Class A is that it is extremely inefficient. The efficiency of this class of amplifier isn't constant but depends on signal input. In the absence of an input signal all the electrical energy delivered to the amplifier by the power supply must be dissipated as heat.

When a signal is supplied to the input of the class A amplifier, both output transistors go to work simultaneously, and some of the dc electrical energy supplied by the power supply becomes part of the audio power output. However, the efficiency of the Class A amplifier is a maximum of 50%. This means that for every 100 watts of dc power supplied, at least half of this must be thrown away as heat. This is a maximum figure since all of the dc energy can be thrown away as heat if there is no input signal.

Class B

Despite its low efficiency some power amps do make use of the Class A amplifier. Keep in mind, though, that Class A isn't a quality or efficiency rating and is simply used to designate a particular class of operation.

Another power amplifier output circuit that is used, and used more extensively than Class A, is the Class B circuit. When a Class B power amplifier delivers its full rated power output power it may be able to convert as much as 60% or 70% of the electrical energy it gets from the power supply into audio energy to be delivered to its load, the speakers. This higher efficiency occurs only when the amplifier is delivering its full rated power.

When you operate the volume control you not only increase or decrease the sound output but you vary the efficiency of the power amplifier as well. At lower outputs, efficiency drops rapidly. If the amplifier delivers 40% of its rated power, the efficiency of a Class B amplifier drops to around 40% to 50%. But as you continue to turn the volume control counterclockwise, thereby decreasing the sound output level, and the Class B power amplifier delivers only 10% of its rated power, that is, 10% of its maximum audio power-delivering capability, the efficiency drops to about 20%.

Efficiency is controllable by the volume control setting, but much more so by the dynamic range of the input signal. Music varies constantly in amplitude. There are some peak values in audio signals, or fortissimo passages in music that demand momentary peak output, but that's exceptional. The best that can be said of the Class B amplifier is that its efficiency is better than Class A, and we can refer to the efficiency in general terms as medium. Efficiencies as high as 75% maximum can be obtained compared to 50% for Class A.

All of the energy for the power amplifier must come from the power supply and so there is a direct relationship between the size of that power supply and the class of amplifier operation. Because of the greater efficiency of Class B amplifiers a smaller power supply can be used.

With Class A the power output can be handled by just a single transistor, although two or more are commonly used. No such option is available in Class B for here we must have a minimum of two transistors since they work alternately.

It would be nice to assume that in Class B the changeover in signal handling is accomplished easily and smoothly. In actual operation that transition isn't that smooth. The signal switches from one transistor to the other and it would be ideal if one transistor immediately picked up where the other transistor leaves off. That isn't the case so at the switchover point we can get what is known as crossover distortion.

Special transistors aren't needed for Class A or Class B, that is, the same type of transistor can be used for either class. The type of operation is controlled by a voltage called bias, a voltage whose strength determines the time of current flow through a transistor. Transistors in Class A are biased so that current flows through the transistor at all times. In Class B the transistors are biased so that one transistor at a time is "cut off," that is, no current flows through it.

The input signal to power transistors is ac and this means that part of the time the signal is positive and part of the time it is negative. If we apply the positive portion of the signal to a power transistor it will overcome the bias; a negative voltage keeping the transistor from conducting in the absence of a signal. The positive half of the input signal takes the transistor out of "current cutoff" and the transistor functions.

The other transistor, though, receives the negative half of the audio signal. This negative half adds to the negative bias and so keeps the transistor from conducting. When the signal changes its polarity, the positive half of that signal will now be applied to the transistor that wasn't conducting, giving it a chance to do some work. The transistor that was working now gets the negative half of the input signal and it stops conducting. Thus, the pair of transistors alternately conduct and do not conduct, constantly switching back and forth between these two conditions.

Class AB

Insofar as possible, manufacturers of receivers, integrated amplifiers and separate power amplifiers try to select power output transistors that are as equally matched as possible. But even if they do there are bound to be some differences, however small, in the operating characteristics of those transistors. Class B operation makes no allowances for such differences.

What we would like to do would be to get the quality of the Class A amplifier and the higher efficiency of Class B and so what we do in practice is to compromise. The result is an amplifier designed as Class AB. To get greater efficiency, class AB amplifiers work in Class A if the input signal is small, switching automatically to Class B for stronger signals. The price is lower efficiency but there is a reduction in crossover distortion.

Class C

A basic difference between Class A and Class B is that in Class A current flows through the transistor at all times. Class B uses a higher amount of bias so that each transistor in the pair has current flow 50% of the time. Since it is current flow that produces heat, restricting that current flow improves efficiency. Obviously, then, a transistor with no current flow through it at all would be 100% efficient, but we would get no sound output.

A Class C amplifier is an attempt to move in this direction. Current flows through the transistor for much less than 50% of the time. Unfortunately, the result of this effort is an extremely high level of distortion. While Class C amplifiers aren't used in audio amplification they are extensively used in transmitters.

Class D

There are several ways of thinking about transistors. In audio, we regard the transistors as amplifiers, but what the transistor is really doing is working as a switch. This is most evident in the Class B amplifier where the transistors are alternately switched on and off by the polarity of the input signal. Transistors as switches is a concept that is widely used in calculators and computers.

When a switch is closed there is no voltage across it because it is really a deliberately designed short circuit. When the switch is open no current flows through it, but full voltage is across the switch.

Instead of using the power transistors as amplifiers we will now use them as switches. When the switch is closed we have a short circuit condition and we get a pulse of audio power. But because the transistor is working as a switch little or no heat is developed. When the transistor working as a switch is open, no current flows, and again no heat is developed. In both operating modes the efficiency is very close to 100%.

Along with the Class D amplifier is a circuit called a pulse width modulator. The switching action of the transistors produces pulses of current in the output. The input signal, of course, doesn't consist of pulses but is an audio waveform. However, we can take a rapid succession of samples and these samples can be in the form of pulses. The pulse width modulator compares the width of the output pulses with the input, making them narrower when the input pulses are narrower, wider when they are wider.

The sampling of the input signal is at least twice the frequency of the highest audio signal and if that signal is 20 kHz, then the sampling rate is a minimum of 40 kHz.

The advantage of Class D is its very high efficiency. This, in turn, means a much smaller power supply. Further, it makes the amplifier independent of nonlinearities in the transistor, that is, there is no crossover distortion. The usefulness of Class D depends on the accuracy of the output pulses, how closely they compare to the samples taken of the input signal. In effect, the responsibility for quality has been transferred from the transistors to the pulse width modulator.

Class G

After considerable research and study, one manufacturer came up with a new kind of power amplifier circuit which they call Class G and which they have included in several of their power amplifiers.

The basic idea behind Class G is quite simple. Suppose you had two amplifiers, one rated at 10 watts per channel, the other at 100 watts per channel. Suppose, also, that you could feed your music signals to both at once and could instruct the low power amplifier to "take over" the music reproducing job for most of the time, while the higher powered amp handled only those peak signals that need to be reproduced more strongly.

Since the low powered amplifier has a rating of only 10 watts per channel and, since most of the time that's how much power the music program demands, that amplifier will be operating at its greatest efficiency for much of the time that it is in use. Then, when an instantaneous 50 or 100 watt peak of audio power is demanded, that small amplifier turns off, so no power is wasted, and the companion higher powered amplifier takes over the job. It, too, is now operating at or near its full rating and therefore in its most efficient operating region.

In Class G we have two sets of output transistor pairs per channel. One set of transistors has a lower output power capability than the other, and is, in fact, powered by a lower voltage supply. The other pair, capable of handling higher power levels, is operated from a much stronger power supply in terms of dc power capability.

While most audio manufacturers state power output ratings in terms of continuous power capability, under actual music listening conditions amplifiers are able to deliver somewhat greater power for the short-term durations of musical signal peaks.

There are no direct conversion formulas by which continuous power can be translated to music power. But for a given Class G amplifier of 75 watts per channel rating and a similarly rated conventional Class B design, the Class G will deliver greater short term peaks before signal clipping is audibly noticeable. This means Class G supplies a greater dynamic range capability (the difference between the softest sounds and the loudest sounds that can be reproduced without distortion or noise).

AUDIO OUTPUT POWER

The amount of audio output power is one of the most important of power amplifier specs and is one that is always given. For some time manufacturers used a variety of power output specs, and since these all supplied different numbers were confusing and quite often misleading. These went under various names such as music power, peak power, or instantaneous peak power. Since it was, and still is, desirable for a power amp to have as much audio power output as possible, power output ratings became a numbers game, but one in which no meaningful audio power comparisons could be made.

Audio power is what lets you hear those satisfying deep down bass tones, without having those tones muddled or distorted. Look for continuous audio power, both channels driven, from 20 Hz to 20 kHz, into 8-ohm speakers, with minimum distortion. Translation: Your receiver has two amplifiers, one for each stereo channel. Continuous power means just that—both amplifiers operating without taking a rest, and both working simultaneously. Twenty hertz is the bottom edge of the bass tone spectrum. Twenty kHz represents the uppermost reaches of the treble. And so, 20 Hz to 20 kHz covers every possible tone you can hear or imagine you can hear. Power output measurements are larger for 4-ohm speakers than for 8-ohm units, but 8 ohms is fairly standard and is most likely what you have at home or what you will buy. Power output should also include a distortion figure, usually a percentage of some kind. Look for 1%, or less; the lower, the better.

Minimum and Maximum Power

The power output of an amplifier is comparable to the power output of an electric light company, but with some important differences. An electric utility will supply all the power you want; the audio power amplifier has limitations.

An audio power amp is intended to work with speakers and these two components, the power amp and the speakers, must be purchased with consideration for each other. Speakers have minimum and maximum power requirements and so the power amp must be capable of supplying enough power to meet the minimum demand, but must not overpower the speakers' maximum needs.

Strapping

In some power amps the two outputs, that of the left channel amp and that of the right channel amp, can be combined or *strapped* to supply a higher audio power output. However, since the outputs are joined, the sound is no longer stereo, but mono only. This is an effective technique when listening to a mono source such as a mono fm broadcast, all am broadcasts, or possibly the input from a single microphone.

RMS Power

The difficulty in measuring an ac voltage, current, or power is that these keep changing, somewhat like trying to weigh a super-active child. Because the waveforms representing these quantities are so changeable, various techniques have been suggested for measuring them. One method is to consider only the maximum amount, the peak value. Another technique is to measure between a positive peak and a following negative peak, and is called peak-to-peak (Fig. 4–3). Another method is called "average" but here we run into some difficulties since an ac wave is positive approximately half the time and negative the other half; arithmetically, these values cancel.

To get around this dilemma the approach most often used is root-mean-square (mentioned earlier in Chapter 2) abbreviated as rms,

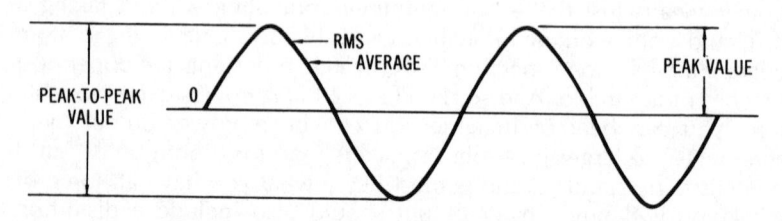

Fig. 4-3. Rms (root-mean-square) or effective value of a regularly recurring wave, compared to average, peak, and peak-to-peak.

and also known as effective value. It is a little more complicated than the other measurement methods, but it works quite well, and is regularly used in connection with voltages, currents, and powers in hi-fi.

In the rms method, a large number of measurements is taken over a complete cycle of a wave, both positive and negative. These numbers are then squared, thus changing all the negative numbers to positive. Any negative number, any minus number, multiplied by itself, supplies a result that is positive. For example, $(-4) \times (-4) = +16$.

All of the numbers obtained are added and then averaged, thus supplying their *mean* value. Assume that we have made four measurements of an ac wave; 4 volts, 4.25volts, 4.5 volts, and 6 volts. If we square each number, multiplying it by itself, we will have 16, 18, 20, and 36 volts. The *mean* or average value of 16, 18, 20, and 36 is 90 divided by 4 or 15. Ninety is the sum of the numbers; and we have four quantities that have been used.

Since all the numbers were squared, the remaining step is to take the square root of the result just obtained. In this example we would have the square root of 90, or approximately 9.49.

An easier way is to find the peak value of a wave, either the positive or negative peak, and multiply it by 0.707. Thus, if the peak value of a wave is 10 volts, the rms value is 7.07 volts. Fig. 4–4 shows the relationship between peak, rms, and average values of a wave.

IMPEDANCE MATCHING

Impedance matching, mentioned briefly in the chapter on preamplifiers, is important in connecting the output of the power amp to its speaker system. For most power amps, the output impedance is 8 ohms and so the speakers connected to such an amplifier should also

GIVEN THIS VALUE	MULTIPLY BY THIS VALUE TO GET			
	AVERAGE	EFFECTIVE	PEAK	P-P
AVERAGE	—	1.11	1.57	1.274
EFFECTIVE	0.9	—	1.414	2.828
PEAK	0.637	0.707	—	2.0
P-P	0.3185	0.3535	0.50	—

Fig. 4-4. Arithmetic relationships among average, effective (rms), peak and peak-to-peak (p-p) values of voltage, current, or power.

have an impedance of 8 ohms. By matching impedances we get the maximum transfer of audio energy from the power amp to the speakers.

Impedance mismatching has its disadvantages. If speakers rated at 4 ohms are connected to the 8-ohm terminals of the power amp they represent a heavier than normal load on that power amp. Because of their lower impedance they demand more current. Whether the power amp can meet the additional demand for current depends on the power output capabilities of that power amp.

Another disadvantage is that the 4-ohm impedance of the speaker system is nominal. This means that on the average it is 4 ohms, but since impedance is frequency sensitive, the impedance can become less than 4 ohms, possibly dropping as low as 2 ohms. This is dangerously close to a short circuit and so there is a possibility of damaging the output power transistors or opening some protector circuit. At the least this action is a nuisance; but it could mean an expensive repair job.

If we have the opposite situation, 8-ohm speakers connected to the 4-ohm speaker terminals of the power amp, the effect is a loss of driving ability on the part of the power amp. No harm will be done to either the power amp or the speakers, but the operating capabilities of the power amp will be reduced. For most of the time during music playback of a hi-fi system, the demand for audio power is modest. But a high peak transient in the bass region can cause a sudden demand for power, and lots of it. Mismatching of impedances means that while the power amp may develop that audio power, it will be unable to deliver it efficiently to the speakers.

The best arrangement, then, is to impedance match and it is easy to do. Connect 8-ohm speaker systems to the 8-ohm terminals of the power amp; connect 4-ohm speaker systems to the 4-ohm terminals.

BIAMPLIFICATION

Since power amplifiers are stereo devices, they are really two amplifiers, not one. But each of these amplifiers is required to handle the full range of audio frequencies, from 20 Hz to 20 kHz. At some point, though, that full range of audio frequencies will need to be separated into bands, possibly bass, midrange, and treble frequencies. The need for dividing the audio spectrum into two, preferably three bands, is a requirement imposed by speakers, since no single speaker is capable of handling all audio frequencies from 20 Hz to 20 kHz. Instead, two or more speakers are used, with one speaker reproducing bass and midrange tones, and the other treble tones. In some speaker systems, one speaker is designed for bass, another for midrange, and still another for treble.

The division of the full audio range into smaller bands usually takes place in the speaker, a job that is handled by a passive crossover network; passive because it consists of parts that do no amplification of the signal. See Fig. 4–5.

In a biamplification system, as shown in Fig. 4–6, the input signal from the preamplifier is brought into an electronic crossover. Here the audio signal is divided into two bands: bass/midrange and treble. The treble tones are brought into a separate treble power amplifier and from there to tweeters—speakers designed for treble tone reproduction. Similarly, the bass/midrange tones are delivered to a separate power amplifier intended to handle just the bass/midrange tones.

The two power amplifiers need not have the same audio output power ratings. Thus, the bass/midrange power amplifier might have an output power rating of 40 watts; the treble power amplifier just 20 watts. The reason for this is the lower audio energy demand of treble tones.

Biamplification is so called since the audio spectrum is divided into two parts. That same spectrum could be separated into three sections: bass, midrange, and treble, and in that case the system could be called triamplification.

Both the electronic crossover as a separate component and the passive frequency divider housed in the speaker enclosure do the same job, but do not do it equally well. The electronic crossover is situated in the component system where the audio signal is weaker than it is at the speakers. At this point it is easier to separate the audio spectrum into bands. But the electronic crossover costs more, requires more space, and needs connecting cables. However, the electronic crossover permits adjustment of the crossover point, something the passive crossover in the speaker does not generally allow. The crossover is the frequency region at which the division of the sound into bands takes place. Adjusting the crossover point is advantageous since it lets you compensate for the particular acoustic conditions of your listening room. You can get a better balance between bass/midrange and treble tones with the electronic crossover.

COMPLEMENTARY SYMMETRY

You may sometimes have seen directly, or via movies or tv, two men using a long saw with handles at each end. In this operation one man pulls while the other pushes, and vice versa. Amplifier circuits that work this way are called push pull, with a pair of transistors taking turns in amplifying the signal. If such circuits are carefully designed, each transistor does its work for exactly 50% of the time. The advantage is that such circuitry results in low steady state or transient distortion.

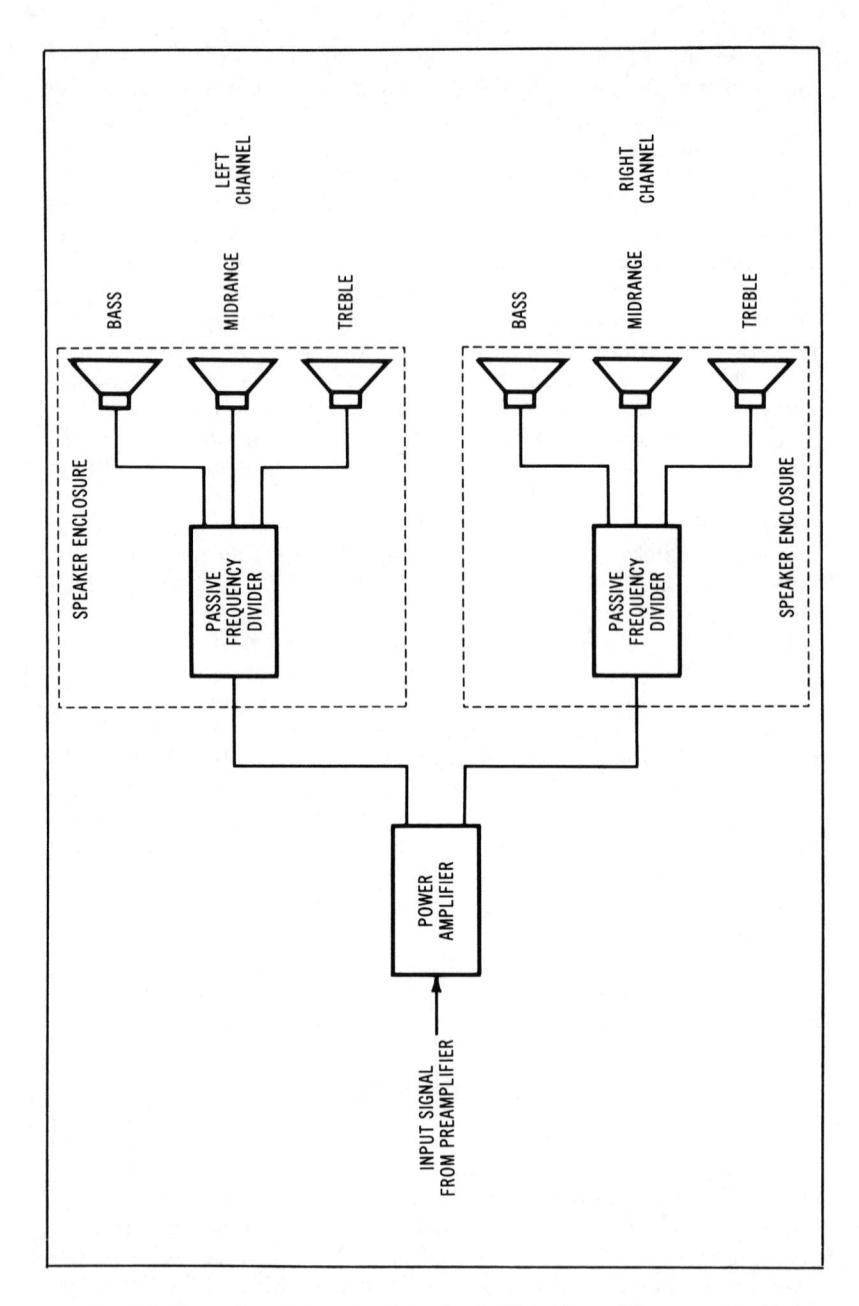

Fig. 4-5. The audio spectrum can be separated into bass, midrange, and treble tones by a passive divider mounted in the speaker enclosure.

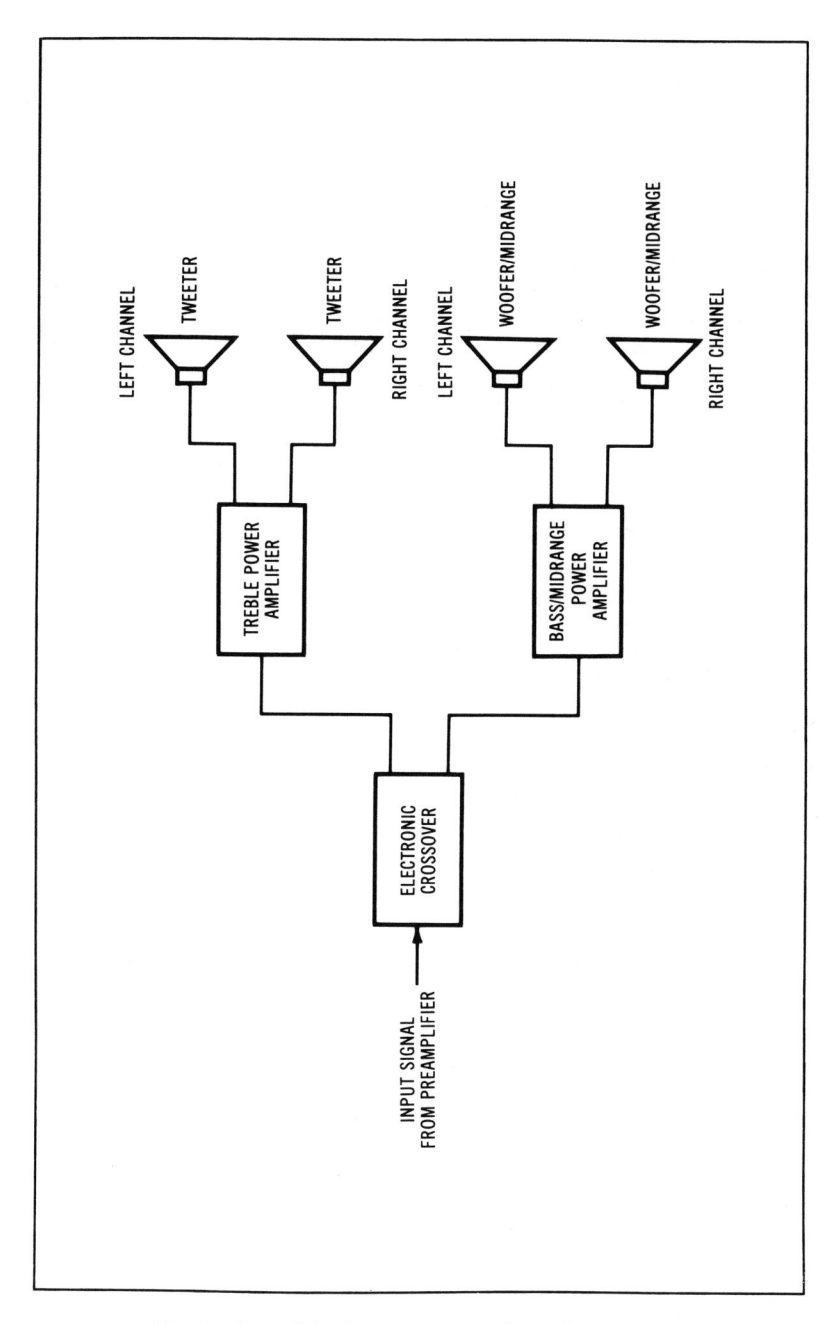

Fig. 4-6. Biamplification system uses electronic crossover.

In some power amplifiers this complementary symmetry arrangement is used only in the output stages. In others, the circuitry is used throughout, from input to output.

NEGATIVE FEEDBACK

Abbreviated as nf, negative feedback is a special circuit arrangement, as shown in Fig. 4–7. Here a small part of the output signal is fed back to the input of the amplifier. It is called negative feedback because the polarity of the output signal is opposite that of the signal at the input. Thus, if at any moment the input signal is positive, then the output signal is negative. As a result the two signals, if brought together, would oppose each other.

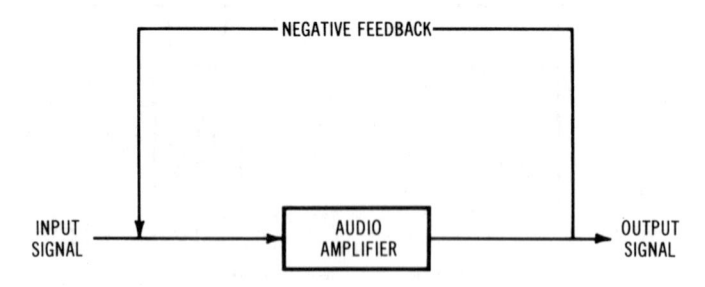

Fig. 4-7. Negative feedback is used to stabilize the audio amplifier.

Because they do oppose each other, the effect of negative feedback is to reduce the gain of the amplifier, but since the amount of feedback signal is small and carefully controlled, this loss of gain is minimal. The benefit is a reduction in distortion.

The input to an amplifier is a very sensitive point. Sometimes stray signals can "leak" over from other amplifier circuits, and these signals may be in step with or have the same polarity as the wanted audio signal. The result is that the amplifier, instead of working strictly as an amplifier, begins to work as a signal generator—an oscillator that produces its own signals. The condition is known as positive feedback. Negative feedback, however, tends to cancel this effect, making the amplifier more stable and less prone to any tendency to work as a generator.

INTEGRATED AMPLIFIER CONTROLS

Since an integrated amplifier consists of a preamp in association with a power amp, some of the controls you will find described in the following paragraphs will be found on preamps, others on power amps; while controls may be partially duplicated on both components (Fig. 4–8).

Volume Control

The volume control adjusts the overall sound level of all the speaker systems connected to the amplifier, whether those are "A" systems, "B" systems, or a combination of the two. The volume control also governs the overall sound levels of left/right speakers front, and left/right speakers, rear. Volume increases with clockwise rotation of the volume control; decreases with counterclockwise rotation. Some volume controls have numbers marked on the panel immediately adjacent to the control knob, but these numbers are relative only, not absolute.

At one time the volume control was incorporated with an on/off power switch. Thus, in shutting off the equipment, the volume control was automatically rotated to its minimum sound level position. However, in hi-fi systems the power switch is an independent part and so turning on a hi-fi system may mean a suddenly large and unexpected blast of sound if the volume control is left turned to a somewhat advanced position. It is good hi-fi operating practice to turn down the volume control and it is even better practice to turn it down before setting the power on/off switch to its off position. This will help eliminate or minimize the possibility of turn-off thumps.

Volume controls are variable resistors and in some amplifiers the control has a logarithmic curve. In this type of control the turning angle is proportional to the attenuation. In other words, the rotation of the control is proportional to the sound volume heard by human ears.

On-Off Control

Since the power amplifier delivers electrical energy to the speakers it must get that energy from somewhere, and that somewhere is the ac power line. Electrical energy from the power line is brought into a power supply that changes the ac of the power line to dc and this dc is used to operate the various solid state units, the transistors, or less often, the vacuum tubes, used in the power amplifier.

On the front panel of the power amp you will find the usual on/off switch and so the line cord of the power amp can be connected directly into any convenient ac power outlet. The disadvantage is that you must now turn the power amp on, and must remember to do so, every time you turn on the preamp or tuner or both. An easier way is to let the hi-fi system do this for you. You can connect the plug of the line cord to a switched outlet on the back of the preamp and so, every time you turn on the preamp, the power amp also turns on, automatically. Many power amps have a glow lamp to show that this component is turned on.

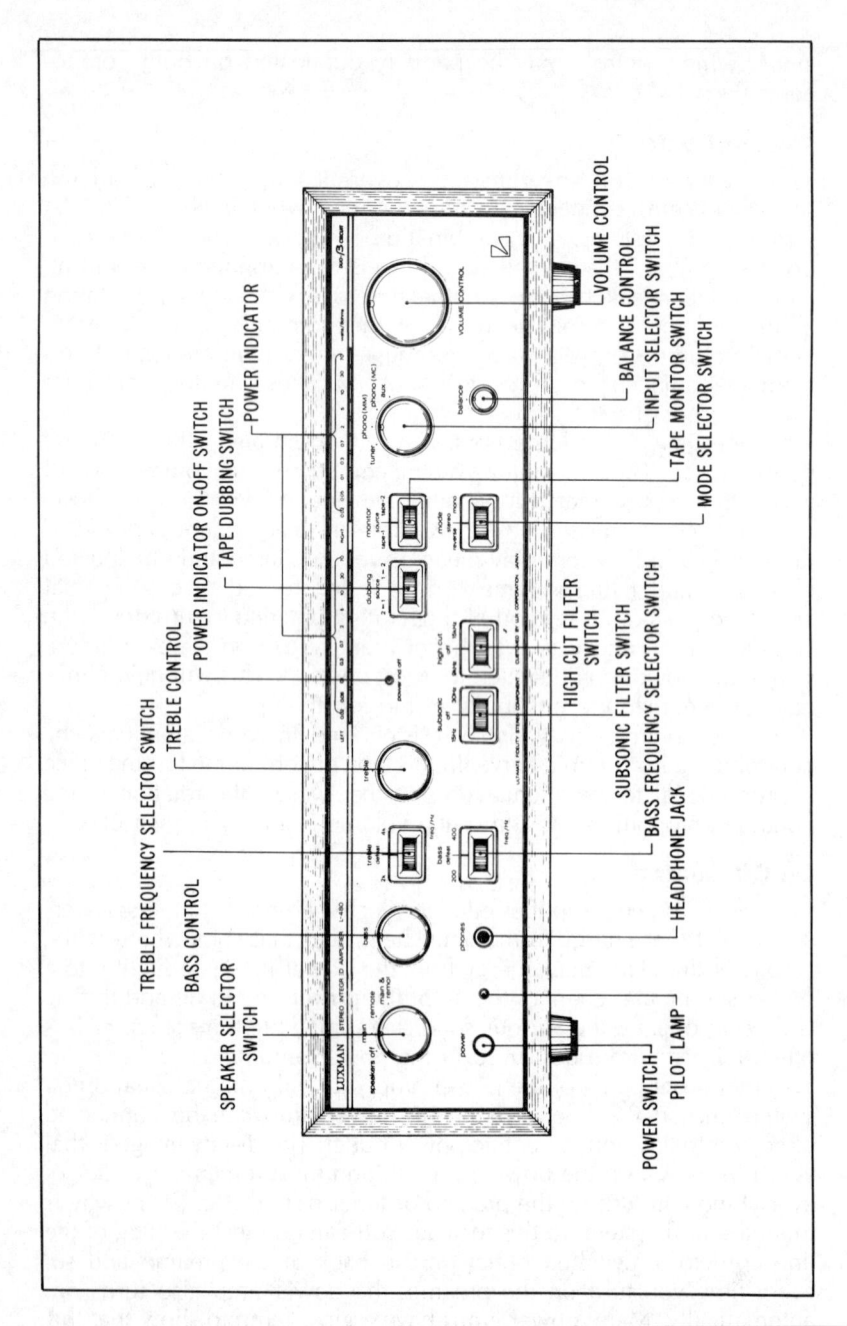

Fig. 4-8. Operating controls of an integrated amplifier.

Outlets used on the rear apron of high-fidelity components should be marked with a power rating. This rating indicates the safe maximum amount of power the outlet can handle. A power amp, for example, rated at 100 watts per channel should not be considered simply as a 100-watt load. Such an amplifier could demand as little as 75 watts of line power with no signal input or very little signal, but could require as much as 600 watts at rated power, that is, with full power output.

Time Delay Muting

In some integrated amplifiers a time delay muting circuit is included which isolates the output for a few seconds after the power is turned on. The reason for the inclusion of such circuitry is to eliminate switching thumps at the time of on/off operation of the power switch.

Balance Control

This control is used to adjust the sound balance between left and right speakers. Turn it clockwise from its center click position and the volume level of the left channel is reduced. Conversely a counterclockwise turn causes a decrease of volume from the right channel.

Input Selector Switch

This switch lets you select the program source you want. The number of choices depends on the integrated amplifier. In a quality component there may be four sound source choices: (1) tuner; (2) moving magnet (mm) phono; (3) moving coil (mc) phono; and (4) auxiliary.

Tape Monitor Switch

In some amplifiers there is provision for the reproduction of sound from two tape decks. These decks can be cassette only, open reel only, or a combination consisting of one cassette deck and one open reel deck.

The tape monitor switch may be marked Tape 1, Tape 2, to let you make a choice between the two tape inputs. When the tape monitor switch is set to either the tape 1 or tape 2 position, playback from other sound sources, such as records or broadcasts, isn't possible. For these sound sources, set the monitor to its "source" position, a click position marked on the panel of the amplifier.

POWER AMPLIFIER CONTROLS

Compared to a preamp or a receiver the front panel of a power amp is relatively free of controls (Fig. 4–9). The number of controls

Fig. 4-9. A power amplifier. Special features include Class A or Class AB operation via a selector switch, 40 watts per channel Class A, 150 watts per channel Class AB, 0.005% thd Class A, 0.008% thd Class AB, s/n better than 115 dB, amplifier and speaker protection circuits. Frequency responses from 10 Hz to 100,000 Hz, minus 1 dB. *(Courtesy Luxman, Div. of Alpine Electronics of America)*

varies from one power amp to the next and you may not even find an on/off switch. In such instances the line plug of the power amp is inserted into a switched outlet on the preamp and is automatically turned on simultaneously with that component.

POWER OUTPUT INDICATORS

A power amplifier may be equipped with a pair of indicators to keep you informed of the amount of audio power being supplied to your speaker systems. These may be meters or a display consisting of a series of light-emitting diodes (LEDs). There may be as many as 15 of these LEDs per channel, indicating the amplifier's average power output, from some nominal value such as a few milliwatts to full rated power. An LED system is easier to read and may be more accurate than meters which are often ballistically slow.

Some power amps are equipped with a pair of meters. The meters that are used are peak reading types and are also used as overload indicators, supplying a visual indication of left and right output levels. A representative pair of meters could cover a range of 0.0032 watt to 100 watts with the output transistors supplying power to 8-ohm speakers. If 4-ohm speakers are used instead, the meter readings must be multiplied by two.

The meters may also be equipped with overload indicators with clipping at about 1% total harmonic distortion. When the clipping indicators glow the input signal level should be turned down somewhat by using input level controls mounted on the rear panel.

SPEAKER SWITCHING

Some power amps are equipped with speaker switches to let you select pairs of speakers. Usually, one pair of stereo speakers is referred to as the "A" group, another pair as "B," and, infrequently, there may be a third or "C" pair.

The speaker switch permits selection of either the "A" or "B" pair of stereo speakers, so you can listen to one pair or to the other, as you wish. Another setting of this switch lets you listen to both pairs of speakers at the same time.

SUBSONIC FILTER

Subsonic filters are sometimes used to eliminate or minimize the sound produced by record warps or turntable rumble. This filter could be capable of supplying an attenuation of -3 dB at 16 Hz. The filter may be switchable, leaving its use as an option.

POWER AMPLIFIER SPECS

There is no standardization on the data supplied in a manufacturer's power amplifier spec sheet, but the most important ones are usually given. The specs supplied in Fig. 4–10 are for one model of one manufacturer and may be helpful as a basis for comparison.

Continuous Power Output

This is the maximum audio power, in rms watts, that can be delivered by both channels with both channels functioning at the same time. The rated total harmonic distortion under these test conditions must be specified. Power output ratings that do not indicate the amount of distortion are meaningless. The load impedance, that is, the voice coil impedance of the speaker system must also be indicated, and is usually 8 ohms, or 4 ohms, or both. The frequency range, generally 20 Hz to 20 kHz, should also be specified.

Total Harmonic Distortion

Ideally, in an amplifier, the component should not only reproduce the fundamental frequency of a tone, but also all its harmonics, not changing them in any way. Further, each harmonic has a different amplitude or strength compared to its fundamental and an amplifier should make no change here either.

RMS Continuous power output per channel, 20Hz to 20kHz (both channels driven) into 4 or 8 Ohms	100 WATTS @ 0.05% Total Harmonic Distortion
Total Harmonic Distortion (THD) from 20Hz to 20kHz at 250 mW to rated power into 4 or 8 Ohms	0.05% Max.
Intermodulation Distortion (IM) from 250 mW to rated power at 4 or 8 Ohms with any 2 mixed frequencies between 20Hz and 20kHz at a 4/1 voltage ratio	0.05% Max.
Frequency Response at rated power	±0.25dB, 20Hz to 20kHz
Noise	Greater than 100dB below rated power
Transient Response of any Square Wave	2.5 microseconds rise and fall time
Slew Rate	40 Volts/microsecond
Stability	Unconditionally stable with any type of load or no load including full range electrostatic loudspeakers
Damping Factor	150 Min (100Hz)
Input Sensitivity	1.5 Volts RMS
Input Impedance	50kOhms
Overload Protection	1. Low impedance electronic-sensing circuit limits with output current below 2 Ohms without limiting with 4 Ohms or higher (or reactive Loads). 2. Thermal sensing of inadequate ventilation.
Loudspeaker Protection	Relay circuit protects loudspeakers from low frequency oscillations and plus or minus DC output. Five second turn on disturbances.
Power Requirements	110-120V, 50Hz/60Hz 75 Watts @ idling to 600 Watts @ rated power

Fig. 4-10. Spec sheet for a power amp.

There should also be no time displacement (phase shift) of any of the harmonics. A musical tone is a composite waveform consisting of the point-to-point summation of the fundamental and its harmonics. A time shift of any of the harmonics means the composite waveform will no longer be an identical replica of the original tone.

If an audio power amplifier modifies any of the overtones or adds overtones that were not present with the original tone, the result is called harmonic distortion, and the sum of such changes is called total harmonic distortion, or thd.

Total harmonic distortion is livable to a certain extent since the new harmonics produced by the amplifier are mathematically related to the fundamental of the tone. It is still distortion however, for we want the amplifier to work only as an amplifier, not as a musical instrument.

How well you can tolerate thd depends on your ears, your musical training, and how well you know the musical composition being amplified at the moment by your power amplifier. Detecting total harmonic distortion is like buying a white shirt. You can only know how white it really is by comparing it with a standard, possibly some other shirt. There are instruments for measuring thd, but if you depend only on your ears, it is better to make a direct comparison between a live and a recorded performance.

Since that is rather impractical, we depend on manufacturers to supply us with thd data in their spec sheets. It is unlikely that you will be able to detect thd if it is less than 1%.

Total harmonic distortion isn't a fixed quantity. It can vary with frequency, so a thd figure at one specific frequency isn't a valuable bit of information. It can also change with the way the amplifier is being driven, tending to be less at low power outputs and more at high power outputs. So look for a thd spec figure for the whole range of 20 Hz to 20 kHz, from minimum to rated power output, with the power amplifier working into 4- or 8-ohm speakers. The spec sheet should also indicate that the thd number, supplied as a percentage, is the maximum. Since this will be the worst case, you can be sure that the thd will be lower since it is unlikely you will be operating your amp at rated output. Near or at rated output total harmonic distortion usually increases dramatically.

Intermodulation Distortion

Another form of distortion, much more annoying than thd, is known as intermodulation distortion and abbreviated as im. The problem is that every amplifier must strengthen not a single frequency but an entire, rather wide band of audio frequencies, from 20 Hz to 20 kHz. If the amplifier is completely linear, that is, if it plays no favorites and does not modify but amplifies only, the tones will pass through unscathed.

However, in the absence of such an ideal situation, musical tones can affect or modulate each other. In this modulation process they may add or subtract. Thus if a 2,200 Hz tone is present at the same time as one having a frequency of 500 Hz, we could possibly get new tones consisting of 2,200 + 500 = 2,700 Hz and 2,200 − 500 or 1,700 Hz. But these newly produced tones aren't harmonically related to either 2,200 Hz or to 500 Hz. Since they aren't harmonically related to the original tones they are dissonances. Further, since the amplifier has generated new tones, it has become a musical instrument. The lack of any harmonic relationship to the original tones makes these new tones musically offensive.

The intermodulation distortion spec, if supplied only at a fixed frequency, is meaningless. It should be given over the full power range of the amplifier, from minimum to rated power output, and this should be done over the full audio range, 20 Hz to 20 kHz. Information about the output load, whether 4- or 8-ohm speakers, should also be supplied. The im spec should also give you some idea of the relative signal levels used by the manufacturer in making the intermodulation test. A test made using one signal that is 20 or more times stronger than the other isn't significant since the amount by which one would modulate the other could possibly not be measured. A 4-to-1 voltage ratio more nearly approaches actual working conditions.

Rated intermodulation distortion figures for power amplifiers are usually less than 1%. Figures ranging from 0.1% to 0.5% are common. A rating such as 0.05% maximum is excellent. The lower the number, the better.

Like harmonic distortion, total harmonic distortion is usually low at minimum output, but climbs extremely rapidly as rated output is approached.

Frequency Response

Frequency response is the complete range of audio frequencies the amplifier can reproduce. A spec without supplying the deviation of that response, in decibels, is meaningless. The spec should also supply the amount of audio output power used when making the test, since the response becomes narrower as power output is decreased. The best spec is one that supplies frequency response at rated power output. Quite often frequency response measurements are made with audio supplied to the aux input and at 1 watt output power.

For many power amplifiers the response is from 20 Hz to 20 kHz, but there are some that are much wider than this, and can be as wide as 0 Hz (dc) to 100 kHz or even more.

Signal-to-Noise Ratio (S/N)

This spec represents a comparison between the wanted signal and unwanted noise supplied by the power amplifier at rated power output and with all tone controls in their defeat or flat position. Expressed in decibels, the higher the ratio the better. Preferably, but not usually, signal-to-noise ratios should be listed separately for each signal input: i.e., Phono (Mag): 65 dB, Aux 70 dB. A ratio of 60 dB means that the amount of signal to noise is 1,000 to 1 while 70 dB indicates a ratio of 3,162 to 1.

Transient Response

With a square wave fed into the input of the power amplifier, this spec indicates the amount of time it takes for the vertical sides of the waveform to reach its maximum value, and is measured in microseconds (μsec or millionths of a second). The spec should indicate both rise and fall times, that is, the amount of time it takes the start of the wave to reach its peak, and the time it takes for the peak to drop to zero. An ideal wave would show zero rise and fall times—2.5 μsec is excellent.

Slew Rate

Slew rate is measured in volts per microsecond and is the speed with which the power amplifier can respond to voltage changes in the signal waveform. An inadequate slew rate is usually a cause of treble tone harmonic distortion. The higher the value of slew rate the better. The amount for a quality power amplifier would be 40 volts/microsecond.

In some specs slew rate is supplied as a plus/minus figure, e.g., 100 volts per microsecond. This does not refer to a deviation of 100 volts above and below the horizontal axis of the waveform. It does mean it is the slew rate on the plus or minus side of the wave.

Input Sensitivity

The input sensitivity (also known as input level), of a power amplifier is the minimum amount of signal supplied to the amp that will permit it to deliver full output power. This means the preamplifier must be able to drive the power amplifier to rated output if the power amp is to do its job of delivering audio energy to the speakers. One of the advantages of using an integrated amp is that this problem is solved for you by the manufacturer. However, if you prefer a separate preamp and power amp make sure the two can work as a team. Preamp outputs can range from a fraction of a volt to one or more volts of audio.

The signal input voltage is in rms, implied or stated. Sometimes the input impedance of the power amp is included in this spec.

Input Impedance

The input impedance of a power amp may be supplied as a separate spec, or, as indicated, this data is included with the input sensitivity spec. A typical value is 47,000 ohms, although you will find others, such as 50,000 ohms, or a much lower amount, such as 20,000 ohms.

Power Bandwidth

Power bandwidth is a relationship between power output and frequency response (Fig. 4–11), and is the frequency range over which the amplifier will deliver half its rated output power without exceeding its rated harmonic distortion. If a power amp can deliver 50 watts per channel with 0.09% harmonic distortion, then a power bandwidth rating of 20 Hz to 20 kHz means that this amp will supply 25 watts between these two frequency points with no more than 0.09% distortion.

In the example just given, the two values of power are 50 watts and 25 watts, a 2 to 1 ratio, or, in terms of decibels, 3 dB. Fig. 4–12 shows the power bandwidth of an amplifier, a measurement that is made 3 dB down or between the half power points.

Frequency response and power bandwidth are quite different, and, if anything, the power bandwidth spec is more significant. Frequency response is the range of audio frequencies the power amp will reproduce from audio signals applied to its auxiliary input with only 1 watt of audio output power.

Damping Factor

Damping factor is another interesting spec and is the ratio of the impedance of the speaker's voice coil to the internal impedance of

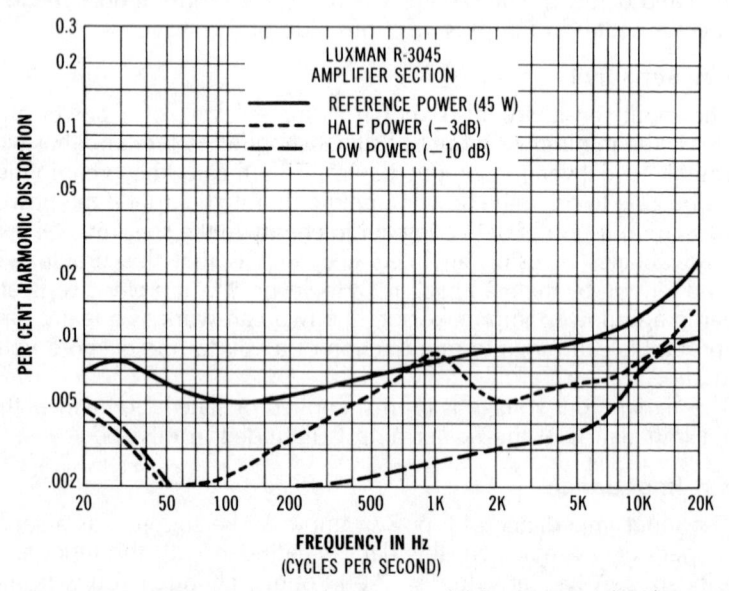

Fig. 4-11. Frequency response (shows as a solid line) compared with power bandwidth in terms of harmonic distortion. For half-power, distortion is substantially lower over most of the audio spectrum except at 1 kHz.

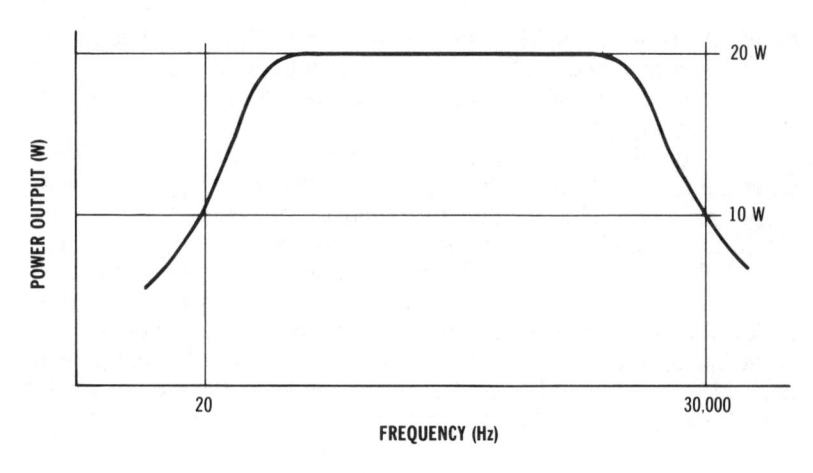

Fig. 4-12. Power bandwidth is measured between the half-power points.

the amplifier. A speaker rated at 8 ohms, used with a power amp whose internal impedance is 0.1 ohm, is 8/0.1 = 80. Damping factor is the amount of restraint of movement of the speaker voice coil by forces other than the audio signal. Typically, power amps have damping factors from 15 to about 100, but anything over 20 isn't meaningful.

EQUALIZERS

The purpose of an equalizer is to compensate for speaker deficiencies and for modification of the particular acoustics of a listening room. To some extent this can be done by the tone controls on a receiver or amplifier, but tone controls are rather coarse devices. If just two tone controls are used, one for bass and the other for treble this means that the entire audio range from 20 Hz to 20 kHz will be divided into approximately two sections, with each tone control responsible for about half.

The advantage of an equalizer is that it divides the audio spectrum into narrow segments and can boost or cut each of these segments. The amount of boost or cut is handled by a control, often a vertical slide lever, one for each audio segment. Unlike tone controls, left and right channel sound can usually be modified separately.

Some equalizers are full octave types. The sound spectrum is divided into 10 octaves with the equalizer supplying a plus or minus boost or cut for each individual octave. The amount of boost or cut varies from one equalizer to the next, but 12 dB is typical.

Some equalizers are half-octave types with the frequency range from 20 Hz to 20 kHz divided into 20 different audio frequency bands. The advantage is that each band to be amplified or cut is a much narrower segment of the audio range and so you can get finer tone control.

Some equalizers are equipped with pink noise generators. This is a wide band of noise you use as a reference to measure your listening response, channel balance, and speaker phasing. A pink noise generator is used since our ears are sensitive to energy rather than amplitude. Consequently, adjustments made using the pink noise generator are more accurate and can be perceived more easily. With pink noise you can also recognize frequency peaks and dips over the audio range in a listening room. Pink noise consists of equal parts of each octave of the audio spectrum. You can get some idea of what it sounds like by listening to the noise between radio stations on the fm band.

There are two types of equalizers: the graphic which divides the audio band into five, ten, twenty, or more bands with each band having boost or cut controlled by its own lever. A parametric equalizer is somewhat different in that it permits control of the width of the selected band, so that it can be narrow or wide. The parametric usually offers a variable frequency control and variable boost or cut.

156

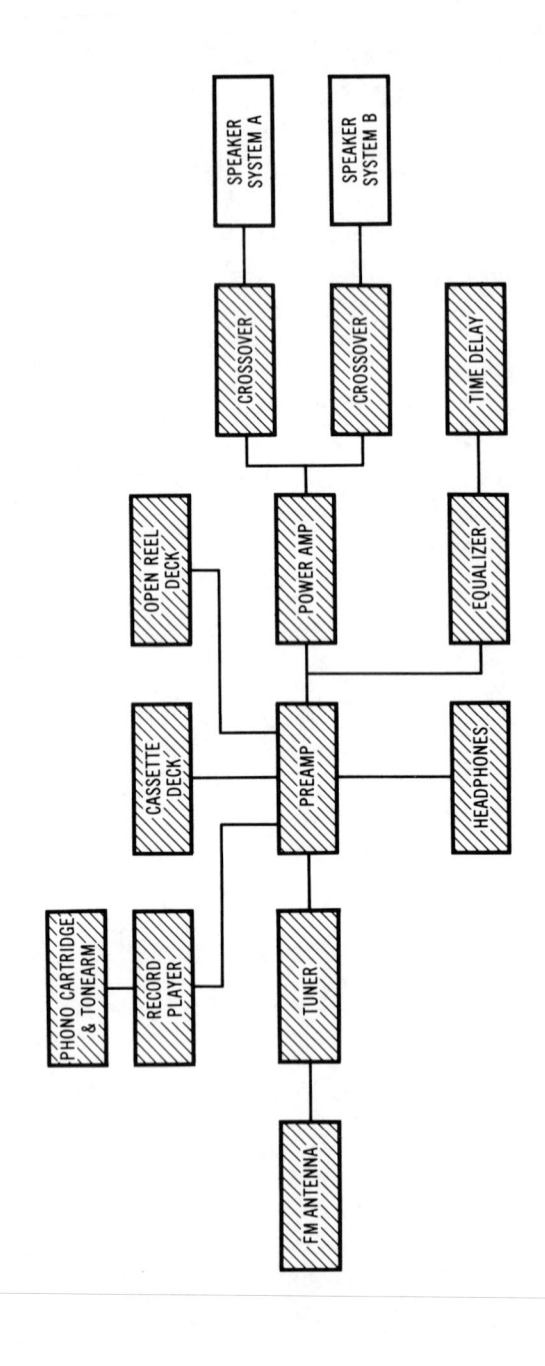

5

Speakers and Headphones

In any high-fidelity system the speakers are the moment of truth. They are still the weakest link in the system, but they do not bear the entire responsibility; for what is heard depends not only on the speakers but on the size of the listening room, its shape, acoustics, speaker positioning, on the physical condition of your ears, musical training, age, and sex.

This means that what is heard from speakers is subjective, tempered by a collection of variables. The overall result is that to get sound that is pleasing it isn't possible to buy just any speakers, to put them anywhere you wish, and then expect superior results. Some luck may be involved, but it plays a very small part.

WHAT DO WE WANT?

Our demands on speakers are reasonable. All we ask of them is that they should not screen the sound, to reproduce it without adding any sounds of their own, and not to emphasize any portion of the musical spectrum. We want the speakers to be present in our listening room, but we do not want them to remind us they are there. We want music to come from the speakers but not to appear as if it is doing so.

We want our speakers to be free of hum and noise. We want them to deliver music as it was recorded or broadcast, with every musical tone remaining in its proper relationship to every other tone. In short, we do not want our speakers to have delusions of grandeur, nor should they consider themselves as prospects for joining any musicians' union.

Perhaps all this is a bit too much to ask, for surprisingly few speakers can cope with these demands.

There are various kinds of speakers but they all have one thing in common and that is they work by pushing air. They do this by the movement of a bit of material, a cone, a flat surface, a cylindrical surface, a diaphragm, or gas. The movement of these substances produces alternate compressions and rarefactions of the air molecules and when this change in the arrangement of air molecules reaches our ears we hear sound. It is these rapidly altering changes in air pressure that give us the sensation of sound.

DYNAMIC SPEAKER

The dynamic speaker has relatively few parts (Fig. 5–1) and these include a permanent magnet, a voice coil, and a cone. The voice coil consists of a number of turns of wire, wound on a thin bobbin made of plastic impregnated paper or aluminum or sometimes not wound on anything at all. It is held in position between the poles of an extremely strong permanent magnet. Lines of magnetism, or magnetic flux, exist between the north and south poles of the permanent magnet and surround the voice coil fastened to the cone at its apex. The cone is held in position by two devices. One is a surround, a flexible material around the outer circumference of the cone, and the other is a spider (Fig. 5–2), a highly flexible support at the voice coil. The spider helps to center the voice coil, holds the voice coil in position, and also has enough spring action to return the voice coil to its original position whenever it stops moving. The outer edge of the surround is attached to a metal structure called the frame.

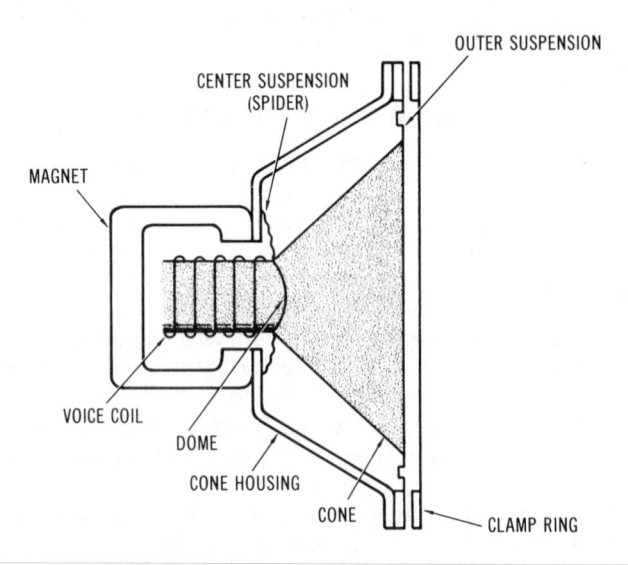

Fig. 5-1. A dynamic speaker's basic structure.

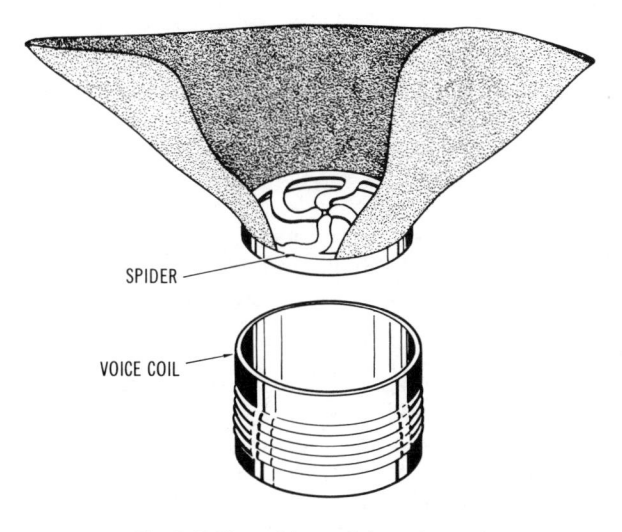

Fig. 5-2. The spider and the voice coil.

Two wires lead from the voice coil to a terminal on the speaker frame, called a basket. From here the voice coil leads are connected to a pair of terminals on the outside of the speaker enclosure. The audio power output of the power amplifier or the receiver is connected to these terminals. In operation, audio currents from the power amp or receiver are led into the voice coil and in so doing produce a magnetic field around the voice coil. This magnetic field, varying at an audio rate, interacts with the magnetic field supplied by the permanent magnet, forcing the voice coil to move back and forth at an audio rate. This motion also drives the cone (hence the name driver for speakers). It is the motion of the cone pushing the air that ultimately results in sound.

Woofers

No single speaker can adequately reproduce the entire spectrum of audio sound from 20 Hz to 20 kHz so this job is shared among a number of speakers. The speaker responsible for handling bass tones is called a woofer. Bass tones carry substantial amounts of energy so this means they produce stronger audio currents that are sent to the voice coil. They require a stronger magnetic field from the permanent magnet whose pole pieces surround the voice coil, so magnets for woofer speakers are more than ordinarily heavy. And, to reproduce bass tones, the cone must also be larger.

Midrange

The speaker designed to handle midrange tones is known as a midrange driver. At one time it was called a squawker, but possibly because this name is so inelegant (and perhaps conveys a wrong impression), it has been discarded.

Tweeter

The tweeter is intended to reproduce treble tones and, because these require comparatively small amounts of audio energy, is small when referenced to the woofer and midrange.

Crossovers

A crossover is a network usually made up of capacitors and coils and sometimes will include resistors (Fig. 5-3). Their function is to divide the audio spectrum into narrower bands so these can be handled by appropriate speakers. Thus, a crossover might divide the audio band into three parts: one section for the woofer, another for the midrange driver, and a third for the tweeter. Since each speaker will be responsible for reproducing only a small part of the audio spectrum it will be able to do a better job.

Crossover Frequencies

Crossover frequencies are borderline between the low, midrange, and high ranges and are generally called crossover points. However, a crossover is not a single spot frequency but rather a small band.

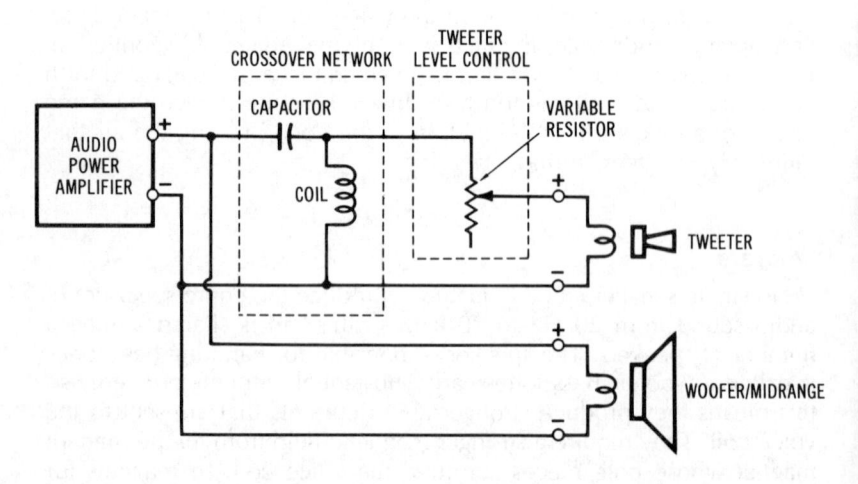

Fig. 5-3. A crossover network is used to channel treble tones to the tweeter; bass and midrange tones to the woofer/midrange driver.

SPEAKER SYSTEMS

A speaker system (Fig. 5-4) may have just one speaker in a vain attempt to cover the entire audio range, or it may have two speakers, three, or more. In some instances a single speaker will be used to handle the bass and midrange tones and another for the treble.

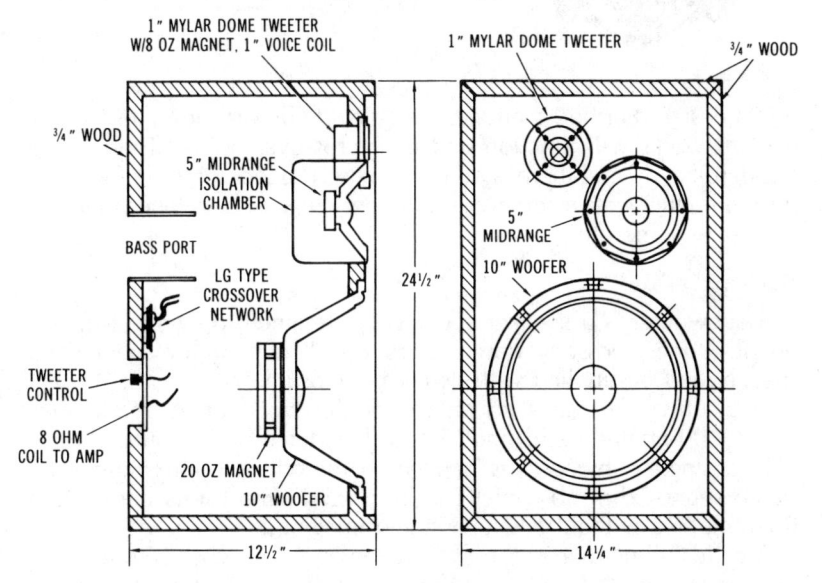

Fig. 5-4. A dynamic speaker system. *(Courtesy Acousti-phase Corp.)*

A speaker system includes any group of two or more speakers (Fig. 5-5) housed in an enclosure. If the audio spectrum is divided into two parts the system is called two way; if into three parts it is known as three-way. A two-way system, then, would consist of a minimum of two speakers, each responsible for reproducing some assigned part of the audio band. A three-way system would comprise a minimum of three speakers. In this arrangement the audio band is divided into three parts.

A system could include one, two or more speakers for each section of the audio band. Thus, a speaker arrangement might include two woofers, a single midrange, and two tweeters. This setup would then be a three-way, five speaker system. The woofer section will produce tones from about 20 Hz to 500 Hz, the midrange from 500 Hz to about 3000 Hz, and the tweeter frequencies from 3000 Hz to 20,000 Hz.

The reproducing range of a woofer, midrange, or tweeter depends

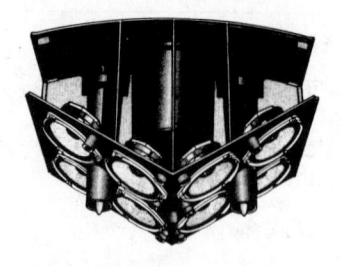

Fig. 5-5. A multispeaker system. (Courtesy Bose Corp.)

on its design. Further, manufacturers decide on just how their crossover networks will function and their crossover points. There is no standardization and there is no uniformity. The woofer in one speaker system might cover a somewhat different range of bass tones than the woofer in some other system.

Subwoofer

A subwoofer is a speaker specifically designed to handle low frequency tones, possibly down to 16 Hz. It is a highly specialized speaker and is supplied with its own enclosure.

The subwoofer takes advantage of the fact that at frequencies below 200 Hz the ear cannot detect the source of the sound, hence only a single subwoofer is needed to reproduce both channels of stereo sound. The subwoofer can be placed almost anywhere in the listening area without affecting stereo imaging.

A typical cone diameter for a subwoofer is 12 inches with a crossover down about 100 Hz. The purpose of the subwoofer is to expand the bass range down to the lowest frequencies, generally in the region of 16 Hz to 28 Hz. One of the reasons for using a subwoofer is that a typical woofer becomes more efficient with higher frequencies, altering the ratio of produced ultra bass tones to the somewhat higher frequency bass tones. The effect is to make the ultra bass tones less audible than they should be. The reason for the increased efficiency of gradually increasing frequency bass tones is that the higher bass tones have a shorter wavelength and so get closer to the actual diameter of the cone itself. This results in better coupling between the cone of the woofer and the air mass it is to move. The subwoofer is one way of extending the cone diameter of the woofer in two steps, rather than using one huge, and quite possibly impractical, cone.

Double Cone

Various techniques have been tried to make a single speaker do the work of a woofer, midrange, and tweeter. One method is the double cone speaker in which the cone is made of two different kinds of materials, one for bass tones and the other for midrange-tweeter use.

This is an interesting speaker but it has no application in high fidelity systems.

Whizzer

Another arrangement is to have two cones fastened to the same voice coil and a whizzer speaker is this type. The speaker has two cones, one of which is larger than the other and, like a usual speaker, has its outer circumference attached to a highly flexible surround. The whizzer cone, usually quite small, is fastened to the voice coil at its apex, but its outer circumference is free. The theory is that each cone will naturally respond to a certain band of frequencies, the larger cone to bass and midrange tones and the whizzer cone to treble, so a crossover for separating these frequencies isn't needed. The whizzer speaker is used in some auto sound systems, but not in in-home hi-fi setups.

Coaxial

A coaxial speaker, as its name implies, has two drivers mounted along the same central horizontal axis. Each speaker has its own voice coil and is operated independently from audio frequencies supplied by a crossover network. The speakers are ordinarily a woofer, and a midrange/treble unit. Sometimes three speakers are mounted this way, using a woofer, midrange, and tweeter. This arrangement is called triaxial.

Horn

The horn speaker (Fig. 5–6) as its name suggests, uses a horn for coupling the output of a speaker to the air. Like other dynamic speakers, the horn speaker uses a voice coil, but instead of being

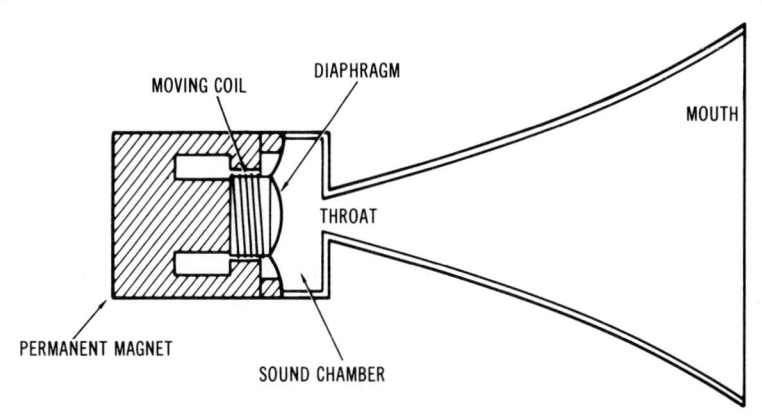

Fig. 5-6. Horn speaker.

connected to a cone is attached to a diaphragm. The vibration of the diaphragm produces changes in sound pressure which are then coupled to the air in the room via the horn. To save space, the horn may be folded back on itself (Fig. 5–7). Some horns, when used as tweeters, are equipped with a number of cells for diffusing the sound. A horn tweeter is sometimes used in the same enclosure with a dynamic cone type (Fig. 5–8).

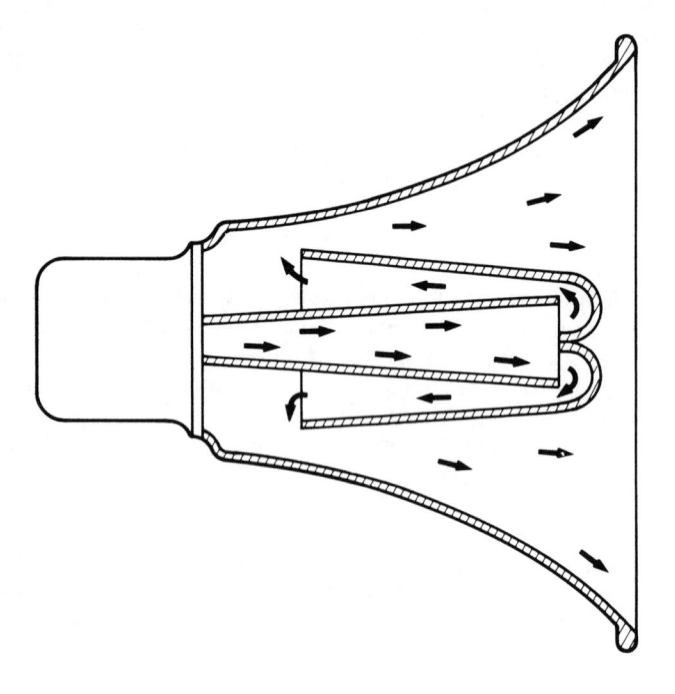

Fig. 5-7. Folded horn speaker.

Multitweeter

Because the sound energy contained in treble tones is less than that present in bass tones but is more directive, listening to treble tones becomes rather critical. Thus, in some instances, depending on the positioning of the tweeters and their number, their effectiveness may be reduced for group listening. To overcome tweeter directional effect, some speaker manufacturers mount a number of these units in the enclosure, forming a tweeter array arranged in an arc, or a diffuser lens may be positioned in front of the driver (Fig. 5–9).

What makes the speaker problem so difficult is that an enclosure with a larger group of speakers doesn't necessarily guarantee better

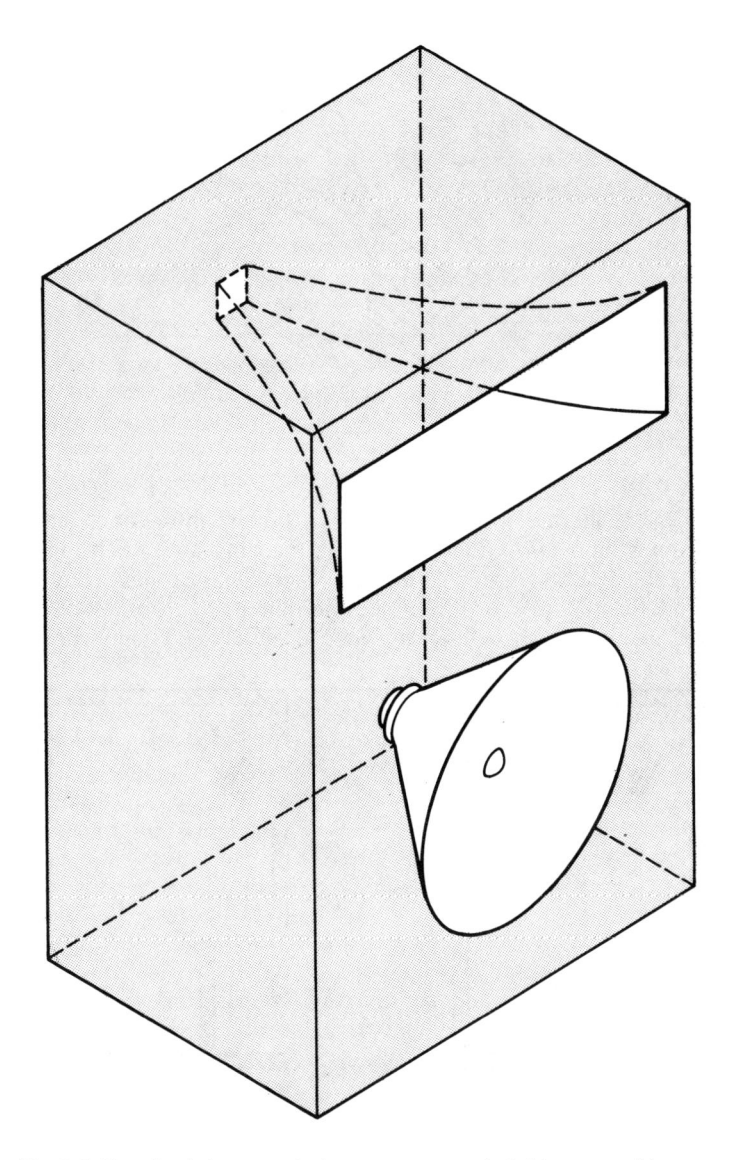

Fig. 5-8. Horn loaded tweeter and a cone type woofer/midrange used in same enclosure.

sound. A number of tweeters used instead of a single one diffuses the treble tones so that more listeners may enjoy the sound simultaneously. It does make treble tone listening much less critical, but this doesn't imply a better treble frequency response.

Fig. 5-9. Diffuser lens for a tweeter.

Ported

Also known as bass reflex speakers or vented port speakers (Fig. 5-10), and, at one time called boom boxes, these enclosure arrangements have fairly good efficiency; their design is such that they are capable of delivering surprisingly good bass tones from small sized enclosures. Because their size permits them to be made small enough to fit on a shelf they are often used as bookshelf types.

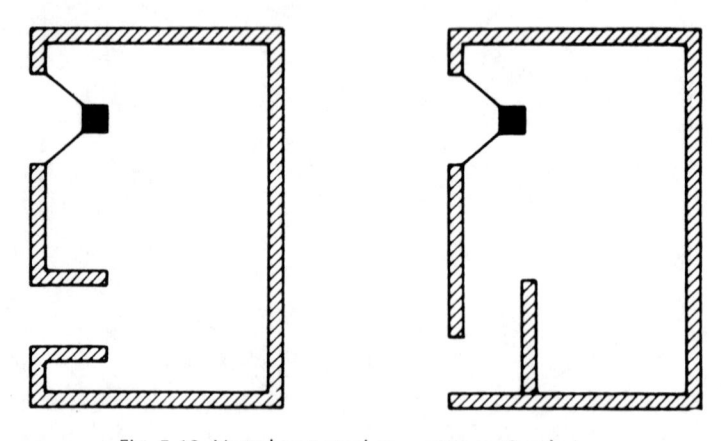

Fig. 5-10. Vented-port speakers—cross-sectional view.

Vented port speakers work by channeling some of the rear sound back to the front through a duct or port. The venting or duct arrangement is such that the phase of the rear produced waves is inverted and thus put in phase with the waves produced by the forward motion of the cone. These speakers have a vent in the front of the speaker enclosure in addition to the hole cut for the speaker itself.

The efficiency of the vented port speaker is based on the fact that the sound produced by the rearward movement of the cone isn't wasted, but rather is utilized by joining with the front movement of the cone. As a result the amount of sound produced by a given watt of audio input power is increased. However, it is important to remember that efficiency is no index to the quality of a speaker but is simply a reflection on its ability to work well—or poorly—as a transducer.

Infinite Baffle

An infinite baffle speaker system uses a sealed enclosure. As a result the sound wave produced by rear movement of the cone is not permitted to escape, but is absorbed by material covering the inner surfaces of the enclosure. The enclosure is often large and the trapped air does not act as an acoustic suspension. See Fig. 5–11.

Acoustic Suspension

The speaker cone, as mentioned earlier, is moved back and forth by the voice coil attached to it. The cone has three main positions: its starting position, its maximum rearward position, and maximum forward. When no current flows through the voice coil the cone should resume its starting position. The surround supporting the outer rim of the cone not only acts as a suspension, but somewhat as a spring restoring the cone to its starting position in the absence of a signal.

This arrangement does present a problem. For deep bass tones the cone must make wide excursions, but the more the cone moves toward its maximum rearward position or maximum forward position the greater the restraint put on the cone by the supporting surround. This restriction distorts the sound and can severely affect bass response.

Fig. 5-11. The infinite baffle speaker uses a sealed enclosure.

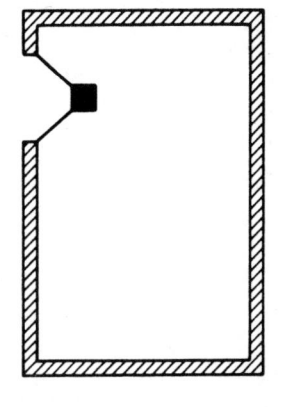

The acoustic suspension speaker (Fig. 5–12) also uses a surround from the basket or frame of the speaker to the outer rim of the cone, and it is made extremely loose. However, the acoustic suspension speaker uses the air inside the enclosure to supply the restoring force for the cone. The enclosure is completely sealed and so the air can be

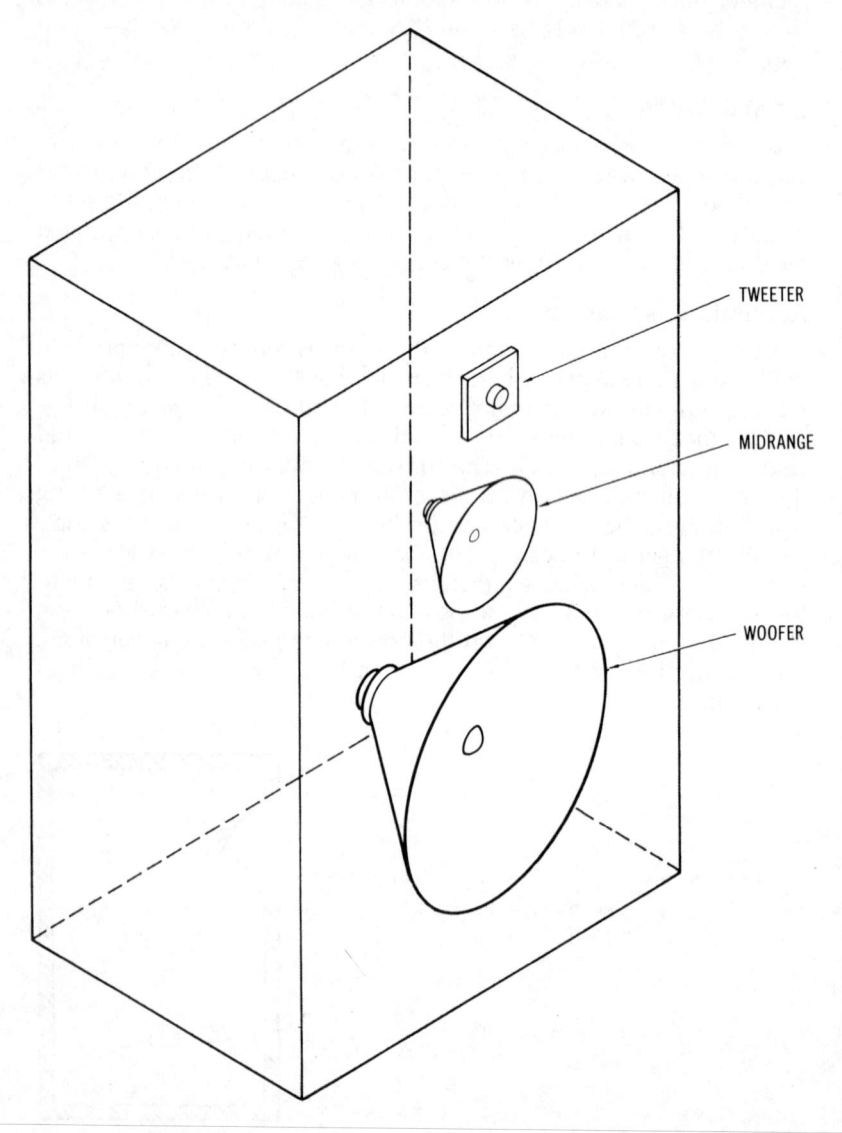

Fig. 5-12. An acoustic suspension system uses a sealed box.

compressed or released. Thus, the acoustic suspension is one type of sealed enclosure technique. The air rests somewhat like a spring against the cone, giving it support, yet allowing it to move the greater distances needed for bass tone reproduction.

An advantage of the acoustic suspension type is that it not only can supply extended bass response but can do so in a small enclosure. Prior to the introduction of the acoustic suspension speaker, enclosures had to be rather large, free-standing floor types to produce clean, deep bass tones. The acoustic suspension speaker can achieve good bass reproduction using a small enclosure. These speaker enclosures can be so small, in fact, that such speakers are often used as bookshelf types.

However, there is a tradeoff, and there usually is. Acoustic suspension speakers are low efficiency types. This is no great problem since there are numerous audio power amplifiers that are fully capable of meeting the minimum audio power input demands of such speakers, but most of that audio power is simply not utilized for the production of sound.

Dome Radiator

High frequency drivers, or tweeters, must be capable of driving their speaker cones as rapidly as 20,000 times per second (20 kHz), so such speakers must be as small and light as possible. This presents a problem because it means that the voice coil must also be small. Physically, this is no difficulty. The problem arises with heat dissipation. A small voice coil means a small radiating surface, and that, in turn, means an inability to dissipate heat rapidly.

Various techniques are used to overcome this problem. In one of the ways being used to overcome this problem, the voice coil of the tweeter is made larger and, instead of using a cone for getting air movement, a dome structure is used instead. The voice coil is attached to the outer edge of the dome. The larger size of the voice coil is an aid in radiating the heat produced by audio currents in the voice coil; the shape of the dome minimizes the beaming effect of high frequency tones. Soft domes are now used to reduce weight and increase power handling capability.

Another advance in tweeter design is the elimination of the spider used to center the voice coil in the space between the poles of the permanent magnet. In place of a spider a magnetic fluid is used to center the voice coil. An advantage in eliminating the spider is that there is more room for a larger voice coil. The magnetic fluid also acts as a heat-conducting medium since it helps carry the heat generated by the voice coil to the basket, the metal frame of the speaker system. The relatively large metal area of the basket provides better heat radiation characteristics.

LZT

Lead zirconate titanate, or LZT, is a semiconductor material that can convert electrical energy to physical motion. In this respect it is analagous to quartz crystals that deform when a voltage is applied.

The driver doesn't have a magnet, nor does it have a voice coil. The semiconductor converts electrical signals directly into sound. Speakers using LZT do not function with the usual crossover network; instead they have a special passband filter that supplies the range of frequencies to be handled by the speaker.

Speakers of this type are also known as compression drivers and are generally equipped with circumferential plugs to force their treble tones through equally spaced circular slots.

High Polymer Film

A high polymer film is a vapor-deposited aluminum coated film having a polyurethane backing with the film expanding and contracting in step with the amplitude of the applied signal. Its behavior has been compared to that of piezoelectric speakers but here the high polymer film moves the air directly. In piezoelectric tweeters a coupling device, such as a horn or cone is needed.

The film, whose thickness is just 0.0012 inch, produces sound that radiates horizontally in a full circle. No voice coil or magnets are used and so the force needed to set the film in motion is much less than in conventional voice coil drivers. No standing waves appear on the surface of the film and the film's expansion and contraction aren't affected by resonances or saturation. The high polymer film is wrapped around a cylinder filled with sound absorbing glass wool and is used for tweeters and supertweeters.

Omni and Dipole Radiators

Although microphones and speakers are at opposite ends of any hi-fi system they are both transducers. In intercom systems the microphone also functions as a speaker so it isn't surprising to find microphone technology applied to drivers.

The two most commonly used microphones are the omnidirectional, or omni, having equal sensitivity to sound from all directions. The other is the cardioid, having a heart-shaped response. Popular, but not as widely used is the figure eight, insensitive to sound from the sides, but having good front and rear responses.

The omni and figure-eight concepts have been applied to speaker design (Fig. 5–13). One speaker system has a full range omni with a response from 30 Hz to 18 kHz and a selectable dispersion angle of 180° or 360°. The enclosure is a truncated sphere, a dodecahedron, that is, a geometric figure having 12 plane surfaces. Consequently the sound is projected toward the listener in all directions.

(A) Omni. (B) Dipole or figure eight.

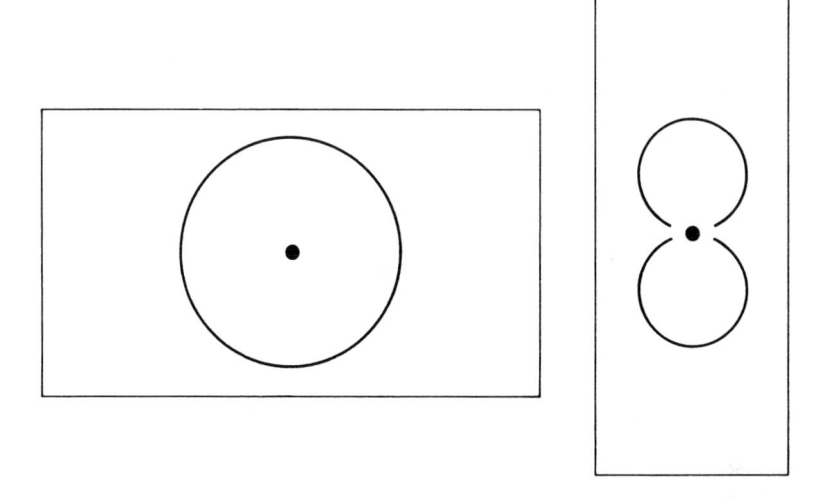

Fig. 5-13. Speaker radiation patterns. Black dot in center represents driver.

One characteristic of an omni is the large number of drivers. One system uses two 8-inch woofers, one 1½-inch dome midrange, two 5-inch cone midrange drivers, two 1-inch dome tweeters and three 1½-inch cone tweeters.

A speaker having a figure-eight radiation pattern is more often known as a dipole radiator. You can get a good idea of this pattern by visualizing a pair of circles tangent to each other with the speaker positioned at the point of tangency.

Plasma

The requirements of a sound producing cone or membrane is that it should be as light as possible, as rigid and nonflexing as possible, and completely air-leakproof. It should also be nonresonant.

There are two ways of meeting these requirements. One is to design a cone or membrane that will have these characteristics. The other is to eliminate the cone or membrane completely, something that is accomplished in the plasma or corona discharge or ion speaker.

A corona is caused by ionization of the air, a process in which atoms comprising air molecules are deprived of electrons. Ionized air has a pungent, sharp odor and usually ionization is an unwanted effect, commonly taking place around the 15.75 kHz horizontal output transformers in tv sets or in high voltage power lines.

When high voltage wiring is terminated in a sharp, needle-like point, the air surrounding this point is exposed to this high potential and becomes ionized. The movement of a cloud of ions in the region near the needle point is known as corona and is accompanied by a glow that is characteristically blue or purple. A corona discharge can occur between any two electrodes, one sharp, the other blunt. It is this corona discharge which is the basis of the plasma speaker.

Typically, a plasma speaker has no mechanical diaphragm. Instead, solid-state circuitry inside the case housing the driver delivers a high-voltage signal to a teflon mounted electrode. This electrode generates a small, blue "flame" of ionized air within a spherical wire mesh ball. The heat of the ion cloud varies with the audio signal. The high voltage oscillator works at a frequency of 27 MHz and is modulated by the audio signal, producing heat variations of the ions. The heat variations result in pressure changes and it is these changes that result in sound. Thus, the unit is a thermodynamic transducer since the sound is generated by changes in the temperature of the ion cloud.

Another speaker along similar lines is one in which a plasma, a mixture of helium and air, is used. Sound is produced by pressure changes as a result of temperature changes in the plasma. The helium, an extremely light and nonflammable gas, is delivered to the speaker from a tank of helium under compression. The tank, of course, requires an occasional refill.

Arc transmitters along the lines of those used circa 1919 for the transmission of signals in Morse code can be modulated by an audio signal to produce sound. While such speakers have been demonstrated, there are none in commercial production.

Flat Plane

Flat plane membranes used for driving air instead of cones aren't new for they have been used for years in electrostatic speakers. But applying the flat plane concept to a dynamic speaker is another matter.

In the flat plane dynamic speakers the drivers are square with the magnets positioned in each of the four corners. The flat planes are aluminum with material in the form of a resin honeycomb between the two surfaces.

Drones

Drones, also known as passive radiators, are well named. They aren't equipped with a voice coil, have no magnets, and have no terminals for connections to a crossover or to the output of a power amp. They work on the same principle as a pair of identical tuning forks placed near each other. Striking one will cause the other to vibrate.

Passive radiators are generally used as woofer supplements. The cone of the drone is made to move by changes in air pressure inside the speaker enclosure. The cone of the passive radiator can be larger than that of the woofer. The advantage of the passive radiator design is that it can yield usable bass response with relatively high efficiency. In tuned port speakers the diaphragm of the drone can be used instead of a port.

Ribbon Transducer

Ribbon speakers (Fig. 5–14) aren't new for various attempts have been made for over a half century to produce one that is practicable, usable. The problem with such speakers lay in the fragility of the

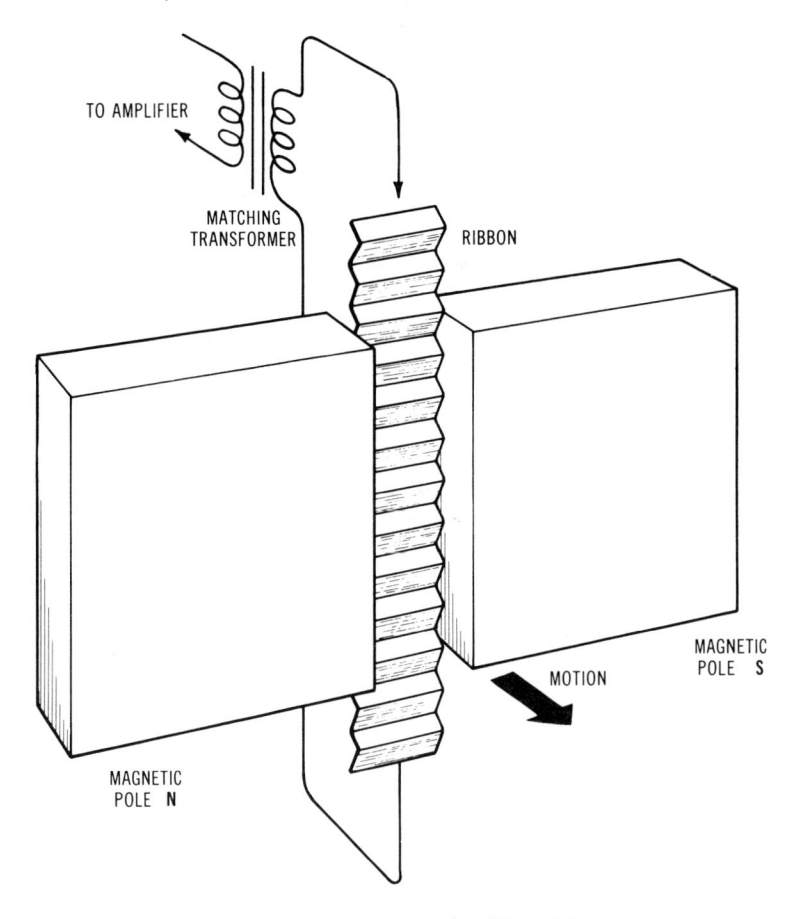

Fig. 5-14. Basic elements of a ribbon driver.
(Courtesy Jumetite Laboratories, Ltd.)

moving diaphragm—a narrow element made of aluminum foil or mylar with an aluminum coating. The efficiency of the ribbon speaker is low and the impedance is generally less than one ohm.

The basic principle of the ribbon speaker is simple. When a current is sent through a conductor placed in the fixed field supplied by a permanent magnet the conductor will move, a principle used in motors. Because the ribbon is completely immersed in the magnetic field it moves as a unit, that is, as a piston. And, since the ribbon is its own suspension, all resonances can be well below the frequency range of the speaker.

Various methods are used to improve the efficiency of ribbon speakers. The ribbon is pleated, transformers are used to match the impedance of the diaphragm, and the speakers may be horn loaded (Fig. 5–15). Ribbon speakers are used as tweeters and can have a wide frequency range, possibly 7 kHz (and lower) to 30 kHz.

Analog Bass Computer

The Analog Bass Computer anticipates speaker cone motion by reading the output of the power amplifier, controlling cone excursion accordingly.

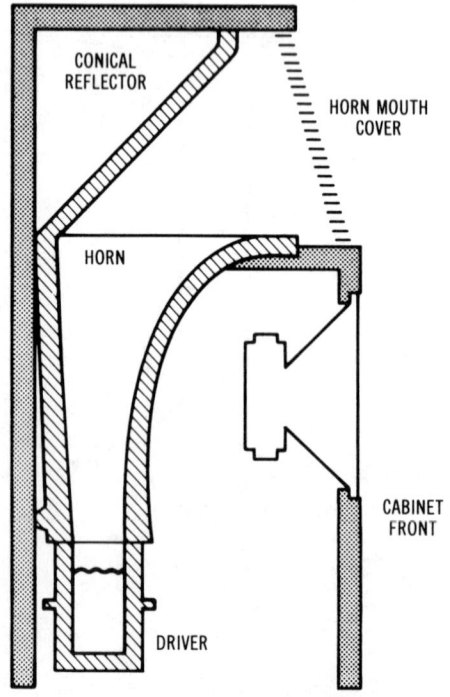

Fig. 5-15. Ribbon speaker is horn loaded.

Fig. 5-16 shows the functioning of the system. Left and right stereo signals are fed into the Analog Bass Computer from a preceding preamplifier and from the computer into a stereo power amp. While the computer does supply equalization at the low end its main function is controlling the voltage going to the speakers, hence the excursion of the voice coil.

The Analog Bass Computer is a separate component connected to and situated near the amplifier or receiver. Optimum performance is obtained on normal music signals in average rooms. When the signal level rises the low frequency input to the speakers is automatically controlled by the Analog Bass Computer according to the demands of the music to prevent mechanical overload. One of its advantages is that it permits the use of smaller diameter woofer cones, resulting in much smaller speaker enclosures.

Motional Feedback

As an alternative to finding new cone materials, or as an adjunct to them, some manufacturers are making use of feedback systems. Feedback of signal voltage from the output of a circuit to the input of that circuit, or some preceding circuit, can be negative or positive. If the feedback is negative, that is, if the output signal is out of phase with the input signal, the effect is a reduction in output signal strength with a reduction in distortion. With positive feedback the output signal is in phase with the input. This increases the gain of the circuit and if the amount of feedback is sufficient the amplifier starts working as an oscillator, generating its own output signal.

Feedback is also used in speaker systems and is part of the motional feedback system (MFB) speakers made by Philips High Fidelity Laboratories, Inc. Feedback is applied only to the woofers, not to the midrange or tweeters (Fig. 5-17). Since the tweeter and midrange drivers have relatively small diameters the chance of cone breakup is reduced. In the case of the woofer its much larger diameter makes unified cone movement more difficult. Further, the bass driver must move larger volumes of air so the load on the cone is more substantial than on smaller diameter drivers. This combination of large cone diameter and heavy load makes the woofer more susceptible to cone deformation. Cone lightness is also essential for making the woofer transient responsive, but cone strength is equally necessary.

In the motional feedback system the audio signal is brought in from a signal source, a preamplifier, and is fed to a 500 Hz electronic filter. Tones ranging from 500 Hz to 20 kHz are sent into a midrange and treble power amp. The output of this amplifier is then fed into a 4 kHz filter where tones from 4 kHz to 20 kHz are routed to the tweeter; those from 500 Hz to 4 kHz to the midrange driver. The filters are

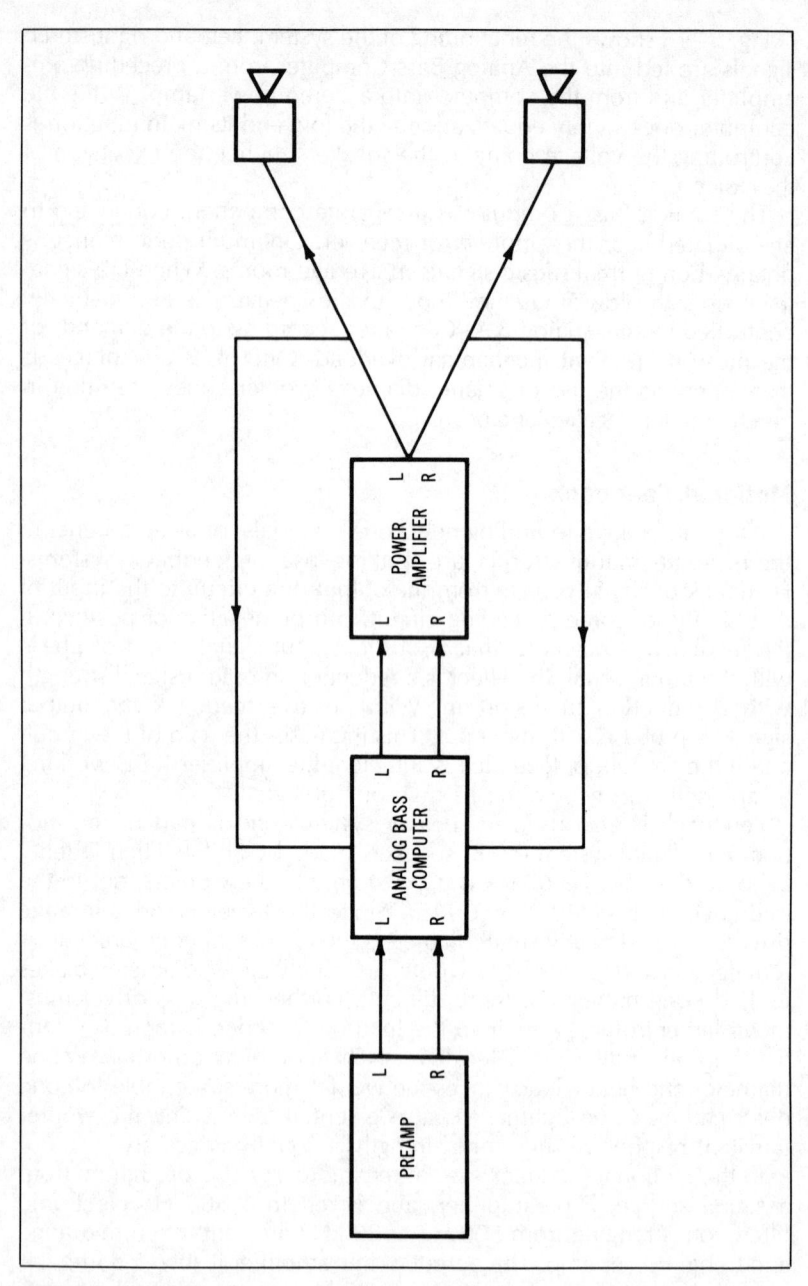

Fig. 5-16. The Analog Bass Computer is inserted between the preamps and the power amps.

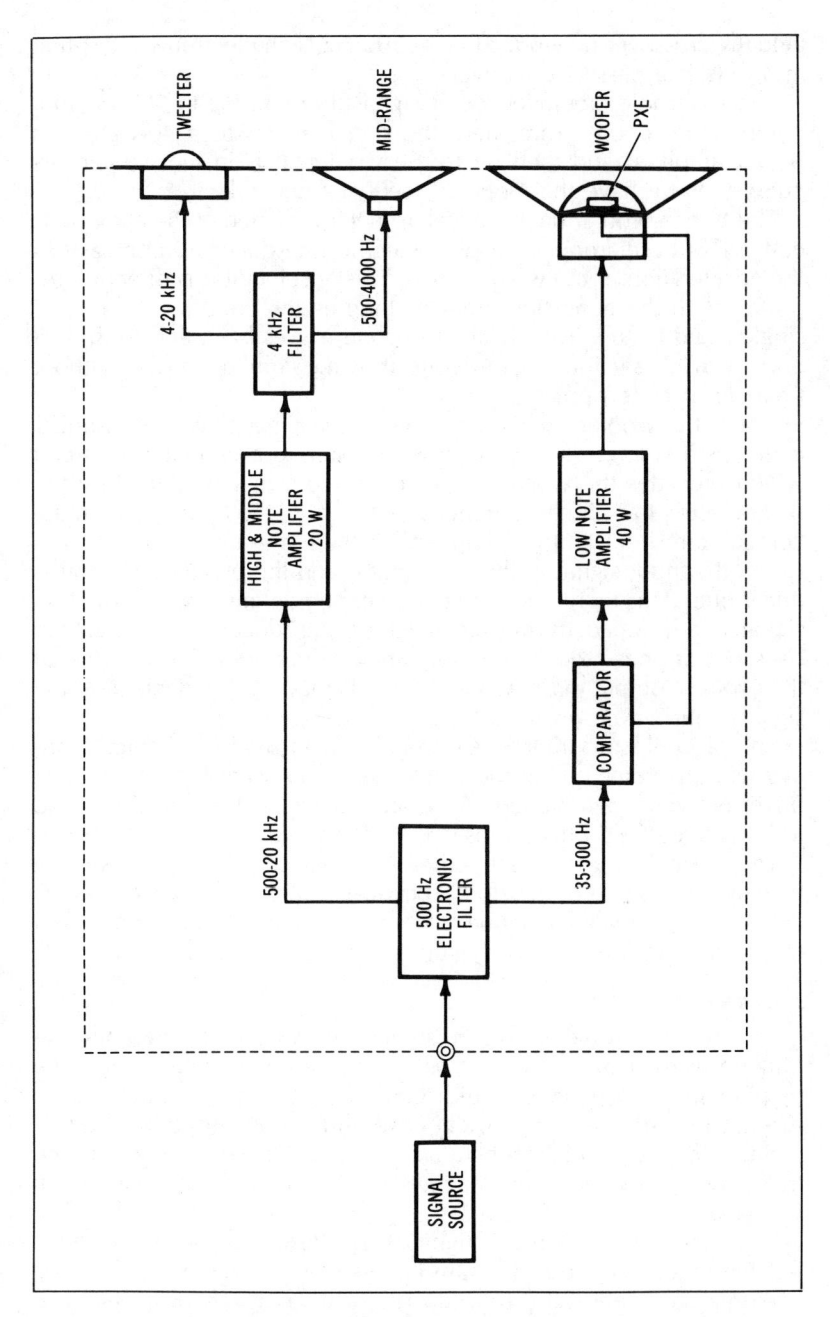

Fig. 5-17. Motional feedback system arrangement.

actually crossover networks; that at 500 Hz is an electronic filter; that at 4 kHz is a passive crossover.

The electronic crossover routes signals from 35 Hz to 500 Hz to a comparator circuit. From there the signal moves to a 40-watt bass power amplifier and then on to the woofer. It is in the woofer circuitry that we have the motional feedback network.

The 8-inch woofer has an accelerometer mounted at the apex of its cone. This accelerometer, a piezoelectric transducer (PXE) measures the acceleration of the woofer cone. This acceleration is directly proportional to the acoustic output as long as the cone functions as a single, rigid piston. This means the cone moves back and forth as a unit, with all sections of the cone moving simultaneously, without breakup or cone ripple.

When the woofer cone is in motion, the piezoelectric transducer develops a voltage. This voltage is fed back to a comparator circuit which precedes the woofer amplifier. If the signal produced by the piezoelectric transducer is identical with that of the input signal in the comparator circuit nothing happens. If there are any differences between the input signal in the comparator and the piezoelectric transducer output, as there could be due to cone nonlinearity, a correction signal is developed by the piezoelectric transducer and is then fed back out of phase to the low frequency power amplifier. As a result the acoustic output of the woofer driver is made to be identical to the signal input waveform.

The piezoelectric transducer assembly is in rigid mechanical contact with the voice coil of the woofer and consists of a printed circuit board on which is mounted the ceramic piezoelectric transducer and its associated field effect transistor (FET) circuitry.

The piezoelectric element senses incorrect cone motion, detecting potential sound distortion before it becomes audible. This sensor, via the FET, sends information on cone movement back to the comparator. As a result, distortion in the bass region is virtually eliminated.

Isotweeter

Another innovative approach is Leak's Isotweeter, an ultrathin diaphragm using a printed circuit pattern voice coil. In the speaker the diaphragm is suspended equidistantly between two barium ferrite magnets. When an audio current flows through the copper pattern, a reaction between the magnetic field around these conductors and that supplied by the fixed magnets takes place and the diaphragm vibrates.

This arrangement is much lighter than the usual voice coil and is less inductive, consequently transient response is improved. The cone is extremely lightweight and consists of a sandwich of polystyrene core material covered on each side with Mylar film.

The Isotweeter gets its name from a magnetic material, Isotropic Barium Ferrite, used in developing the unit. To obtain optimum performance the tension on the diaphragm must be adjusted precisely. This is done by putting the speaker assembly in a micrometer-accurate jig and stretching the diaphragm by small amounts prior to mounting in the magnetic assembly. The Isotweeter is a treble tone reproducer only.

ELECTROSTATIC SPEAKERS

Dynamic speakers are current operated devices; electrostatic speakers are voltage operated. In its simplest form, the single-ended electrostatic, the unit consists of two plates, one of which is a large metal plate that is fixed in position. The other is a film diaphragm coated to make it conductive. The fixed plate has a pattern of holes to permit sound to pass through. There must also be an air gap between the fixed plate and the diaphragm to permit diaphragm motion.

If a fixed dc voltage, known as a polarizing voltage, is applied to the two plates, the electrostatic forces generated between the plates will cause them to be attracted to each other. If a signal voltage is mixed in with the polarizing voltage, the signal will alternately add to and subtract from the polarizing voltage, in effect making it vary at an audio rate. This means the electrostatic force between the plates will also vary at an audio rate, and the diaphragm will then vibrate, producing sound.

The push-pull arrangement shown in Fig. 5-18 is an improvement. The signal from the output of the power amp is brought into a step-up transformer to match the high impedance input of the speaker. The signal is delivered to each of a pair of acoustically transparent plates. The signal, made into a high voltage by the step-up transformer, is recitfied and filtered, and the resulting high voltage dc is delivered to the centrally located movable membrane.

During the time no signal is received the unit remains charged because it behaves like a capacitor, with the charge dissipating gradually if the power amp is turned off. During signal time the charge on the plates is varied by the audio signal and as a result the membrane vibrates at an audio rate. The perforated plates are rather large and may be flat or curved. Panel size can be 3-to-4-feet square and instead of being a single driver may consist of a group of as many as 10 smaller electrostatic panels.

Electrostatics are sometimes used to supply mid and treble tones and work in conjunction with a dynamic type woofer. In the push-pull electrostatic, maximum sound is radiated front and rear, minimum at right angles to the speaker, and so the unit behaves like a dipole speaker. The electrostatic doesn't require an enclosure, (but it

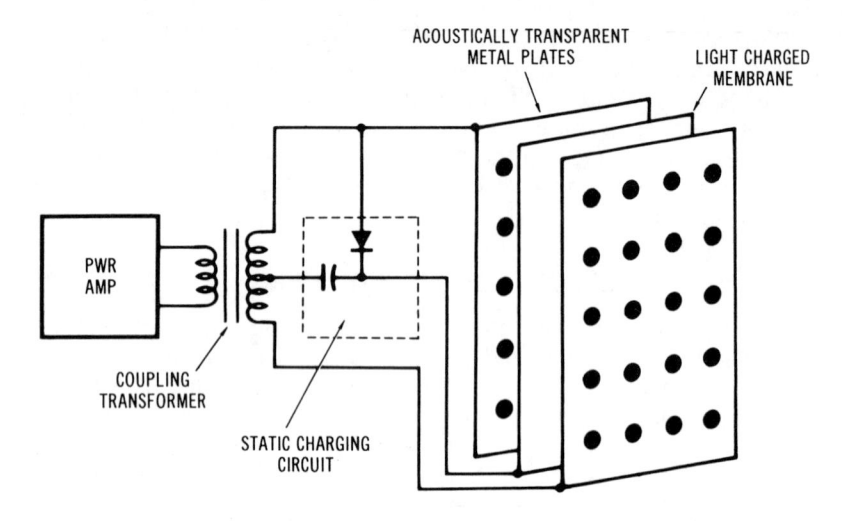

Fig. 5-18. Basic push-pull electrostatic speaker. A single-ended electrostatic would consist of a single fixed plate and a moving plate.

does need a frame or housing of some kind), harmonic distortion is low, and transient response is good.

LABYRINTH SPEAKERS

The labyrinth or tuned-pipe speaker, shown in cross section in Fig. 5-19, provides a long path for the rear sound wave produced by the speaker's diaphragm. Part of the rear wave is absorbed by acoustic material mounted against the inside walls of the labyrinth, while a duct, port, or vent is cut in the front of the enclosure to permit exiting of rear produced sound. The vent can also be put at the side or rear of the speaker. The rear-produced sound emerges in phase with that produced by forward motion of the driver.

TRANSMISSION LINE SYSTEM

The ideal situation in the case of a cone driver is that all parts of the cone should move simultaneously, with the cone functioning as a piston. The mechanical energy initiated by the moving voice coil first becomes evident in the small cone area nearest the voice coil. This energy travels outward until it reaches the rim of the cone, that part attached to the surround suspension. It is true that the time elapse of this energy movement is extremely small, but it still means that not all parts of the cone are set in motion at the same time. As a consequence, the sound produced by that part of the cone nearer the voice

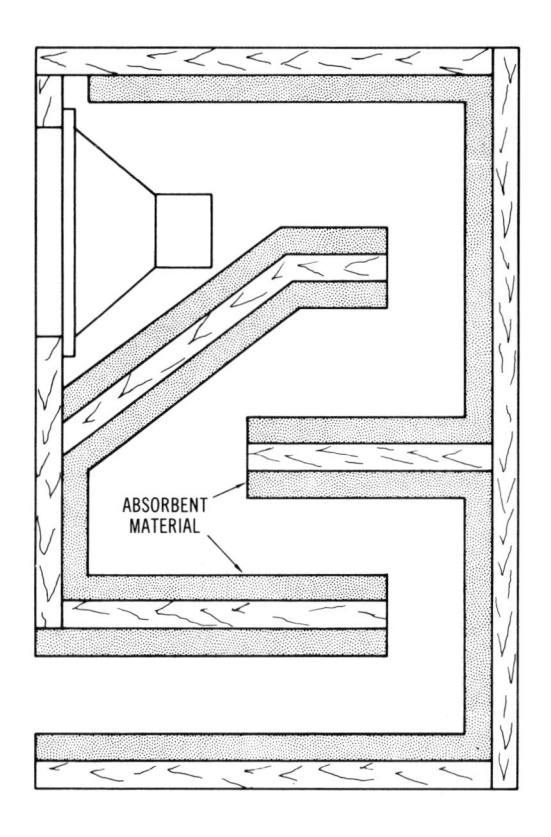

Fig. 5-19. Cross-section of a labyrinth speaker.

coil will be heard earlier than that generated by the section of the cone closer to the surround.

The transmission line tweeter uses a metallic cone positioned vertically, with each conic section producing sound at a precise time, so that the acoustic output from the top of the cone, compared to the bottom, is in phase. The vertical arrangement supplies sound dispersion that is omnidirectional.

Transmission line speakers can be woofers, woofer/midrange, or tweeters.

HEIL DRIVER

This treble speaker makes use of an unusual driver, one that is a rectangularly shaped membrane made of polyethylene and only 0.5 mil thick. The plastic film is covered on both sides with conducting strips. This combination is pleated and put into a plastic frame measuring about 2 inches × 5 inches. The mass of the driver constructed in this manner is very small, a highly desirable feature for tweeter drivers.

The pleated film and its conductors, forming the diaphragm, are surrounded by the strong field furnished by a ceramic magnet. One end each of the connected conducting strips receives the audio current input from a power amp. This causes the strips to move, changing the spacing between the folds of the diaphragm. The alternate opening and closing of the folds produces a fairly large total air movement even with there is a relatively small movement of the individual folds.

This speaker falls into the range of moderate efficiency, about halfway between poor and high efficiency drivers, and is in the region of about 4%. Because sound is moved from the front and rear of the diaphragm, with little or no movement at the sides, it is a bipolar type.

PIEZOELECTRIC TWEETERS

Piezoelectric effect is a property of various crystalline minerals such as quartz, tourmaline, and ceramic substances. Quartz crystals have been used by broadcasting stations for many years for maintaining close adherence to their assigned frequencies, and more recently in hi-fi components such as receivers. Ceramic materials are used in phono cartridges and microphones.

By applying a mechanical pressure along one of the axes of a piezoelectric material, a voltage is produced across another axis. Conversely, it is possible to put a voltage across one axis, and get a mechanical expansion or contraction along another axis. If this voltage is ac, the expansion and contraction will be directly related to the frequency of that voltage.

Consequently, when a crystal is connected to the audio output of an amplifier, the crystal will expand and contract vibrating at a rate determined by the frequency of the audio signal. However, the amount of air displacement produced directly by this vibration is inadequate to permit the crystal to work as a speaker without assistance. This is supplied by a horn or cone made of metal, or plastic, or paper coupled to the crystal. The piezoelectric tweeter is sometimes slot loaded to improve sound dispersion, possibly using a T-configuration.

Piezoelectric tweeters can supply excellent transient response and are capable of extending the reproduced range into the super high audio regions. Such speakers can have problems of temperature and humidity sensitivity.

MIRROR-IMAGE SPEAKER CONSTRUCTION

Mirror imaging is a method of mounting drivers in their enclosures now used by many speaker manufacturers.

With mirror-imaging, each speaker (left and right channels) has its drivers mounted so each of the two speakers is literally a mirror image of the other. In other words, a tweeter will be mounted in the upper right corner of the enclosure of one speaker system and in the upper left corner of the other.

This improves imaging, discussed in Chapter 1, and provides better control of frequency response in the listening room. Mirror imaged speaker systems are generally sold only in pairs, for obvious reasons.

DIRECTIONALITY

Ideally, a speaker should radiate a spherical sound wave regardless of the frequency of the tone being produced. But with rising frequency the sound tends to have beam-like characteristics. Although often not considered as such, bass sounds also beam, although to a much smaller extent than treble. This doesn't present a problem if the treble devices are mounted in their enclosure so there is a fairly direct path to the ears.

Not all tweeters beam to the same extent, depending on the way the cone is constructed and its diameter. The smaller the diameter the higher the frequency the speaker can reach giving uniform dispersion of sound. The top range of a one-inch tweeter before beaming is about 12 kHz. If the diameter of this tweeter is reduced to ½-inch, circular dispersion can extend up to about 20 kHz. A woofer, based on its diameter, will typically begin to beam at about 800 Hz.

Various methods are used to overcome directionality effect. One technique is to use a diffuser — a set of parallel plates placed in front of the tweeter driver. Another is to use a multicellular horn design. Dome structures are also found with tweeters to ensure uniform dispersion, but no such effort is made with woofers or midrange drivers. And, because of the fact that a large part of the audio spectrum can beam, listener position can be critical.

EFFICIENCY

Speaker efficiency is the ratio of audio electrical power input to sound power output. The greater the amount of sound power output from the speaker for a given amount of audio power input, the higher the efficiency of the speaker. Speaker efficiencies aren't supplied as a number, a unit, or a percentage, but in relative terms only. Thus a speaker may have low efficiency, a moderate, or a high efficiency.

To say that a speaker is inefficient does not mean to imply that it isn't a good speaker. Here there is a tradeoff. Efficiency is swapped for sound quality reproduction. But if a speaker is one that has poor efficiency, it can mean that only five watts out of every hundred watts of audio power delivered to it manages to get converted to sound. If we use numbers, just for this example, it means the speaker has an efficiency of only 5%. And so, for such speakers, power amplifiers with high driving power are essential. As the efficiency of a speaker is increased, the power demands on the amplifiers are lowered.

MINIMUM AND MAXIMUM SPEAKER POWER

In spec sheets associated with speakers you will generally find a listing of minimum and maximum audio power. Minimum is the smallest amount of audio power the speaker requires to be able to deliver suitable sound. Maximum is the largest amount of power the speaker's voice coil can handle before it burns out.

A voice coil carrying audio currents is like any other wire carrying current. Heat is developed when a current of electricity passes through a wire and this current produces heat. Any current carrying wire, and that includes a voice coil, must be able to dissipate that heat, to get rid of it. If it cannot do so, the heat will accumulate, and eventually the coil will burn out.

Because speakers and audio power amplifiers must work together, you should not buy either without reference to the other. A power amplifier must be capable of driving the speakers; the speakers must be capable of handling the audio power delivered to them. If an amplifier can only "underdrive" a speaker system, never delivering as much power as the speakers should have for proper audio power-to-sound power conversion, then the resulting sound will be weak and washed out. Conversely, if the speakers do not have adequate power handling ability, high levels of audio power will drive the speakers into distortion, with the always-present possibility of speaker burnout.

ENCLOSURES

A speaker will produce sound whether it is enclosed in a box or not. The box, more elegantly known as an enclosure, is an important and essential part of the speaker system.

Roughly, all enclosures can be placed under two general headings: closed boxes and bass reflex types. A closed box is exactly what its name implies: an enclosure sealed on all sides, and that includes the back, so that the only escape for the sound is from a hole or holes cut in the front of the enclosure.

When a cone or membrane moves, it displaces air and it is this air displacement which we interpret as sound. But the cone or membrane not only moves forward, but backward as well, and so the sound from the speaker is produced in two directions. When a speaker is positioned out in open air, the sound produced by the rear movement of the speaker cone can come around to join the sound produced by the forward movement of that cone. We thus have two sound waves produced by the forward and rearward movement of the cone, but it would be most unlikely for these waves to be in phase. They are usually out of phase to a greater or lesser degree, and the amount of out of phase condition keeps varying. The result is that we get sound cancellation and this sound cancellation could also be frequency selective. The result is sound that is distorted and is really not listenable.

CONE MATERIALS

One of the main reasons speakers sound like speakers rather than live or real is that the movement of the drivers doesn't reflect accurately the electrical input. It is the mechanical characteristics of the cone material which help determine how correctly the electrical signal from the output of the power amp is translated into sound. As a result speaker manufacturers try every conceivable material for use as cones. Paper cones have a typical sound of their own which gets in the way of the sound of music. Early plastic cones also tended to add their own acoustic flavor, sometimes called "quack." Cone sound is caused by the uncontrolled movement or breakup of the cone from moment to moment in the music and results in loss of clarity, definition, detail, and depth.

Various materials other than paper are used for cones, including beryllium, an extremely light, hard material that is difficult to machine, treated wood pulp, polypropylene, and bextrene.

SPEAKER SIZE VERSUS POWER

The size of a speaker or its enclosure has little to do with its power handling ability. Power handling is determined by the size, shape, and heat dissipating ability of the voice coil. You could get a bookshelf type speaker that could easily handle 100 watts while a larger floor model might burn out at half that power input.

It doesn't necessarily follow that big speakers are for big rooms; small speakers for small rooms. If you have a small room you might pick bookshelf speakers, not out of choice, but necessity. And if you have a fairly good size listening room speaker size may no longer be a forced decision.

FERROFLUIDS

One of the requirements of dynamic speakers subject to high energy inputs is the need to dissipate the heat produced in the voice coils because of the large currents flowing through them. Another is the problem of centering the voice coil precisely between the pole pieces of the surrounding magnet. Mechanical centering, mentioned earlier, is often used; a flexible material, known as a spider, holds the voice coil in position and also returns the coil to its starting position, working like a spring. The difficulty with this arrangement is that this device has a stronger restriction against voice coil movement as the voice coil moves toward the limits of its excursion.

To overcome these difficulties, ferrofluids are used in some midrange and tweeter drivers. Ferrofluids are thick, oil-like syrupy liquids that are colloidal suspensions of magnetite, the stable inert oxide of iron, Fe_3O_4. This liquid has a number of advantages. It increases a speaker's power handling capabilities since it conducts heat more efficiently than air by a factor of about 500%. Ferrofluids are also used to center voice coils eliminating the need for spiders in cone type speakers.

Various ferrofluid formulations are available to meet the needs of different speakers. Those with a higher viscosity increase damping, reduce Q, and lower the impedance peak at resonance.

DISTORTION

Since the speaker cone is a mechanically moving mass it can have a resonance point or range—a frequency or small band of frequencies at which it enjoys vibrating more than at other frequencies. If permitted to do so the result becomes an emphasis of a small section of the audio range. We hear it and it sounds unpleasant since it has little relationship to the original sound.

Manufacturers attack resonance distortion in various ways: by special design of the moving element of the speaker, or the enclosure, or by feedback control.

TIME DELAY DISTORTION

Time delay distortion works by delaying different parts of the audio spectrum. The harmonics of a tone are related to its fundamental both in frequency and time. If the frequency relationship remains unchanged, but the harmonics are heard a few thousandths of a second later the result is sonic smear. The transient isn't sharp; it is blurred and the sound becomes indistinct. The word *muddy* is sometimes applied to this kind of sound.

One of the techniques used in speaker design for lowering time delay distortion is to stagger the woofers and tweeters so they aren't arranged in an on-axis configuration on a panel. Woofers, midrange units, and tweeters may not only be staggered on the supporting panel, but may not be arranged in the same plane, that is, some of the drivers are positioned farther forward than others.

The edges of the enclosure also cause sound reflections. The result is that the speaker system may behave as though it had a large number of drivers located both in and out of the enclosure. The result is that the sound has a vague, indefinite character, affecting the stereo image and making sound localization impossible.

Various techniques are used to keep enclosures from producing sound reflections including rounded corners and enclosures that are a departure from the typical rectangular shape. One arrangement uses a truncated pyramid in addition to the rounding of cabinet edges.

ACOUSTICS

In its travels through a hi-fi system the audio voltage waveform is carefully supervised electronically to make certain it is as close to a true representation of the input signal as possible and that it is a true analog of the original sound. This has been made possible by the use of an interesting variety of circuits to keep the noise floor as low as possible, to make sure that unwanted harmonics are not introduced, and at the same time shielding the signal from extraneous voltages.

Thus, by the time the audio signal reaches the voice coil it has been fairly well coddled. Ultimately, though, that signal must emerge into the real world, away from the electronic cocoon supplied by the tuner, receiver, record player, tape deck, preamps, and power amps.

In the voice coil of the speaker, though, we are not concerned so much with voltage as we are with the current flowing through the voice coil. It is this audio current, interacting with the magnetic field supplied by the surrounding permanent magnet, that forces the voice coil and its attached cone to move. It is at this juncture that our problems begin.

The alternate compressions and rarefactions of the air molecules around the driver ultimately produce corresponding compressions and rarefactions in the vicinity of our ears, resulting in the sensation of sound. These air pressure changes, though, interact with anything and everything. The sound produced by the various drivers can bounce back and forth inside the enclosure, they can be refracted and reflected by the hardware and various surfaces inside the enclosure, and outside it as well. The sound can and does strike the floor, walls, and ceiling of the listening room, so what we hear is not only direct sound, sound that moves in a straight path from the speakers to our ears, but reflected sound as well. Since the reflected sound takes a longer path, there is a time difference between the arrival of directed and reflected sounds.

What we hear, then, is this combination. But the reflected sound is affected by the sound absorption and reflective properties of everything in the listening room, and that includes listeners as well. Further, we do not hear equally well in both ears and what we hear is also affected by where we sit with respect to the speakers and how we adjust various controls: the balance, loudness, volume, and tone controls. What we hear is also determined by age, sex, and previous musical training. And so, because of the large number and variety of these variables, listening to electronically reproduced music is a subjective experience.

There are various ways in which we can modify the acoustics of a listening room. One method, somewhat impractical for the average home listener, is to add or remove various acoustic materials: acoustic tile can be used to cover the ceiling and or walls; cushions can be added or removed; rugs can be added or removed. What you hear is also determined by speaker positioning and here more dramatic changes can be made easily. Speakers can be separated or brought closer together, placed catty-cornered against a pair of adjoining walls, mounted above the floor by the use of speaker supports, or put on a bookshelf.

The easiest and most convenient way is through the use of tone controls. The problem here is that this isn't a very effective method since most receivers use only two: one for bass, the other for treble, thus dividing the audio spectrum into just two parts. Some receivers, a few, have added a third control, a midrange, and while this is some improvement, the method is still coarse. If tone controls are used, it is better to have separate controls for left and right channel speakers, but often this is not the case and so bass and treble are adjusted simultaneously ignoring the fact that each of our ears might require a different response.

A better method is through the use of a graphic or parametric equalizer described in the preceding chapter. Such units divide the audio spectrum into a number of octaves and boost or attenuate the signal level in each of these octaves by a certain number of decibels. Further, the equalizer has separate controls for left and right channel sound. Thus, an equalizer is actually an extension of the tone control system, but while it is incorporated into a few receivers, it is, for the most part, a separate component.

Still another acoustic technique is the use of a reverberation (reverb) unit, a device for adding reverberation to recorded sound. Reverb units are time delay devices and their purpose is to create something of a replica of the original sound field. A reverb or time delay unit creates the feeling of presence at a live performance by creating reflected sound to arrive at the ears later than direct sound waves.

Reverberation time is generally considered to be the amount of time it takes for the sound to decrease by 60 dB, a reduction to one-millionth of the original sound level. Reverberation time depends on the volume of the listening room — the sound absorbing properties of the walls, ceiling, and all objects in the room. The larger the room in terms of cubic size, the greater the distance the sound will have to travel from the surfaces it will strike, and thus the greater time required for a given decrease in intensity.

The greater the amount of sound absorption, the smaller the reverb time. A hall having a volume of 10,000 cubic feet will have reverb times ranging from 0.7 to almost 1.2 seconds, depending on the contents of that hall. A larger hall of 80,000 cubic feet will have reverb times from a little more than 1 second to less than 1.6 seconds. A listening room in the home, averaging about 1500 cubic feet, will have a much smaller reverb time. Sound heard directly from a source is called *dry* sound; reflected sound is sometimes called *wet* sound.

STANDING WAVES

Standing waves are produced in a listening room when reflected sound waves are in phase with the dry sound, and in such instances we get sound reinforcement. We can also get the opposite effect, an attenuation of sound when dry and reflected sounds are out of phase. Between these two extremes we have a variety of conditions, and so what you will hear depends on where you sit. This is as applicable to listening "live" as it is to sound reproduced in your listening room.

As you walk through a listening room you may note certain places where the sound is weak; others where it appears to be stronger. The point at which standing waves cancel is called a node; where they reinforce is known as an anti-node or loop.

SPEAKER SPECIFICATIONS

Speakers are relatively simple devices when compared to receivers, preamps, or power amps and so you will find the specs aren't as detailed. What a speaker will do depends on the room in which it will be placed, the positioning of the speakers, and personal listening preferences. Speaker specs are important (Fig. 5–20) for the component must interface with the preceding power amplifier.

Voltage Sensitivity

This is the amount of input signal voltage to produce a sound pressure level, generally measured on axis at a distance of 1 meter from the drivers.

> **FREQUENCY RESPONSE:**
> 32-20,000 Hz ±3db
> **DRIVER COMPLEMENT:**
> 1 – 1" tweeter with low coloration diaphragm and Acoustical Loading Sphere. Voice coil gap is filled with ferro fluid.
> 1 – 6" midrange with special magnetic structure and high viscosity ferro fluid damping.
> 1 – 10" bass driver with special low distortion, high power capacity magnetic structure.
> **CROSSOVER:**
> Bass/midrange – 475 Hz 12db/octave low pass 6db/octave high pass.
> Midrange/tweeter 2,000 Hz 18db/octave; constant resistance type. Has three position L-pad tweeter attenuator control.
> **ENCLOSURE:**
> Low resonance composite with a veneer of oiled walnut. Unique cabinet shape eliminates diffraction. Bass and midrange drivers mounted with vibration isolating fasteners. Inside of cabinet is lined with several layers of asphalt and lead damping material.
> **GRILLE:**
> Acoustically transparent black foam.
> **SIZE:**
> 16½" square at bottom, 8½" square at top, 41⅜" high.
> **RECOMMENDED AMPLIFIER POWER:**
> 30 to 125 watts
> **WEIGHT:**
> 62 lbs.

Fig. 5-20. Speaker specifications. *(Courtesy Epicure Products, Inc.)*

Number of Drivers

This spec shows the number of speakers used in the system. For tweeters, the spec may contain some information about the type, whether dome or cone or any other. The same is true for midrange units, but for woofers the usual data concerns only the diameter. The spec may also indicate the kind of enclosure, *i.e.*, vented port or acoustic suspension. The spec might include the division of the audio spectrum, such as two-way, three way.

Power Handling Ability

This is the minimum and maximum power rating of the speaker, in watts. In this case, the reference is to continuous power per channel, throughout the entire audio frequency range, from 20 Hz to 20 kHz, and is for a composite signal, such as music, and not a single frequency test tone. There is also an assumption that the audio signal is undistorted and that there is no clipping.

Efficiency

Efficiency is the amount of sound pressure level obtained for an input of 1 watt across the input terminals of the speaker using a test frequency of 400 Hz.

Impedance

This refers to the impedance of the speaker's voice coil. Impedance is the vector sum of the resistance and inductive reactance of this coil, with the reactance making the greater contribution to the overall impedance. The dc resistance of a voice coil is usually quite low in comparison to its impedance.

Most speakers have a voice coil impedance of either 4 ohms or 8 ohms, with the latter much more common. A few speakers are made that have an impedance of 16 ohms. However, most power amplifiers are intended to work into impedances of 4 or 8 ohms.

Every speaker is a load on its power amplifier and a speaker having a voice coil impedance of 4 ohms is twice the load, that is, it will draw twice as much current as an 8-ohm speaker. Because of the heavier current demand made by 4-ohm speakers, the voice coils are made of heavier—that is, thicker—conductors to carry the current.

Impedance figures supplied in specs are nominal and, since the reactance of a coil is frequency sensitive, impedance will change with frequency. As a result the actual working impedance may drop to a fairly low value at a particular frequency. Minimum impedance figures are sometimes supplied, e.g., 3.2 ohms for a nominal impedance of 4 ohms.

Frequency Response

This indicates the audio range supplied and should be accompanied by a deviation number, usually 3 dB, although you may find some with variations of as much as plus/minus 6 dB. If the audio range is considered to be 20 Hz to 20 kHz, most speaker systems appear to do better in the treble region, often showing a response well beyond 20 kHz. At the low frequency end speakers rarely approach 20 Hz, unless supplemented by a special woofer type driver, such as a subwoofer. With such a driver response can get down below 20 Hz; without it the lower end of a woofer is generally between 28 Hz and 65 Hz.

Crossover Frequency

Crossover frequencies are considered as the points at which various segments of the audio band are routed to the different drivers and vary considerably from one speaker system to the next. Sometimes this is accompanied by an attenuation range. Actually, the crossover frequencies indicated in the spec are of no great value to the user since they are dictated by the requirements of the drivers. Possibly more significant is the attenuation rate per octave, with the best having an attenuation of 18 dB/octave. However, this information is not usually supplied.

CONNECTING SPEAKERS

Speaker connections are polarized as are the speaker terminals on the output of a power amplifier or receiver. Polarization may be indicated by symbols such as (+) and (−) or the plus terminal may be marked red; the minus terminal black. A plus terminal on the speaker should be connected to a plus terminal on the receiver or power amp, and minus terminal to minus terminal.

Although speaker wire is sold in various stores, any type of insulated wire can be used provided it is the correct gauge. The standard for wire sizes is the American Wire Gauge (AWG). Wire gauge is a number that refers to the diameter or thickness of the wire and the smaller the gauge number the thicker the wire. Wire is available as single conductor or stranded. To select the correct gauge of wire, measure the distance from the output terminals of the receiver or power amplifier to the farthest speaker. You can then use the following table as a guide to AWG size.

Speaker Distance	Wire Gauge
25 feet or less	18
25 to 40 feet	16
40 to 60 feet	14
60 feet or more	12

The advantage of speaker wire is that it is marked as to polarity. If you do use stranded wire make sure none of the strands reach over to touch an adjacent terminal, producing a short. After stripping stranded wire, count the strands to be certain you haven't accidentally cut away one or more.

Always use a thicker wire, if possible. Thus, if your budget permits, use 16-gauge in preference to 18, or 14 in preference to 16.

Speaker cables are an important part of the hi-fi system so a better option is to use cables made specifically for connection to speakers rather than lamp cord. This is especially true when using high power amps. True speaker cable is more expensive than lamp cord, but percentage wise, it is a small part of the total cost of a hi-fi system.

Do not connect speakers in parallel unless specifically recommended by the manufacturer. A parallel connection consists of two speakers wired plus-to-plus terminal, minus-to-minus terminal. If the speakers have identical voice coil impedances, the effect will be to cut the impedance in half. A pair of 8-ohm speakers in parallel will have the equivalent impedance of a single 4-ohm speaker. A pair of 4-ohm speakers in parallel will be 2 ohms.

Putting speakers in parallel increases the load on the receiver or power amplifier. If these components can meet the heavier current demand of the speakers in parallel then overload distortion will be no problem. There will be distortion if the power amp or receiver cannot meet the power demands of the speakers.

Impedance figures supplied by speaker manufacturers are always nominal values, and may be higher or lower depending on frequency. A pair of 4-ohm speakers in parallel will be the equivalent of 2 ohms or less, and at times may get dangerously close to a short circuit.

HEADPHONES

Speakers and headphones are closely related in the sense that both are transducers, devices that change one form of energy into another. The input to speakers and headphones is electrical energy supplied by audio currents. The output is sound energy. In the process some energy is lost in the form of heat.

Headphones and speakers are alike in the sense that they are both air pushers. The device could be in the shape of a cone, a flat surface, one that is cylindrical, or circular. But whatever the shape, the objective is always the same and that is to move air in step with the frequencies of the audio currents being supplied.

At one time a headphone was a strictly utilitarian component following the puritan ethic of work without comfort. The headbands were of spring steel so that the earpieces literally clamped the ears, flattening them to the head. Actually, this arrangement was a necessity since in early radio sets, known as crystal receivers, the signal output was so small that sound could not be permitted to escape from the earphone reproducers. In those early headphones the moving element was a circular disc of metal and so all the reproduced sound was tinny.

Modern headsets are luxurious by comparison. They are comfortable. Some come equipped with individual tone and volume controls, eliminating the need for getting up out of a comfortable chair to make any needed adjustments. The ear cups are padded and you can get a pair of headphones with liquid-filled padding so that the ear cups follow each tiny contour of your ears. The headbands on the headsets are also padded to make sure there is no unnecessary pressure on your head.

A common statement is that if you have speakers you don't need headphones. Don't believe it. That's like saying if you have feet you don't need a car.

Headphone sound isn't speaker sound. It isn't that headphones are better than speakers, or vice versa. Headphones and speakers aren't competitive. Headphones do not supplant speakers, nor do speakers replace headphones. Rather, they supplement each other. And as far as saying that speaker sound is the same as headphone sound, that is just so much nonsense.

Headphone listening produces a sound sensation that is quite different. When sound is radiated by your speakers the distance between your ears and the reproducers is measured in feet and so, in making that trip, anything can happen to that sound. It mixes with ambient sound, sound produced by sources other than the speakers—voices, a telephone, someone walking, the slamming of a door—you name it. And the sound from the speakers does get some bouncing around before you hear all of it, reflected from the floor, walls and ceiling.

Headphone listening is another matter. The distance between the drivers or sound reproducing units, the ear pieces, is about an inch or so. As a result, depending on earphone design, you may hear little or no outside sound. The acoustics of your listening room are practically eliminated and so what you hear is more nearly the sound that is actually delivered by your hi-fi system. Further, headphones require very little audio power compared to speakers and so your headphones practically float, electronically speaking, on the output of your audio amplifier system. See Fig. 5–21 for typical headphone specs.

Headphone Comfort

If you intend wearing headphones for several hours, and there's no reason why you shouldn't be able to, try them on first for comfort. You want headphones that are reasonably light, that are padded, and that have a coiled connecting cord that is sufficiently long so you can get up and move around should you want to do so. The coiled headphone cord is like a spring, accommodating its length to your movements.

> **Transducer Types:** Dynamic moving coil low-frequency system; Fixed-charge electrostat high-frequency system
> **Frequency Range:** 16 Hz to 25,000 Hz
> **Crossover:** 4000 Hz, 6 dB/octave
> **Nominal Impedance:** 400 ohms, each channel
> **Sensitivity:** 1.6 Pa/volt*
> **Normal listening level requirements:** 0.63V (1 mW) for 94 dB SPL (approx. 1 Pa). "IHF Sensitivity" rating. 5.0 V (63 mW) for 112 dB SPL (approx. 8 Pa).
> **Sound Pressure Level for ≤ 1.0 THD:** 104 dB (200 mW), approx. 117 dB SPL.
> **Cable:** 3 meter (9.8 feet) 4 conductor cable with 3-conductor (stereo) 1/4" phone plug.
> **Weight:** 385 grams (13.5 oz.) less cable.
> **Contact force:** Approx. 3 newtons (10.5 oz).
>
> *The Pascal (Pa) is now the standard unit for sound pressure level. 1 Pa ≅ 10 microbar (μb) ≅ dynes/square cm. ≅ 94 dB SPL.

Fig. 5-21. Headphone specifications. *(Courtesy AKG Acoustics, Inc.)*

Connecting Headphones

Headphones come equipped with connecting plugs for easy insertion into jacks located on the front panel of a receiver or preamplifier. If there is only one such jack and you want his and her listening you can get an adapter which will let you do just that. Generally, males aren't fussy about getting hair mussed by using headphones, but women are different. Females like headphones which are attractive and stylish and want to get away from those which are solidly black or brown. They also prefer lightweight units. And if they are concerned about hair-dos, they can wear the headphones upside down with the headstrap under the chin or resting on it. There's no possibility of leakage or damage, and headphones work well in any position.

Inside the Headphones

The units inside modern headphones are equivalent to speaker drivers and you can regard them as miniature speakers. They will contain woofers, units designed for the reproduction of bass tones, and tweeters for reproducing treble. You will find them with woofers having a controlled acoustic suspension and ceramic tweeters with printed circuit crossover networks. Crossovers are used to divide the audio spectrum of 20 Hz to 20 kHz into narrower bands, each of which can be handled separately. Some have a polyester film diaphragm and aluminum voice coil or else Mylar cones, resulting in greatly reduced moving mass. You'll also find headphones with removable and washable earpads.

Most headphones are dynamic types, but you can get electrostatic units as well. Inside each earcup of the electrostatic are a pair of adjacent mesh electrodes, acoustically transparent metal plates separated by an ultrathin metallized Mylar diaphragm. Dynamic types, like their speaker counterparts, use a copper voice coil attached to one side of a miniature cone or diaphragm.

One manufacturer has a headphone that is quite unconventional since it uses a planar moving coil to reduce size and weight. The coil is a flat spiral of conductive aluminum, photoetched on both sides of the diaphragm. Instead of a magnet slug the unit uses two perforated disc magnets on each side of the diaphragm.

Circumaural and Supra-Aural Headphones

Headphones can be fully closed types (also called circumaural) with the cushions encircling the ear, excluding all (or almost all) outside sound. With headphones of this kind the user can retreat from the real world into a world of sound only. The open air (or supra-aural) type has a structure that lets the wearer hear ambient sound; a telephone in another room, a door bell, the sound of voices, but these are all muted.

A good feature of headphones is that they supply listening privacy. Sound levels cannot disturb other members of the household or neighbors. They are also useful for evaluating records and tapes, for with headphones the ears are about an inch or so away from the sound reproducing element, not five feet or more as in the case of speakers. Faults in records and tapes, with the possible opportunity to correct them, are more evident with headphones.

Passive Drivers

In one type of headphone the unit has a centrally located active transducer surrounded by six passive diaphragms. These encircle the driver and are spaced at 30° intervals. These passive radiators or "slaves" are activated by the sound pressure waves produced by the active driver. The passive diaphragms eliminate unwanted mid- and high-frequency cavity self-resonance of conventional circumaural headphones.

Headphone Controls

Some headphones are equipped with volume controls, with individual controls on each earpiece. Our ears aren't perfect twins and so the volume controls can compensate for hearing deficiencies. The controls also help produce stereo-related effects and can cause the sound to appear to come from the left, from the right, or from somewhere between.

Headphone Spl

It's a good idea to keep sound levels reasonably moderate when wearing headphones since the sound source is so close. Sound pressure varies inversely as the square of the distance, but with headphones, unlike speakers, distance cannot work well as a protective cushion. It's the sound pressure level (spl) at your ears that can do some damage.

Headphone Weight

For some audio listeners light weight is of prime importance. Most headphones are in the 10- to 20-ounce region, including the weight of the cord, but some are down to about 7 ounces. However, the total weight of headphones can be misleading for several reasons. Headphone bands and earpieces are usually generously cushioned and padded. Further, the weight is supported not only by the head, but by the ears as well, so the weight is distributed.

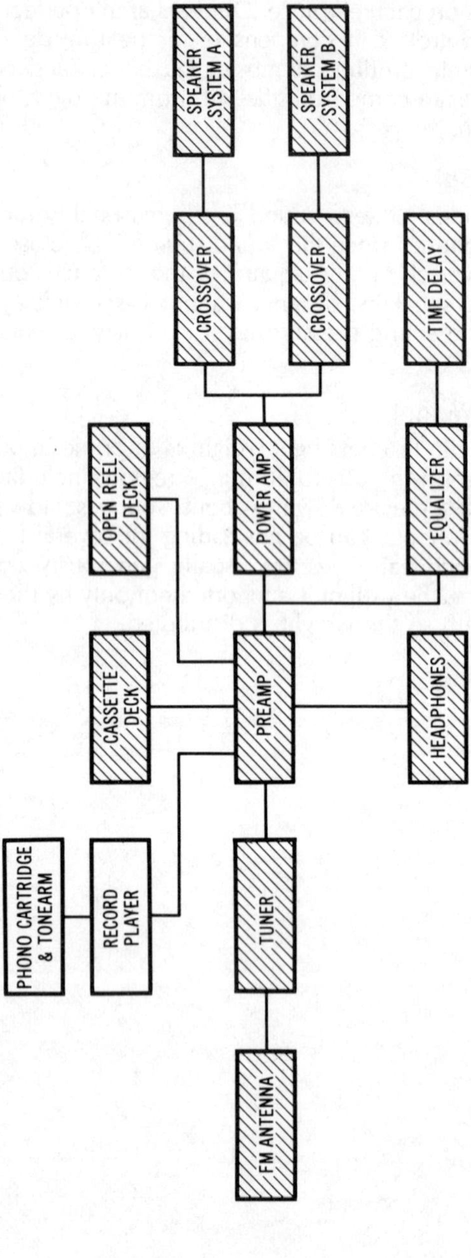
SPEAKER SYSTEM A
SPEAKER SYSTEM B
CROSSOVER
CROSSOVER
TIME DELAY
OPEN REEL DECK
POWER AMP
EQUALIZER
CASSETTE DECK
PREAMP
HEADPHONES
PHONO CARTRIDGE & TONEARM
RECORD PLAYER
TUNER
FM ANTENNA

6

Record Players, Phono Cartridges, and Records

Electromechanically, a record player, with its turntable motor (or motors), stylus, phono cartridge, tonearm, and assorted tracking force and antiskating adjustments, is an oddball contraption. But it works quite well and gives us as much music as we want each day. See Fig. 6-1.

To say that a record player works is damning it with faint praise. Its performance can be extremely impressive. The motor is designed to run at constant speed, regardless of load and line voltage fluctuations.

Fig. 6-1. A two-speed vacuum disc stabilizer turntable. Special features include: minimum resonance construction, automatic operation with quartz-locked direct-drive motor, 0.03% wow and flutter, straight-line low mass tonearm, 33- and 45-rpm with plus/minus 3% speed adjustment, 12-inch aluminum die-cast turntable platter, integral dust cover.
(Courtesy Luxman, Div. of Alpine Electronics of America)

The heavy platter, often die cast and machined, is now so well made that its uniformly distributed mass keeps wow percentages to an astonishing minimum, particularly when it gets an electronic assist from a speed controlling sensor.

RECORD PLAYER SPEEDS

For a while record players were offered as four-speed units, supplying 16-, 33⅓-, 45- and 78-revolutions per minute (rpm), but the 16-rpm disc is an idea whose time has come and gone, while the 78-rpm record is a collector's item.

AUTOMATIC RECORD PLAYER

An automatic record player indicates that cuing, the action of positioning the stylus in the start groove, and shutoff, the lifting of the arm when the last groove is completed, is machine handled.

There are a number of variations in automatic play. A basic automatic will start and stop by itself. With some, the tonearm will swing back to its starting position, but with others the tonearm will hover near the spindle and depend on you to return it to its cradle. With some automatics the tonearm will be returned to its starting position and the motor will turn off. With others there will be the same tonearm return but the motor will continue operating. And with a few, the tonearm will move down to the starting groove again so you get automatic replay.

RECORD CHANGER

A single play automatic has some or all the automatic features but is capable of being loaded with just one record at a time. If you can put a stack of records on the platter and the machine can play these sequentially you have a record changer. And, of course, record changers can be automatic, semi-automatic, or manual.

TURNTABLES

As a high-fidelity component, turntable refers to the platter and its driving motor and associated parts.

Manual

Various names are used in connection with devices for playing records. A manual turntable means you must use your hands to put the cartridge in its starting position in the first groove and to remove it from the last groove upon completion of play. A single-play manual means just that: all you can do is put one disc on the turntable at a time and you also need to take care of positioning the stylus on the lead-in groove of the record.

Transcription Turntable

The word *transcription turntable* may be used by some manufacturers who want you to understand immediately that their component is high quality. For many years the transcription turntable was used by some broadcast stations to indicate single play units, but with the type of players now designed for home-entertainment use, the phrase transcription turntable is applicable.

Automatic Versus Semi-Automatic

A fully automatic turntable is one in which no manual assist is needed. A semi-automatic indicates that some function, such as lifting the tonearm at the end of play, or returning the arm to its starting position is needed.

DRIVE MOTORS

The task of the record player's drive system seems simple at first—to do nothing but rotate the disc at a specified, constant speed. Any mechanical system with moving parts, however, generates vibration and wobble, and in a turntable these can produce unevenness of pitch, referred to as wow and rumble.

Platter

Obviously, the turntable platter must be large enough to support fully the largest phono record, but there's another reason for making this part of the record player as big and heavy as the motor torque will allow. The large mass of the platter supplies a flywheel effect, helping to maintain constant rotation and, through its inertia, tends to smooth any slight speed variations that might occur. However, we want the start up time of the platter to be as short as possible, so what we have here are two requirements which are in conflict.

At one time turntables were made of steel plate, but die-cast aluminum alloys are preferable. Such platters can be machined with greater precision and they aren't subject to magnetization by stray magnetic fields.

The Motor

A phono motor in a turntable must meet several basic conditions: it must rotate at constant speed; it must reach that speed as quickly as possible; it must have adequate torque (turning power) to rotate the platter and phono disc with the stylus traveling the grooves, and to do so without strain and without vibration. To these you can add that the motor must not permit magnetic flux leakage since this could produce hum in the pickup system. And it must meet these conditions on a continuously operating, rather than an intermittently, working condition.

Induction Motor The induction motor is generally used with low-cost turntables. Its speed can vary with load and so would require some sort of speed regulation technique.

Synchronous Motor The synchronous motor, similar in type to those used in ac clocks for the home, is locked in to the frequency of the ac power line, and since line frequency is remarkably constant, helps maintain uniform speed of platter rotation. A variation is the hysteresis synchronous motor, notable for its freedom from vibration and its noise-free operation.

Induction and hysteresis synchronous motors are fast running types and usually operate at 30 times the frequency of the power line (60 Hz) or 1800 revolutions per minute (rpm). But since turntable platters are required to rotate at 33⅓-rpm or 45-rpm, there must be some arrangement for speed stepdown.

Some record players with automatic features have two motors instead of one. This could include a hysteresis synchronous type to rotate the platter and another low speed gear motor for handling automatic features. Single motor operation, usually found in cheaper models, puts a heavy burden on the solo motor. The division of labor concept is followed even more in some record players that have three motors: one for the platter drive system, another for cuing, and still another for handling the tonearm.

Other Motors There are many other motor types including the dc servo, the ac servo, quartz phase lock loop (pll), outer rotor hysteresis synchronous, 120-pole linear ac servo, 120-pole linear quartz phase lock loop, brushless Hall effect, brushless, slotless dc servo, coreless dc servo 20-pole, 30-slot dc servo, and quartz dc servo. Some motors have names coined by record player manufacturers. The quartz type locks in the rotation of the platter to the extremely accurate and stable frequency generated by a quartz crystal.

The problem with conventional dc motors is the noise and friction produced by their brush and slot elements, and cogging. The number of poles in a motor can be as few as 4, and some are 24-pole equipped. Cogging is the tendency of a motor to stop slightly at each pole.

THE DRIVE SYSTEM

In the early days of the phonograph, the platter was driven by a spring type motor which was hand wound by a crank handle protruding from the side of the phonograph. The spring had its maximum tension at the start of the record but much less by the time the last groove was reached. The result was that the platter turned with a constantly decreasing speed, but the change was gradual, so much so, that the change in pitch caused by the gradually reduced speed wasn't always noticed.

The modern record player is motor driven, but even today with some motors that are marvels of speed constancy, there are various ways in which the platter is made to turn.

Rim Drive

One method, once widely but no longer used, is called rim or idler drive (Fig. 6–2). With this technique a wheel, mounted on the motor shaft, makes contact with another wheel, an idler, positioned on the inside rim of the platter. This sounds sort of devious, but its great advantage is that it was easy to change platter speed and its cost was low. The problem with rim drive is that the platter wheel tended to get "out of round," that is, it developed "flats" on parts of its surface.

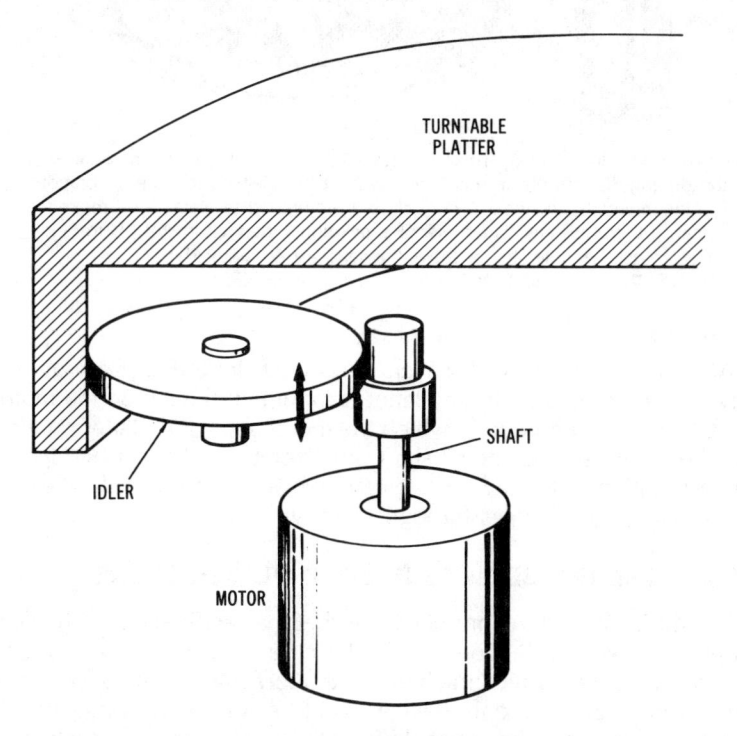

Fig. 6-2. Rim drive uses an idler wheel to couple the rotation of the motor shaft to the platter.

Further, with the wearing action on the idler a certain amount of slippage developed.

Belt Drive

Belt drive (Fig. 6-3) is probably the most widely used method for revolving a platter. This technique uses a belt made of synthetic rubber but more commonly polyurethane because of its resistance to heat, humidity, and oil, plus low elasticity.

In the belt drive system the motor wheel is kept, but a belt is used to connect a pulley on the motor and turntable platter. When the wheel turns, so does the platter. Since the diameter of the platter is so much larger than that of the wheel, the motor must drive it very rapidly. An advantage of belt drive is that it isolates the motor from the platter, so vibration problems are minimized.

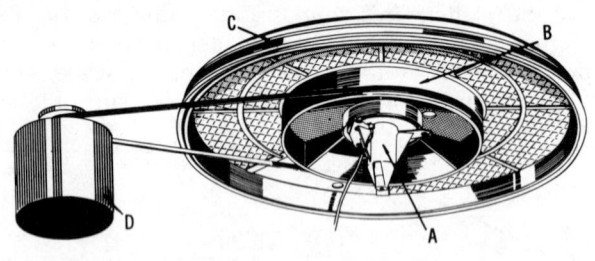

Fig. 6-3. Belt drive system. This unit uses a 160-pole tachometer generator (A) at the driving disc (B) electronically monitoring the platter's (C) rate of rotation. The tachometer's dc signal is continuously compared to a stable dc reference signal. If there are any variations, plus or minus, the tachometer (A) instantly accelerates or slows the separate dc motor (D).
(Courtesy North American Philips Corp.)

Direct Drive

Almost as many record players use direct drive (Fig. 6-4) as belt drive, a system in which the platter is mounted directly on the motor shaft. This is a technique in which the motor shaft and platter rotation are the same, so the motor shaft must rotate at the correct speed. Since the platter, a heavy mass, can now act more as a flywheel, it can resist speed changes through its inertia.

THE PROBLEM OF RECORD PLAYER SPECS

The difficulty with record players is that they are electromechanical rather than purely electronic as in the case of tuners, preamps, and power amps. It's easier to achieve extremely high levels of performance with electronics-only devices. And because record players are electromechanical it becomes difficult to set up specs by which such components can be fairly compared.

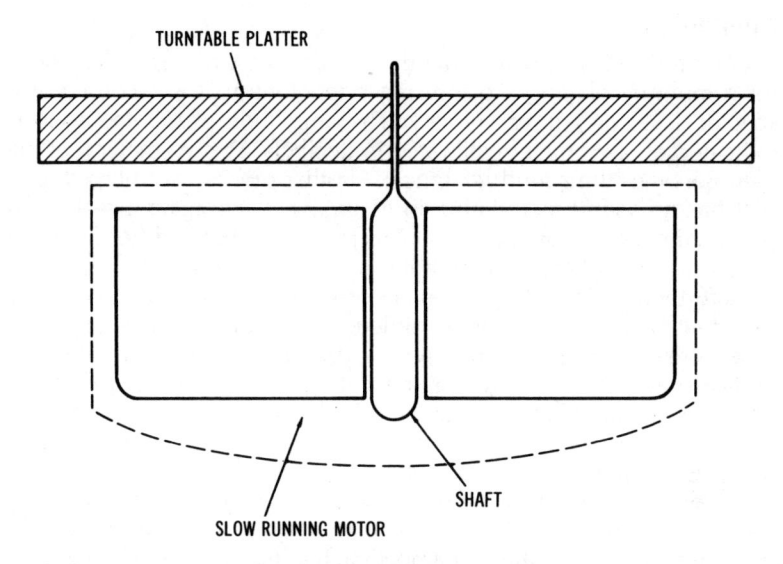

Fig. 6-4. Direct drive turntable.

The demands on the turntable impose certain requirements and it is these requirements that distinguish a quality unit from one that is so-so. The turntable must revolve at a constant speed, not influenced by the drag of the phono cartridge or by changes in line voltage, and, in the case of multiple play units, remain unaffected in speed by the weight of the discs. The platter must turn as uniformly as possible in a plane that does not deviate from a horizontal level. The turntable must reach its operating speed in the least amount of time without overshooting that speed and without hunting for it. The unit should have minimum rumble, wow, and flutter.

RECORD PLAYER SPECS

Record players not only produce music, but noise as well. There are three types of noise associated with turntables: rumble, wow, and flutter. Fig. 6–5 shows the specs for the phono motor section of a turntable.

Driving System:	Direct-Drive System
Motor:	DC-servo brushless & slotless motor
Turntable Platter:	30cm aluminium die-cast (weight 1.8kgs including platter-mat)
Rotation:	33-1/3 rpm, 45 rpm (2-speed)
Adjustable Range of Rotation:	±4%
S/N Ratio:	No less than 60dB (IEC-B)
Wow & Flutter:	no more than 0.03% W.R.M.S.

Fig. 6-5. Specs for the phono motor section of a turntable.

Rumble

Rumble is a low pitched sound. Manufacturers of turntables check for rumble by playing a disc using a test tone that is recorded at a specific level or groove velocity. The amplitude of this test tone is indicated on a vu meter calibrated in decibels. This vu meter reading then becomes the sound reference. Another disc is then played, one that has no sound modulation in its grooves, and again a reading is observed on the same vu meter. This reading, subracted from the first or test signal reading, is the rumble spec, and is supplied in decibels.

Unfortunately, there are so many possible variables and so many test standards that it becomes difficult to evaluate rumble properly. One of these variables is the phono cartridge used in making the test and since it is the cartridge that translates groove modulations into an electrical signal, what you may hear in the way of rumble is partially dependent on the cartridge you use. Consequently, rumble data is often not included in the spec sheet.

Since rumble is a low-frequency sound, the amount of it you may hear also depends on the preamp that follows the record player. In this case, a high quality preamp that has excellent low frequency response may supply more rumble output than a unit that has a higher low-frequency rolloff. With a poorer quality preamp, that is, with one that does not have a good extended low-frequency response, you will hear less rumble, but you will also fail to get good bass response.

Another variable depends on the test tone used for making a rumble evaluation. The frequency of this test tone and its strength will influence the final rumble spec. Still another factor is the kind of "silent" record used in making the rumble test. How silent is silent? In this case the word "silent" can be a bit of wishful thinking since some such records are more silent than others. Every phono record, even if unmodulated, produces sound in the form of noise, varying from one record to another.

Finally, the amount of rumble is determined by the weighting network used. A weighted system emphasizes those frequencies which are most audible.

Rumble Standards

A number of standards are used in measuring rumble and these include NAB (National Association of Broadcasters) supplied as weighted and unweighted. Another is DIN (Deutsche Industrie Normen) furnished by the German Standards Committee. These are available as Din B or unweighted or Din A unweighted. Din B is often used by manufacturers for it supplies higher decibel figures than other methods. There is also a standard set up by CBS Laboratories and known as the ARLL or Audible Relative Loudness Level. In the case of Din standards a manufacturer may sometimes supply both Din A and Din B figures; thus, −50 dB (Din A); −70 dB (Din B).

These different standards are not the same for they use different amounts of level and frequency for the reference tone of the test record and they also differ in the weighting networks used.

Rumble not only can be produced by a playback turntable but is modulated into every phono record. A phono record has recorded wow and flutter because the record is made on a turntable lathe that can produce these characteristics. Further, we assume that the center hole of a record is perfect when in manufacturing practice it may be eccentric, however slight. While all recordings from a particular master will have the same amount of rumble, this spec can vary with the records produced by different labels.

Rumble figures are supplied in decibels and should be preceded by a minus sign, although you will find rumble specs without it. The larger the decibel number, the smaller the amount of rumble as measured by the manufacturer. You will find figures as low as −30 dB to as much as −90 dB. The minus sign is an indication of the amount by which the rumble is less than the recorded material. Since rumble figures used in spec sheets do not indicate the type of test that was used and also do not indicate whether the numbers represent a weighted or unweighted figure, this spec shouldn't be considered rigorous.

Wow and Flutter

Wow is a sound variation caused by a low speed change. Flutter is a higher speed quiver and is particularly noticeable in notes which are sustained.

Wow and flutter are caused by deficiencies in the drive system and are expressed as a percentage. Wow and flutter are alike in the sense that they are deviations in speed of rotation of the platter and so they are generally combined.

Weighted and unweighted figures are supplied for rumble, flutter, and wow. Sometimes the percentage figure for wow and rumble will be followed by the letters wrms. The "w" indicates that the measurement is weighted; "rms" is an abbreviation for root mean square which means that the wow and flutter figures have been averaged.

For wow and flutter, expressed as a percentage, the lower the figure the better. The actual percentage depends on the standard used. Wow and flutter specs, when stated without any reference to a test standard, are meaningless when used for making comparisons between record players. However, when supplied, wow and flutter figures range from about 0.025% to 0.15%.

As far as rumble, wow, and flutter are concerned, specs on modern turntables indicate these are now down to low levels. How, then, is it possible to hear these nuisances when manufacturers' specs indicate they should be inaudible? The disc you play may be the direct cause of rumble you can hear. Before you condemn your player, try a different record. A good recording will have rumble of about −55 dB, DIN unweighted. But your record player may well have a rumble figure that is much better than this. The disc you have selected may not, technologically speaking, be on a par with your record player. With the wonderful progress turntable manufacturers are making, it would seem as though records will be the limiting factor on the reproduction of high-fidelity sound.

RECORD WARP

A record warp means more than a record that isn't absolutely flat. All records are warped to a greater or lesser degree, but many warps aren't noticeable since they are tiny bumps that are manufactured into the record. There are also so-called eccentricities due to the fact that the center hole of a record may be imprecise.

A stylus not only responds to music, but to warp as well. When a stylus moves upward toward the peak of a warp, the stylus is pushed toward the cartridge body. The opposite effect takes place when the stylus reaches the bottom end of the warp, for then the stylus moves away from the cartridge body. This stylus motion generates unwanted signals and these are reproduced by the hi-fi system.

VACUUM DISC STABILIZER

Nearly every record exhibits a certain eccentricity and/or warp. Because of the prevalence of record warp, the bearing pivot point should ideally lie in the same plane as the playing surface. The longitudinal displacements of the stylus caused by warp will thus be kept to a minimum and the resultant wow-and-flutter components will be lower. Fig. 6-6 illustrates this effect as it occurs with a tonearm of normal length. Of course, it isn't easy to lower the pivot point completely to record level. However, the closer it is, the lower will be the wow and flutter produced by warped records.

High levels of wow and flutter occur with short-arm designs using a relatively high bearing pivot point. If you assume a quality turntable exhibits a basic wow and flutter of 0.05%, then with a record warp of just 1 mm, the total wow and flutter will rise to 0.17%. The DIN standard specifies a wow and flutter of less than 0.20% for high fidelity reproduction: 0.20% can be detected by an untrained ear.

Various techniques are used to minimize the effects of warp, including stabilizers and special platter maps. The vacuum disc stabilizer, illustrated in Fig. 6-7, is, in a sense, an audio stabilizer by

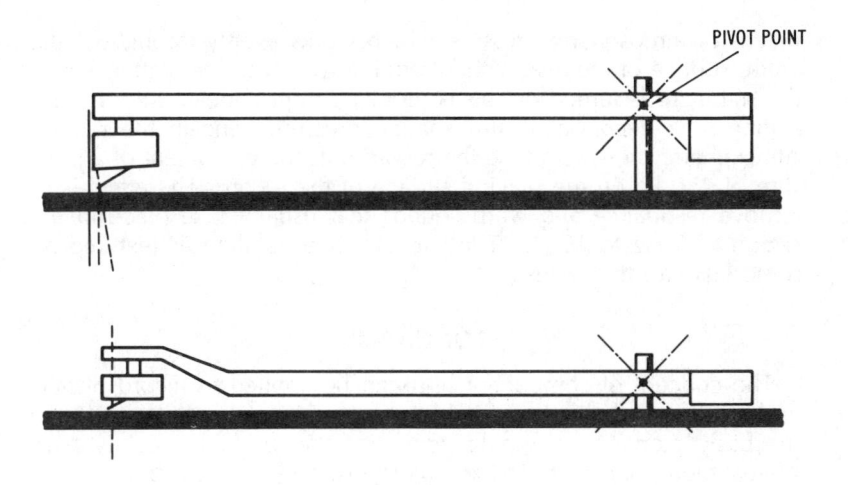

Fig. 6-6. Lowering pivot point, as shown in the bottom drawing, reduces effect of record warp. *(Courtesy Thorens Corp.)*

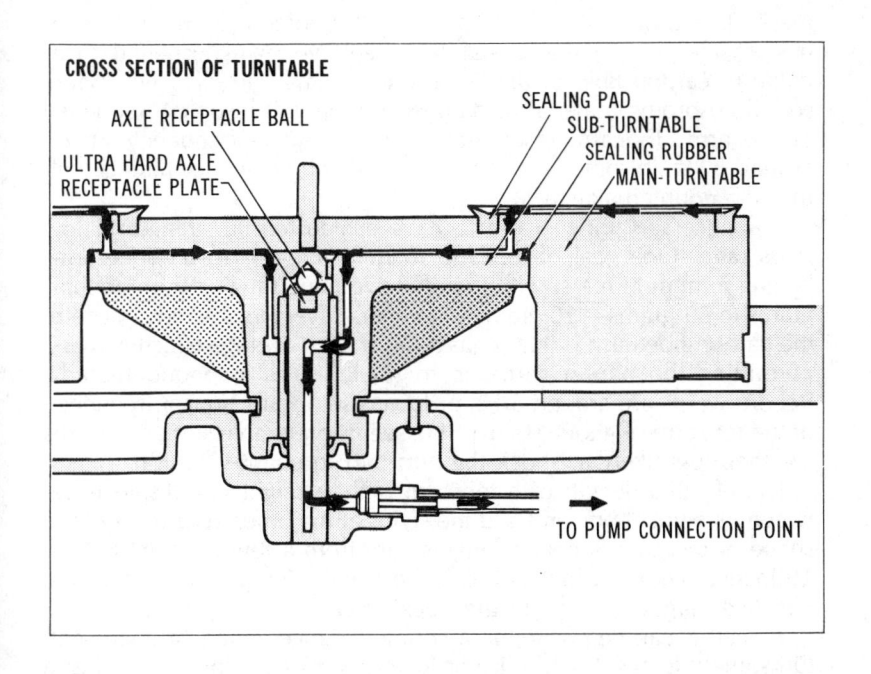

Fig. 6-7. Vacuum disc stabilizer.

means of atmospheric pressure which works evenly throughout the entire surface of the disc without adding weight to the platter. When the stabilizer is turned on, air is removed from beneath the disc, resulting in a decrease of atmospheric pressure beneath the record. Atmospheric pressure above the record puts the equivalent of a pressure of 250 kg on the playing surface of the record. The effect is to remove resonance and warp sounds that usually take place in the region of 5 Hz to 10 Hz. The vacuum disc stabilizer is featured on some Luxman turntables.

TONEARMS

The concept of component hi-fi can be applied to record players and some record players are sold this way. For a record player system the basic components would consist of the turntable and its drive motor, the phono cartridge and its stylus, and the tonearm.

The function of a tonearm is to hold the phono cartridge in place while it moves across the record. It also holds the wires which carry the audio signal from the cartridge. The tonearm is usually made of tubular metal, often of a uniform diameter but occasionally tapered toward the cartridge end to increase rigidity and reduce effective mass. The tonearm metal is generally anodized aluminum, magnesium alloy, or seamless stainless steel; you will also find them made of carbon fiber or titanium. The arm may be filled with some sound absorbing material to dampen the natural resonant frequency of the arm. Tonearm resonance is quite low in frequency, down around 8 Hz. In some cases a mechanical anti-resonance filter is built into the counterbalance.

It is necessary for the arm to have a high tensile strength to low mass ratio. One goal of tonearm manufacturers is to produce arms having minimum mass, with the effective mass ranging from about 4 grams to 33 grams—10 grams is a general average. The advantage of the carbon fiber arm is that it has a lower than average effective mass, permitting the phono cartridge to track warps in phono records. Record warps are around 5 Hz or lower, so a self-resonant frequency of the tonearm of about 10 to 11 Hz seems preferable since the warps are then less likely to shock the arm into resonant vibration.

The effective length of a tonearm is the straight line distance between its pivot at one end and the stylus at the other, regardless of the shape of the arm. Tonearm lengths vary from a low of about 8¼- to 10-inches, with 9½-inches being the usual average. The longer the effective length the smaller the tracking error.

Tonearms can be pivoted in a counterbalance gimbal arrangement. On some it is possible to raise or lower the pivot point, usually just a few millimeters, to make sure that the arm is parallel to the flat surface of the disc. The gimbal supports the tonearm at one end and has either jewel bearings or ball bearings, or a combination of the two to permit easy movement of the arm, vertically or horizontally. At the rear of the tonearm a rotatable cylindrical counterweight can be adjusted to permit changes in tracking force, usually in the range of 0-to 3-grams.

In some units the headshell of the phono cartridge is integrated to the tonearm, that is, the headshell and arm form a single, nonseparable unit. The design concept is that this minimizes mass at the critical point of the tonearm and also removes possible poor contact of the cartridge terminals.

Tonearm Shape

Tonearms can be grouped into four categories: straight, *J*, *S*, and tangential. The straight (or *I* shaped), *J*, and *S* types are shown in Fig. 6–8. The drawings show the effective lengths of each of the arms. The letter *a* represents the offset angle, the angle the cartridge makes with the straight line portion of the tonearm. The purpose of the offset angle is to minimize tracking error. Greater control is also possible through the use of a sensor and a microcomputer.

Tangential Tonearms

Instead of swinging across the record from a fixed pivot, tangential tonearms, also called linear tracking tonearms, cross the record in a straight line and in this way follow the action used by the cutting stylus in making a master disc.

For this reason tangential tonearms can trace the grooves in the phono record more accurately. But because their bases must also move they lack the solidity of the more firmly anchored pivoted arms and tend to vibrate in response to low-frequency musical signals. One advantage of the linear tracking arm is that it is easy to operate.

Tracking Error

Tracking error, measured in degrees, represents the deviation of the cartridge axis from the record tangent. Tracking error varies from the outer to the inner grooves of a record and can be from as little as less than one-half of a degree to as much as two and one-half degrees. Typical values supplied by manufacturers are $+1.8°$, $-1.0°$ and $+2.5°$, $-0.8°$ of record radius.

Cartridge Interchangeability

Except for integrated tonearms in which the phono cartridge and the arm are a single unit, tonearms have standard plug-in connections permitting interchangeability of cartridges.

STRAIGHT TONEARM

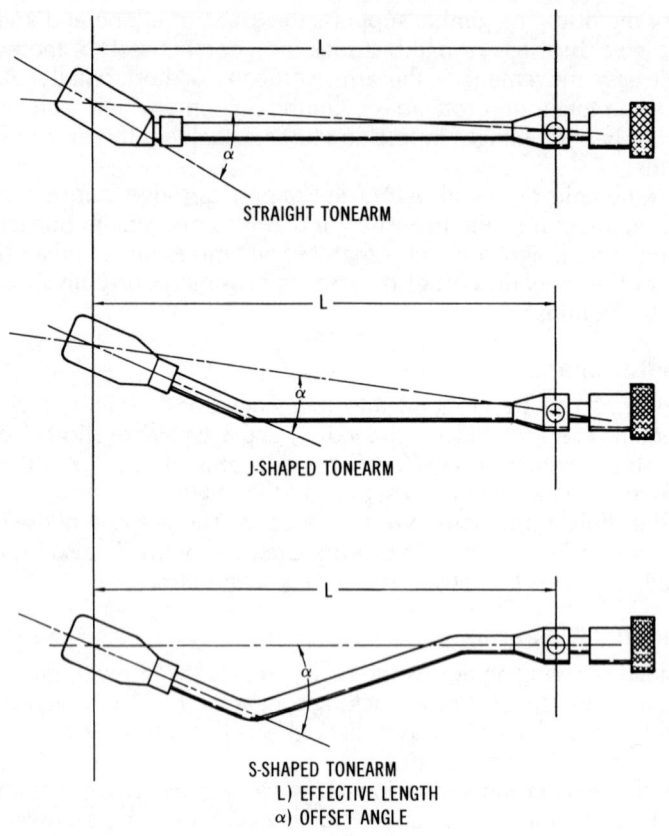

J-SHAPED TONEARM

S-SHAPED TONEARM
L) EFFECTIVE LENGTH
α) OFFSET ANGLE

Fig. 6-8. Tonearm shapes.

RECORD PLAYER MAINTENANCE

You can get better results from your record player if you follow some simple precautions.

The platter surface should be flat, and to make sure it is, you can check it with a bubble glass (level), available in hardware stores. Get a small bubble glass so you can make a number of checks on the surface of the platter since you will get uneven turning if the player is tilted, even slightly. Some record players are supplied with adjustable feet, so with the help of the bubble glass you can adjust these until the platter is absolutely horizontal. If the record player doesn't have this feature, use shims made of small pieces of thin cardboard.

The fault may not be in the player but in the table on which it rests, so start with that first. Your floor may be uneven or warped or the table legs may not be of equal length. Check the table top for flatness, moving the bubble glass to a number of different spots to make sure. Put the record player back in its operating position and then check it also for flatness.

Use the most solid support for your record player you can get. The ideal arrangement, but impractical for the home, would be a solid block of concrete. Broadcast stations have been known to use cinder block surfaced by cement. The idea here is to keep floor vibrations from reaching the record player and to discourage self-induced vibrations. A bare wooden floor has excellent vibrational characteristics and your walking across it belongs to the same family as wall thumping. Use carpeting or a rug. This will come between your footsteps and the floor, either preventing acoustic vibration or absorbing it.

Your record player probably comes equipped with rubber feet, one at each corner. In older models these feet are rather skimpy; in newer units the feet are quite wide, supplying more contact surface area. If you've had your record player for a long time, you may find that the feet have become hard. If so, replace them.

If you find you can't get replacements for the feet, take them off completely and then put a rubber pad between the record player and the cabinet top on which the record player rests. You can get such pads in a stationery or department store. If you need to-use shims to get the player flat, put the shims between the player base and the rubber pad, not between the pad and the top of the cabinet.

Unless the center hole of your record is worn, there will be a snug fit between it and the spindle. When you put a record on for play, press down on the record label all around the spindle to make sure the disc is flat against the mat of the platter. It's easy to assume records are warped, especially since they are much thinner than they used to be. To check, turn the record player on and move your head down so your eyes are a few inches away from the rotating edge of the platter and the record. You should be able to see if the record travel is flat.

If the dust cover is a removable type you may find it easier to clean it if it is off and away from the turntable. Return the cover to the record player as soon as possible.

Don't oil your record player unless the manufacturer's instruction book advises you to do so. Most record player bearings are oilless types. If you do use oil, do so sparingly; just a single drop is usually enough. Don't let oil drop on the belt (or belts) used in belt-drive systems.

THE PHONO CARTRIDGE

Of all the components associated with a high-fidelity system the phono cartridge is probably the least understood and the least appreciated. Perhaps this is due to its small size, for it is nothing more than a somewhat rectangular box sitting at one end of the tonearm. However, the cartridge is the beginning of a long chain of events that takes place in a high-fidelity system, but what that system can do for you in the way of quality sound reproduction from phonograph records depends directly on the cartridge.

Aside from the modest modifications in the frequency response curve to correspond to RIAA requirements, the preamplifier that follows the cartridge can do nothing to improve the sound. If the cartridge signal output is distorted, then the preamplifier has no choice but to amplify that distortion. If the stereo separation of the cartridge is poor, the stereo amplifiers in the preamp will not, and cannot, improve matters. If the cartridge frequency range is poor, the preamp will not extend it. If cartridge crosstalk between the left and right channels of sound is excessive, the preamp will pick up and continue that fault.

In short, what you get in the way of sound is what you put in, and the sound signal you put into your preamplifier is what the phono cartridge supplies. Since phono records are our most important sound source, there is no logic in building an elaborate, and quite often expensive, high-fidelity system without having some prime consideration for the cartridge.

At one time all record players and changers came equipped with integrated tonearms—that is, the tonearm and its associated cartridge were single units, available only as a joined pair. All the user was required to do was to change the stylus, then referred to as a needle. With these cartridges record wear was severe, frequency range was very limited, noise and distortion were high. But as the job complexity of the cartridge became known and appreciated, it started to occupy a category all its own and so it is now most often sold as a separate item.

What It Does

A phono cartridge is a miniature electric generator. Its function is to change the mechanical energy supplied by a stylus following the sound modulated grooves in a record into an equivalent electrical voltage.

Types Of Cartridges

There are several basic types of cartridges, depending on the techniques they use for producing an output voltage. In some the transducing element is a bit of ceramic or other piezoelectric material but cartridges using these aren't generally regarded as suitable for high fidelity. All quality cartridges, also called pickups, make use of a coil (or coils). The difference among them is how this part is used.

Moving Magnet There are various ways in which you can generate a voltage across the turns of a coil. You can take a permanent magnet and move it in and out of the coil. As long as that magnet is in motion the lines of magnetic force surrounding the magnet will move across the turns of the wire of the coil inducing a voltage that can be measured across the ends of the coil. This action is that of a transducer, a device for changing one form of energy to another. In this case the mechanical energy involved in pushing the magnet in and out of the coil appears as electrical energy obtainable from the coil. This technique is used in the moving magnet (mm) type and is the most common method in phono cartridges.

Moving Coil An alternative method is to have the magnet fastened in position and to move the coils in and through the magnetic field existing between the poles of a permanent magnet (Fig. 6–9). Again, we have relative motion between the coils and the magnet and so a voltage will be induced in the coils and can be measured across its ends. This is the moving coil (mc) type.

Fig. 6-9. Elements of a moving coil phono cartridge. The moving stylus moves the cantilever which moves a coil of wire closer to or farther from a permanent magnet. This motion through the magnetic field of the permanent magnet induces a voltage across the moving coil.

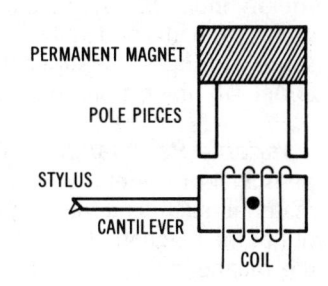

Generally, in moving coil cartridges (there are exceptions) the stylus cannot be replaced by the user but is a job for a trained technician who has the tools and expertise required.

Some audiophiles claim that moving coil cartridges supply better sound and there are technical reasons for supporting this belief. One is that the recording cutter head for making phono records also uses the principle of the moving coil and so the moving coil in a phono cartridge duplicates that technique.

The advantage of the moving coil type of cartridge is that it generates the signal with a minimum of distortion. To keep the moving mass as low as possible the coils usually have just 15 or 20 turns, but this small number means that the output signal voltage is very small. This is the reason most moving coil cartridges need the assistance of a pre-preamplifier (also known as a head amplifier) or step-up transformer at the output. In the moving magnet type, by contrast, the coils which are fixed in position, may have as many as 3,000 turns of wire. The larger the number of turns, the greater the induced voltage, and so moving magnet cartridges have no need of output voltage assistance, do not require head amplifiers, or step-up transformers.

One of the ways of getting higher voltage output in a moving coil cartridge is to use more turns of wire in the moving coils. This has been achieved with an extremely thin aluminum ribbon measuring 10 microns × 100 microns which permits more turns of wire in a limited space. It also lowers the effective tip mass and supplies better transient characteristics since the weight of the coil and the mass to be moved is at a minimum.

In the moving coil there are two coils, one for left channel sound, the other for the right. They are shaped to fit into the gap between the pole pieces of the fixed position permanent magnet. This is the region of maximum magnetic flux density and so is a contributing factor in the production of a higher than usual voltage output.

Another technique for getting more output signal voltage is to use a magnet with a strong magnetic field. In a typical cartridge the magnet has a strength of only 2 K gauss, that is, 2,000 magnetic lines per square inch, but some cartridges have more than seven times as much strength, or 15,000 lines per square inch. The magnet used in some cartridges is a super high energy anisotropic Alnico 9. This has about 1.8-times more magnetic energy than the normally used Alnico 5.

Variable Reluctance In this next method, both the coils and the permanent magnet are kept in a fixed position. A small bit of ferrous material, such as a very light, hollow piece of iron, is moved in and out of the magnetic field of the magnet. The motion of the iron causes the magnetic field to increase and decrease, cutting across the turns of wire of the coils and inducing a voltage. This is also known as the moving iron type. See Fig. 6–10.

Record Groove Signal Amplitude

The maximum amplitude of a signal in the record groove is 25 microns and the amplitude of noise is 1/1000 (−60dB) of this amplitude, or 0.025 micron. (A micron is a millionth of a meter.) A signal of only 1/10 to 1/30 noise level can easily be recognized by a listener. Roughly, this is equal to about the size of 10 atoms, as the separation between atoms is about 0.0001 micron. The noise level in the record groove, 0.025 micron amplitude, is almost equal to 250 times the separation between atoms.

The Cantilever

The basic technique for moving a magnet, or a coil (or coils) or a bit of ferrous material is to attach these to a small bar, rod, or thin wall tubing made of aluminum beryllium, or other substances such as carbon fiber, titanium, boron, or precious stones such as ruby, sapphire,

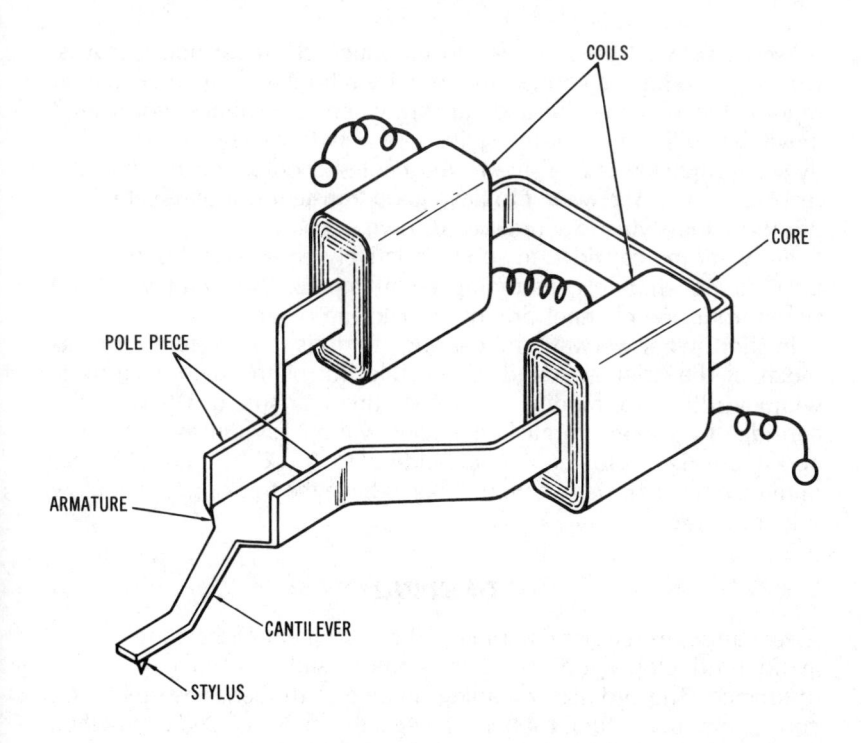

Fig. 6-10. Variable reluctance phono cartridge.

or diamond. Known as a cantilever, the other end holds the stylus (Fig. 6–11). As the stylus moves in the record groove, the cantilever moves accordingly, and then so does the coil or magnet or iron fastened to its other end.

CARTRIDGE REQUIREMENTS

All of this sounds simple enough, and it is. Up to now the cartridge is simply a device for generating a voltage. It is when we start making other demands on it that we encounter some problems.

We want the cartridge to develop enough voltage so that the voltage will be strong enough to drive the preamplifier to which it is connected, but not so strong that the preamplifier will be overloaded with input signal voltage and will distort.

Fig. 6–11. The stylus is fastened to one end of the cantilever.

We don't want the cartridge to introduce electrical noise, that is, we want it to have a good signal-to-noise ratio. We want to be able to replace the stylus easily and quickly when it becomes worn, as it inevitably will. We want the stylus assembly to be able to move easily, to be quickly responsive to the faintest modulation cut into the record grooves. We want it to have good transient response yet we do not want the stylus to vibrate of its own accord.

We want the cartridge to supply relatively equal signal outputs for both stereo channels, assuming equal inputs. We don't want the signal from one channel drifting over to the other.

In short, what we want the cartridge to do is to recapture for us as much of the original sound as possible out of the record grooves, without adding to, modifying, or distorting that sound. We want the cartridge to respond to all frequencies without favoritism, starting at about 20 Hz or lower and extending to 25 kHz, or higher. To minimize record wear we would like to have the tracking force below 2 grams.

DAMPING

All cartridges require damping of the movement of the cantilever to avoid producing spurious signal outputs and unwanted cartridge resonance. The ordinary damping method is rubber. It is quick and convenient, but natural rubber is a vegetable product, and even when made with anti-oxidants, gradually loses its flexibility and stiffens.

THE STYLUS

Sitting right at the end of the cartridge mounted in the tonearm of the record player or changer is a tiny bit of mineral——a diamond (Fig. 6–11). Because it is so small it is barely visible, yet it is this part that decides whether you will hear music just the way it was recorded on discs, and whether the records you buy will have a decent life span before you throw them away.

A stylus presents problems not found in other hi-fi components. We want the stylus to stay in the record grooves (Fig. 6–12); the ability of the stylus to do so is called trackability. This means we must put a pressure on the stylus to keep it in those grooves, but as the record revolves it tends to throw the stylus up and away. However, if we put pressure on the stylus, at least two things can happen, both bad. The first is that record wear increases. The other is that the stylus will wear and need replacement. Since records and styli represent the biggest cost of upkeep of a high-fidelity system, anything that can be done to prolong record and stylus life is a move in the right direction (Fig. 6–13).

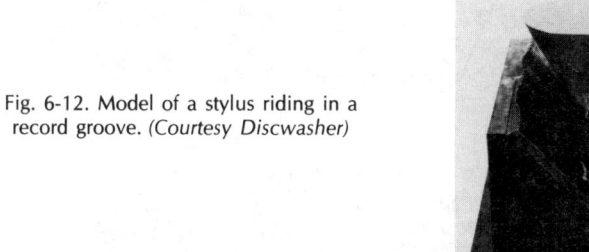

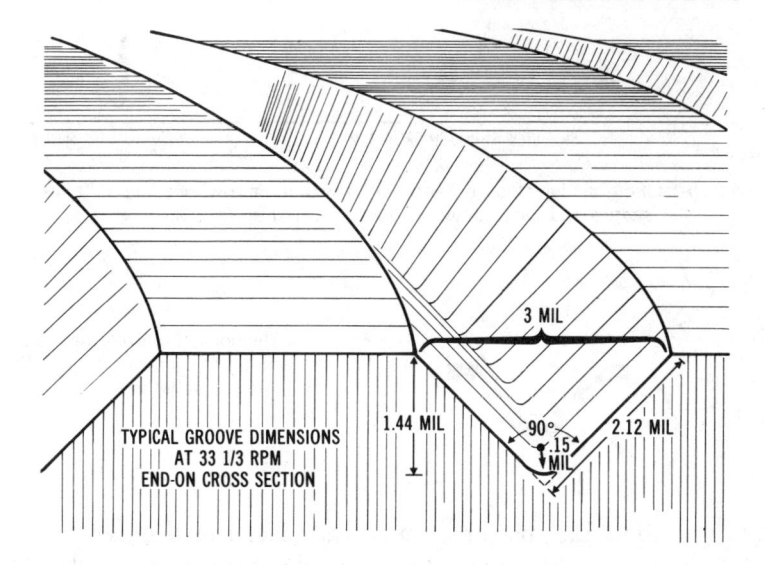

Fig. 6-12. Model of a stylus riding in a record groove. *(Courtesy Discwasher)*

Fig. 6-13. Typical groove dimensions. *(Courtesy Discwasher)*

There are various types of styli: conical (also called radial or spherical), elliptical, and Shibata (Fig. 6–14). There are also some variations of the Shibata stylus, produced by various cartridge manufacturers, but these need not be categorized separately.

Conical Stylus

Of these different types, the conical (Fig. 6–15) at one time was the most widely used. Easier to manufacture than the others, it is less expensive and its shape could be compared to the writing end of a ball point pen. The tip is machined to extreme precision with a tip

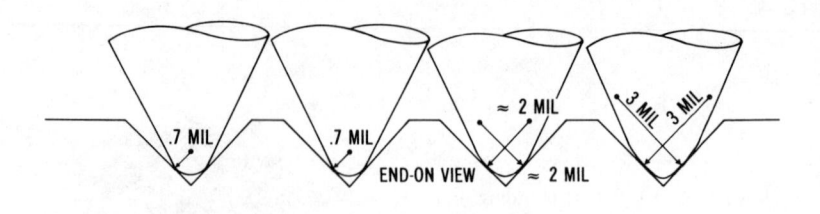

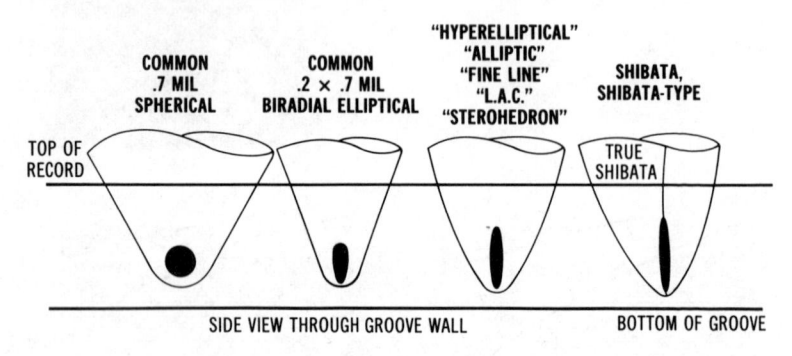

Fig. 6-14. End-on view (top row) and side view (bottom row) of common stylus shapes and their contact geometry. *(Courtesy Discwasher)*

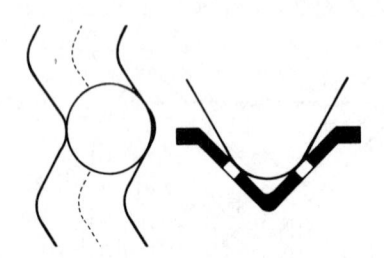

Fig. 6-15. The conical stylus is the easiest to manufacture but is limited in its ability to trace treble tones.

radius of about 0.5 mil for a stereo cartridge. However, with such dimensions this stylus cannot follow the high frequency modulations in the groove walls, especially near the end grooves at the center of the disc. The result is distortion due to poor tracking.

Elliptical Stylus

Elliptical styli (Fig. 6–16), having a narrower profile, can deliver better response to high frequencies since such styli can follow fine groove tracings with more precision. Stylus force adjustment for an elliptical is more critical than for the conical. You can also do more damage more quickly by using too much or too little tracking force with the elliptical. And, since the elliptical stylus is more difficult to manufacture and install correctly, it costs more than the conical.

Fig. 6-16. The elliptical stylus is better able to follow treble tones, hence results in lower distortion. The amount of stylus wear is about the same for spherical and elliptical types. Wear is more affected by the quality of the polish put on the stylus.

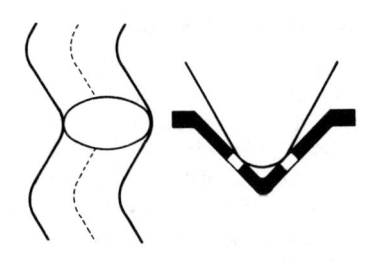

Shibata Stylus

The Shibata is a modified form of elliptical stylus whose structure permits more contact with the record grooves. The Shibata stylus was originally designed to be used with four-channel records but it is also highly desirable for stereo discs.

Made of diamond, the durability of a stylus depends largely on tracking force, the weight with which the stylus is pushed up against the record groove walls. The shape of the stylus and the mass of the stylus tip are also important. Ideally, the mass of the tip should be as low as possible, since the lower the mass the less force required for tracking loud passages.

With the elliptical stylus, the cross section that touches the groove walls has a radius of about 0.4 mil. (A mil is one-thousandth of an inch.) The advantage is that this smaller radius enables the stylus to respond to higher frequencies. The long radial axis of the elliptical is about 0.07 mil and is used to keep the stylus at the right height in the groove.

Because of its construction, the elliptical stylus has less surface area in contact with the record groove walls than the conical. If the tracking force for a conical and an elliptical are the same, then the pressure of the elliptical on the groove wall is greater. Consequently, tracking force with an elliptical must be made less than with the conical to prevent groove wall damage.

The Shibata stylus has a tip with a large supporting radius having about four times as much contact area with record groove walls compared to the elliptical. This means the tracking force is spread over a much greater area and so the tracking force can be made lower than that of either the conical or elliptical styli. An advantage of the Shibata is that it can respond to groove modulations at frequencies as high as 50 kHz.

The advantage of the elliptical stylus over the conical is that it requires less tracking force. Since the pressure on the record surface is lower, there is less record wear. For the same reason the Shibata stylus, using less record pressure, results in less record wear, even though a particular cartridge using a Shibata stylus might require the same tracking force as a comparable elliptical stylus.

van den Hul Stylus

One of the more recent of the stylus entries has been the van den Hul, named after its Dutch inventor. The design of this stylus is such that it fits the record grooves fairly precisely, with a recommended tracking force in the range of 2 to 2½ grams. The stylus, whose shape closely resembles that of the cutting stylus used at the start of the record making process, must be precisely positioned. An alignment mark on the phono cartridge permits easy azimuth adjustment. However, if this adjustment is not done or done improperly, it is possible for the stylus to deform the record grooves or to cut away sound modulations.

Nude Diamond

Diamonds are expensive and so to cut costs some styli aren't made just of diamond but of diamond fastened to some other, less expensive, material. These may use a bonded metal shank, but its effect is to increase effective tip mass.

A better, but costlier, arrangement, is a stylus made completely of diamond, attached to the cantilever. Known as a nude diamond this type of stylus supplies the lowest possible moving mass and can deliver a more accurate frequency response, especially in the treble range.

EFFECTIVE TIP MASS

While we ordinarily think of the stylus as pressing down into the record groove, we can consider it in a different way, regarding the modulated groove as a way of moving the stylus and the cantilever attached to it, laterally and vertically.

The stylus, the cantilever and whatever is attached to it comprise the effective tip mass (etm). The etm is mostly made up of stylus tip mass, plus about 40% of the mass of the cantilever and about 10% of the mass of the generating element. Ideally, we should get etm down as close to zero as possible, hence it is desirable to make the cantilever as short as we can. If we disregard this requirement and opt for a higher output voltage with the help of a longer cantilever, we increase etm, causing serious record wear in the frequency range between 12 kHz and 16 kHz, particularly on the inner walls of the record grooves.

CARTRIDGE SPECS

Cartridge specs (Fig. 6-17) are important since they permit an evaluation prior to purchasing. However, consider that even if a pair of cartridges made by different manufacturers have almost identical

[Tonearm Section]

Tonearm:	Straight Arm of static balance type
Effective Length:	240mm
Tracking Error:	+2°13′, −1°08′
Overhang:	15mm
Cartridge Weight:	4g – 11g
Cartridge Height:	16mm ~19mm (by use of spacers)
Stylus Pressure:	0 – 3g (direct reading)
Accessories:	Anti-skate Adjustment, Tonearm Elevation

Fig. 6-17. Tonearm and cartridge specs.

specs they may well sound different because of variations in construction techniques and materials.

Stereo Separation

Separation refers to the amount of separation, specified in decibels, between left and right channels and is the ability of the cartridge to maintain them distinctly. The greater the stereo separation, the better.

Separation at spot frequencies is inadequate, but is sometimes indicated at 1 kHz. It should be described for the entire frequency range claimed for the cartridge and should be given for both channels independently. Thus: left channel more than 25 dB from 35 Hz to 5 kHz; more than 18 dB from 20 Hz to 20 kHz. Right channel: more than 25 dB from 20 Hz to 5 kHz; more than 20 dB from 20 Hz to 20 kHz.

Channel separation is especially important in the treble range, but it generally decreases with increasing frequency. A typical cartridge could have a separation of 23 dB minimum at 1 kHz and 17 dB minimum at 10 kHz. Preferably, the output voltage supplied by the cartridge for each channel should be fairly close over the entire frequency range within a fraction of a millivolt.

Compliance

Compliance is the ability of the stylus assembly to move when a force, measured in dynes, is applied. High compliance is desirable since it means more accurate reproduction of the groove modulations and also less record wear.

Compliance figures are supplied based on the movement of the stylus in millionths of a centimeter. Mathematically this is expressed as 10^{-6} cm. A representative compliance figure is given in a cartridge spec could be: 20×10^{-6} cm/dyne. This means that for a force of 1 dyne the stylus has moved a distance of 20 millionths of a centimeter. Since compliance is always specified in terms of cm/dyne just the first number is often supplied. In the example just given the compliance would simply be referred to as 20. The higher this number the more compliant the cartridge.

In stereo discs, the grooves are modulated vertically as well as laterally. This means that compliance figures in specs should be for both. Good vertical compliance is an indication of good stereo separation.

While lateral and vertical compliance figures are often fairly similar, lateral compliance is often larger in some cartridges. In some it is possible for the lateral compliance to be twice as great as vertical compliance.

Tracking Force

Tracking force, also referred to as vertical tracking force, or vtf, is the amount of downward pressure applied to a stylus to make sure it remains in the record grooves. If tracking force is too small it is possible for a strong bass tone to send the stylus skittering across the face of the record. Insufficient stylus force also means inadequate contact with the groove walls, thus losing some of the signal. If tracking force is too large, the stylus will cause unnecessary groove wear and may even shave some of the weaker signals right out of the grooves. Tracking force is dependent on the type of stylus used and the tonearm: it is lower for a Shibata than for an elliptical (Fig. 6-18).

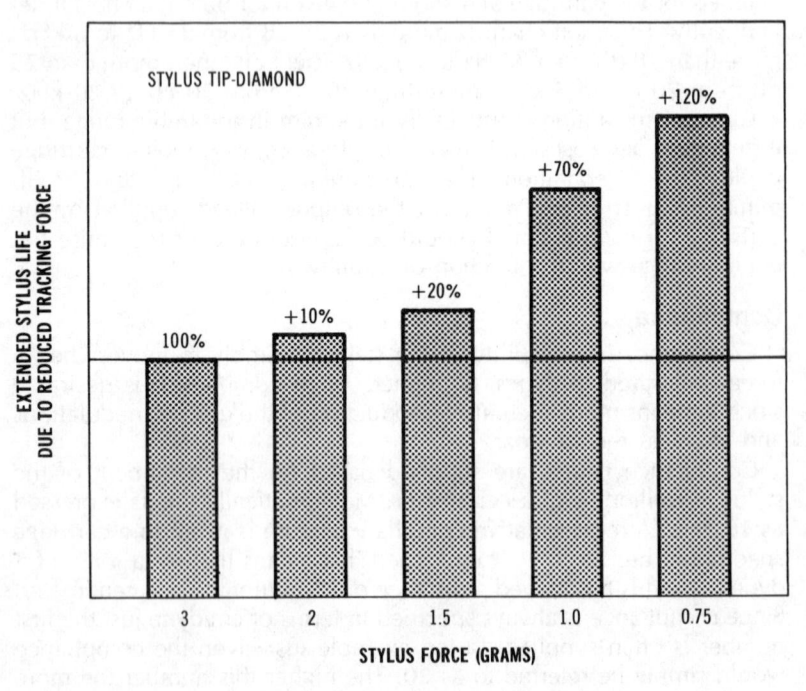

Fig. 6-18. Effect of tracking force on tip life.

Some record players come equipped with a phono cartridge, but in most cases the phono cartridge is a separate component. At the time of purchase the manufacturer will supply you with information about the recommended tracking force, in grams. Translated from the metric to the English system, a gram is 0.0353-ounce avoirdupois. That's just a little more than three one-hundredths of an ounce.

The smaller the tracking force the more careful you must be. Typical values range from about ¾ gram to 2 grams. Since manufacturers supply minimum and maximum tracking force figures, a safe method is to use a tracking force that comes right in the center. If the manufacturer recommends ¾ gram to 1¼ grams, set your tracking for 1 gram. The counterweight on your tonearm may be calibrated to let you do this or you can get an inexpensive device that will let you measure the tracking force.

Once you have the tracking force selected, play a record that has some strong bass tones. The stylus should track the record without being pushed out of the grooves. If it is, increase the amount of tracking force until the stylus stays in the groove.

Record player tonearms are equipped with counterweights for adjusting tracking force. Sometimes these are two section types: one for coarse, the other for fine adjustment.

Tracking is a measure of the velocity of the stylus tip in cm/sec while maintaining good groove contact at a stated tracking force at a specific frequency. Trackability is a refinement of the compliance figure, taking into account both stylus tip mass and tracking force.

Because of the extremely tiny area of contact between the stylus and record groove walls, fantastic pressures occur. At a tracking force of a few grams the pressure is several thousand atmospheres, in other words, thousands of times higher than the air pressure in an automobile tire.

Load Impedance

This is the preamp's input impedance as seen from the cartridge. For moving-magnet and induced-magnet pickups the standard value is about 47,000 to 50,000 ohms, which is what most amplifier phono inputs are designed for. Moving coil cartridges have low load impedances, usually about 2 to 40 ohms.

The sound of moving magnet and variable reluctance by the amount of capacitive loading they "see" at the input to the preamplifier. A Shure V-15 IV, for example, specifies a load of 275 picofarads for optimum performance. If the cartridge is loaded at a value that is less, it will sound bright and tinny. A higher value will make it sound dull, with the midrange emphasized. The better the equipment, the more these anomalies will be noticed. A variation of just 25 picofards is audible under ideal conditions.

Frequency Response

This is the useful response range with the cartridge terminated in its recommended load. It is quite often given in a 10 octave range from about 20 Hz to 20 kHz. The variation, in decibels, should also be specified. Severe peaks will produce tonal coloration. Look for two things: frequency range and flatness of response.

Output

This is generally supplied as a measurement in thousandths of a volt (millivolts) and abbreviated as mV. It should be specified separately for left and right channels and is often given as a number such as left channel, 5.3 mV; right channel, 4.8 mV. This information is inadequate since it does not include the amount of input.

With magnetic pickups the amplitude of the electrical signal output varies in proportion to the velocity of the displacement of the cutting stylus, hence the need for a velocity reference.

The standard reference recording level is 5 cm/sec. A more accurate listing would be: left channel, 5.3 mV at a velocity of 5 cm/sec; right channel, 4.8 mV at 5 cm/sec.

Additional Specs

Specs for record players may also include a listing of options, special features, or may supply further details. Fig. 6-19 is typical.

[Additional Features]

Dust Cover:	Detachable with semi-freestop hinge
Stroboscope:	Mirror-reflex type.
Automatic Function:	Auto-Start, Auto-Repeate Cut-Off

[General]

Power Consumption:	12W
Dimensions:	472(W) x 152(H) x 348(D)mm (18-37/64" x 6" x 13-45/64")
Weight:	Net 10kgs (22 lbs.) Gross 12kgs (26.4 lbs.)

Fig. 6-19. Supplementary information included in a spec sheet.

HOW RECORDS ARE MADE

Recording is the first step in the record-making process. In the recording studio the sound of music, an orchestra, a jazz or rock group, a western group, an instrumentalist or vocalist, is recorded on magnetic tape. Quite often each musician or vocalist is supplied with an individual microphone and a separate sound channel. Unlike tape recorders used in the home, professional units can have as many as 32 individual channels. Further, the musicians and or vocalists need not be all together at the same time, for each sound channel can be recorded separately. The advantage of such an arrangement is that the sound of individual instruments or vocalist(s) can be adjusted, balanced, weakened, or enhanced.

After all the channels of sound have been recorded, they are mixed down into a single stereo channel consisting of a pair of tracks holding left channel and right channel sound. It is during the mixing stage that decisions are made that affect the quality of the final performance.

The audio signal is then transferred to a *lacquer* (Fig. 6–20), the industry name for a mastering disc, by means of a precision cutting lathe. Once the master has been cut it is sent to a record pressing plant.

Since the lacquer or mastering disc will not withstand pressing, a metal mold must be made (Fig. 6–21). Silver is used as the first agent because it will adhere to the lacquer surface. It will then harden and

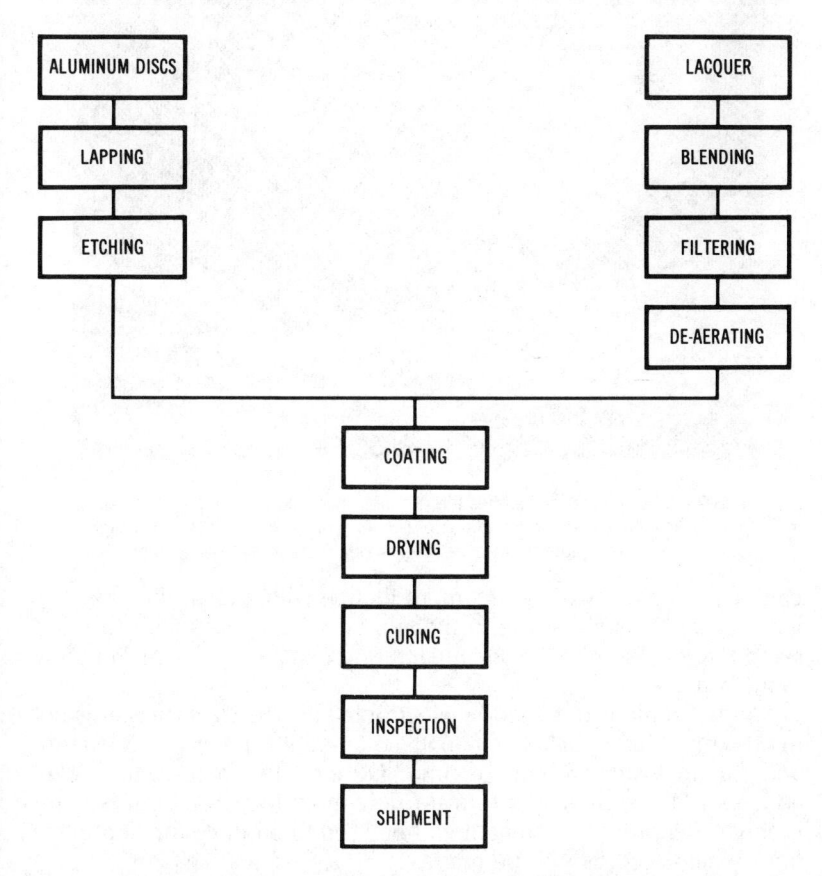

Fig. 6-20. Flow chart of phono record manufacturing process.

Fig. 6-21. After a lacquer reaches the pressing machine it is sprayed with silver, forming the first of several metal molds that will be made from the original master disc. Lacquers themselves are too delicate to withstand pressing.

can be peeled away in perfect reproduction without harming the original master. This silver mold, though, is not tough enough to withstand large volume record manufacturing demands, so a nickel alloy mold is made.

A nickel mold is obtained by electroplating. The silver master is put in a highly diluted nickel sulfamate solution at a precise temperature and for an exact amount of time. During this process the nickel adheres to the master and forms a new mold (Fig. 6-22). It is commonly called a *mother* since it will be used to produce the final plate from which records will be made.

Each mold formed from the original lacquer master must be

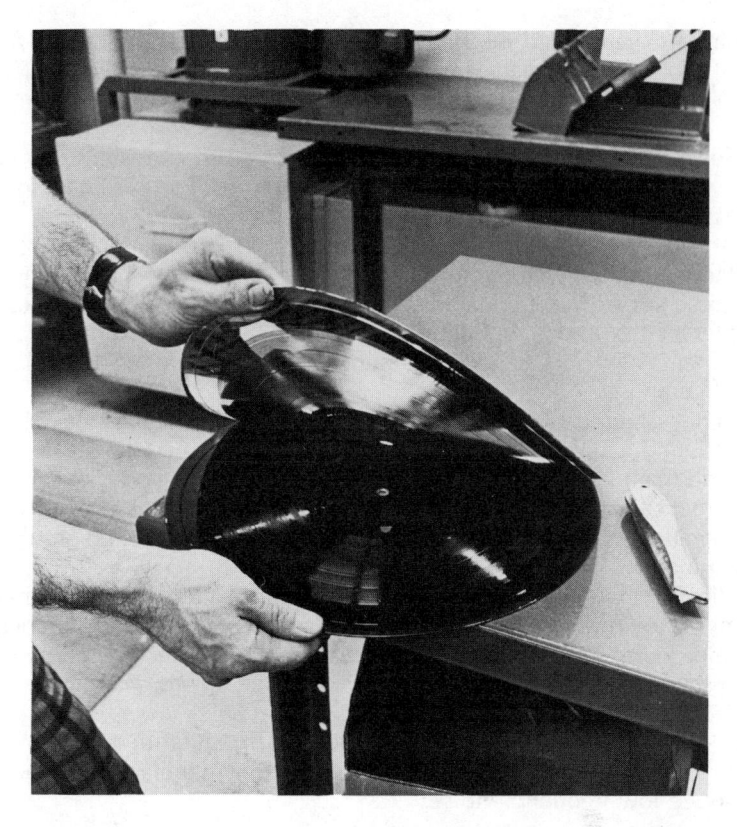

Fig. 6-22. The silver master is stripped away from the lacquer, forming the first mold made from the cutting surface. This metal master will now go through electroplating to form the next mold, called a "mother."

stripped away from the previous mold, cleaned, and trimmed before the next process can take place. In each case the mold is put on a revolving turntable and carefully cleaned to remove any stray particles which might have lodged on the surface or in the mold's grooves. Alcohol is used to clean silver masters; jewelers' rouge is used for nickel molds.

When the mother is finished it is sent to a testing area to be checked for sound quality. Using special playback equipment the quality control inspector listens to the audio signal. If any pops or groove damage are evident, the inspector either rejects the mother outright, or, if the problem is minor, inspects the mother under a special microscope and then makes the repair.

Since molds must be exact copies, but in reverse, of the final product, one step remains. The lacquer master was a positive; the silver mold is negative and the nickel mother is then a positive. Since the final mold must be negative, the nickel mother is sent back to electroplating for a final mold, called a *stamper* because it is later inserted in a pressing machine and used to stamp out records.

Several stampers are usually made from each mother since stampers wear down during the manufacture of records. For extremely long runs as many as 1,000 or more stampers may be used.

After a stamper is made, trimmed, and cleaned, it is taken to a centering machine where its record spindle hole is punched. Using a microscope with a graduated screen, a worker determines the exact center of the mold. At this point in the manufacturing process the stamper is also back sanded, die punched, and formed to fit the mold configuration.

While stampers are being made in the electroplating area, vinyl compound is being mixed for the final pressing of records. Bulk resin for the manufacture of vinyl is stored in giant holding silos and then automatically pumped into the compound area where it is mixed automatically via weight monitoring and then pumped to the pressing machine area.

Vinyl compound is released through the pressing machine in small lumps known as *biscuits*. Labels are attached and then the stampers clamp down, pressing out a record. Each pressing machine can hold two stampers at a time, one for each side of the record. The process is fully automated.

A phono record may have tiny surface irregularities known as *orange peel* produced at the time the record is manufactured. These produce low-frequency noise.

VIRGIN VINYL

Vinyl that is 100% pure and that does not contain previously used records is known as virgin vinyl. Most phono records are not made of virgin vinyl in an effort to cut down on costs, and so a typical record is an intermix of virgin vinyl and reground or used vinyl. However, since used vinyl contains dirt, grease, and other impurities, an intermix inevitably produces record noise which we hear as pops, clicks, and hiss. If a new record sounds noisy the first time it is played it is quite possible it wasn't made with virgin vinyl. And, since the dirt particles are molded into the vinyl, it is unlikely they will be removed by record cleaners.

Not all virgin vinyl is of equal quality. Here again a record company may economize by using virgin vinyl of a lower grade. All virgin vinyl is different in its chemical composition and so each type imparts its own sonic subtlety to the finished disc.

Vinyl will also break down as a result of repeated playing and in turn this increases pops and clicks. Mistracking, using an improperly balanced phono cartridge, a dirty or worn stylus, or failure to clean phono records regularly will significantly reduce the useful life span of the record.

RECORD CLEANING

We live at the bottom of an ocean of billions upon billions of dust particles, visible only when a beam of sunlight strikes them. For the most part these particles are so light they float and relatively few fall onto the exposed surface of phono records. Instead, they are pulled down by electrostatic forces.

Electrostatic forces are almost always high voltage and consist of electric charges built up on nonconducting surfaces when they are in motion. A charge of thousands of volts develops across your skin area when you walk across a rug in a room that has dry air. Just removing a phono record from its dust jacket can easily produce a thousand volts or more along its face. When that record rotates the charge climbs to several thousand or more volts.

It is practically impossible to reduce that electric charge to zero and to keep it at zero. The best we can do is to minimize it. Cleaning records (Fig. 6-23) is a battle that can never be won permanently for it is unending, but we can fight a good rearguard action. If you smoke in your listening room you give aid and comfort to the enemy for tobacco smoke may have an even higher particle density than dust.

Record cleanliness can be deceptive. A brand new record, nice and shiny when first removed from its dust jacket, would seem to be a model of cleanliness. The fact is that many new records are dirty from the chemicals used in the manufacturing process, so a new disc isn't exempt from the need for prior treatment.

Phono records not only contend with dust but with the oil and

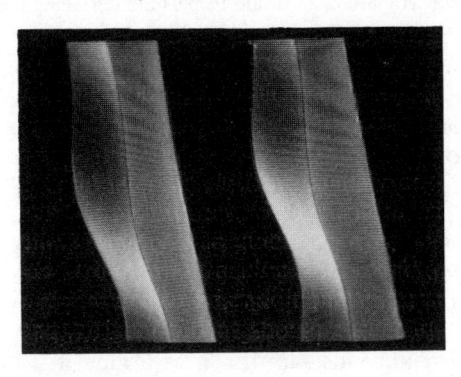

Fig. 6-23. Clean record grooves magnified 500 times. *(Courtesy Discwasher)*

acidic perspiration from fingers. A low tracking force stylus will actually "play" these deposits. A good rule is always to hold discs by the edge only.

Washing a record with soap and water or with laundry detergent will make records look new and shiny but also leave soap deposits in the grooves. Oddly enough, one of the problems of water is that it is not wet enough, and so cannot get down, deep down, into the grooves to be effective. Records should be kept as dry as possible and while a high level of humidity in a listening room will lower electrostatic charges it helps age the vinyl and encourages microorganism growth.

Dirt cannot only damage record groove modulations, but can cause pops and crackles during playback. To the swiftly traveling stylus the dust may not appear as microscopic barriers but as boulders (Fig. 6-24). The collisions between the stylus and dust particles can cause the stylus to bounce against groove walls, removing some of the wall modulated signals.

Fig. 6-24. Dirt in a phono record after 200 plays. Note the large collection of dirt particles at the bottom of the V, plus the dirt along the sides.
(Courtesy Discwasher)

There are any number of commercial record cleaning devices available but a good first step is to develop the proper hi-fi housecleaning attitude. One way to do that is to become conscious of the fact that records require constant protection. The dust jacket that comes with discs is precisely that—a device for keeping dust out. Always keep records in their jackets until the records are to be used for play. As soon as disc play is completed, put it back in its jacket.

If, after you finish playing a record, you note that the record and/or stylus have accumulated some lint or dirt, don't blow across either. It is futile and all it does is prove you can breathe, something that requires no confirmation. Instead, use it as a clue that both the stylus and record should be cleaned before returning the record to its jacket. To clean the stylus use a soft camel-hair brush. Always bring the brush toward you, from the back of the stylus to the front, never from front to back or side to side (Fig. 6-25).

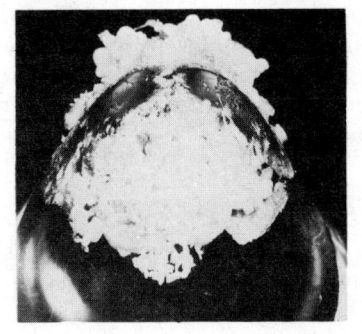

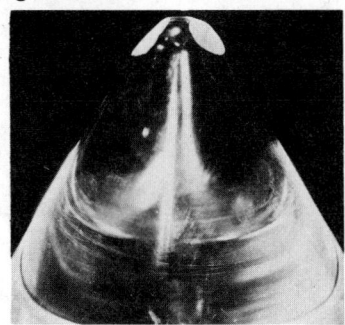

Fig. 6-25. Dirt collected by a stylus (left), and clean stylus (right).

One of the problems involved in using liquid record cleaners is that they may leave unwanted residues in the grooves after the cleaning liquid has completely evaporated, akin to putting dirt back after you've removed it. This undesirable swap, mentioned earlier in connection with soap and detergents, isn't true of denatured alcohol. However, alcohol has a strong affinity for the stabilizers used in making records and extracts them, reducing record life.

Record cleaning isn't a one-step process since cleaning a record doesn't automatically eliminate the cause of record dirt. This means that the static charge on records must also be reduced. Record cleaning, if done as a separate operation, should precede the use of a static discharge device or liquid, since the action of cleaning can increase the charge on a record.

If the record cleaner is a brush type, and these are common, the brush should be cleaned before using it on another record; for it is possible for the brush to act as a transfer agent and will move dust from one record to the next.

No liquid can clean a record by itself but requires a brush to get the contaminants out of the grooves. The grooves are microfine and it is necessary for the brush bristles to reach down into the bottom. Nylon, sometimes used for this purpose, will have fine points and a tip radius of about 0.00025 inch.

Records can also pick up dirt from the supporting mat. Clean the mat using detergent and warm water, followed by a number of cold water rinses. Wipe both sides dry with a lint-free cloth.

Keep the dust cover of your record player closed at all times except when you put on or remove a record. Do not try to clean the cover while playing a record. When cleaning the dust cover, use a moist cloth or you will find you are simply pushing the dust from one part of the cover to another, but not removing it. Don't assume that the dust cover is clean, just because it looks that way. Don't use any liquids on the dust cover that will scratch or discolor the surface.

If the dust cover is a removable type, put it in a solution of warm water and detergent. Rinse a number of times with clean water and then wipe dry with a lint-free cloth. Return the cover to the record player as soon as possible. You want that record player to remain uncovered for a minimum amount of time.

STYLI CARE

Styli don't last forever. Useful life depends on record cleanliness, tracking force, correct antiskating, and how often you play. Recommended figures run from 400 to 800 hours which means it's anybody's guess. One way to compare is to examine the stylus, when brand new, under a magnifying lens so you can become familiar with its original shape.

RECORD PLAYER CARE

Make your record player off limits to anyone but yourself. Keep it as a one man or one woman device. It's bad enough that you must become familiar with the care and feeding of record players—don't expect anyone else to have your interest or concern. And this caution applies particularly to children. A record player is not a toy.

If you have a hum problem, even if it is slight, start by disconnecting the ground wire coming out of your record player. Usually this lead is attached to the ground terminal of your tuner or receiver. With this lead disconnected, the hum level will increase. With the volume turned up but with no record playing, try transposing the male plug of the player in its power outlet socket. You will hear hum both ways, but use the position with less hum.

And now, with the volume turned up a bit, and with the platter rotating (but with the stylus not touching the grooves), try touching the ground lead of the record player to different grounding points to see if you can get the hum to disappear completely. Try the ground terminal of your tuner or receiver. Also try touching this lead to the metal chassis of other components, if you can get at them. Your power outlet has a center screw holding the cover plate of the outlet in position. Try attaching the ground lead to this screw.

The usual record player is a two-speed device. Some of these have speed change controls near all the other operating controls. However, in some belt drive units you will need to remove the platter. You will see a polyurethane belt connected to a small drive wheel. This drive wheel is a two-section affair. The larger wheel is for 45 rpm; the smaller for 33⅓ rpm. Look at the belt as it turns the platter. There should be no slippage. If there is, get a belt replacement. Slippage is always a possibility with belt or rim drive, but not with direct drive.

Sometimes a record player may seem to have problems that are caused externally—that is, they are not due to an inherent fault in the player. If your ground lead is loosely connected you can have a hum level that is annoying. Make sure the signal leads coming out of the record player aren't near any power line cords or near the power transformer of other high-fidelity components. To check, move these leads with the record player turned on, but without playing a record.

The tonearm contains wires which bring the audio signal from the cartridge to the input of the preamplifier. If the arm is made of metal and is grounded this is usually quite effective shielding. However, if you have fluorescent fixtures, you may find they radiate a strong signal and that this signal is being picked up by the tonearm wiring. It's easy to check. If you hear a buzzing sound with the lights turned on, only to have the buzz disappear when the lights are turned off, you have a clear case of incompatibility. Move the player and the lights away from each other.

If your high-fidelity system starts to howl every time you put on a record, or seems to be on the edge of howling, you may have acoustic feedback from your speakers to the turntable. Don't use speakers as a support for your turntable. Separate these components to get rid of the feedback problem.

SPEED CONTROL

Strobe patterns for 33⅓- and 45-rpm around the edge of the platter are now commonly used. If you neglect adjusting the fine speed control, the pitch of the sound will be either higher or lower than it should be, but since this higher or lower pitch will remain fairly constant throughout record play, you may not be aurally aware something is wrong. You will, though, if the pitch deviation is high or low enough. Adjust the fine speed controls, following the instructions in the operating manual. If your player doesn't have strobe markings you can get a low-cost strobe disc to take its place. These are small and fit right over the spindle. Don't try to slow the speed of the platter by putting your finger against the edge. This could throw the belt.

If you are planning to replace your cartridge, and you have decided you do not want a unit identical to the original, check with the manufacturer of the record player. The combination of cartridge weight and effective tonearm mass results in a resonant arrangement. That resonance should be below the frequency of the lowest musical tone you expect to get, preferably somewhere between about 5 and 15 Hz. In short, the cartridge should be compatible with the arm.

236

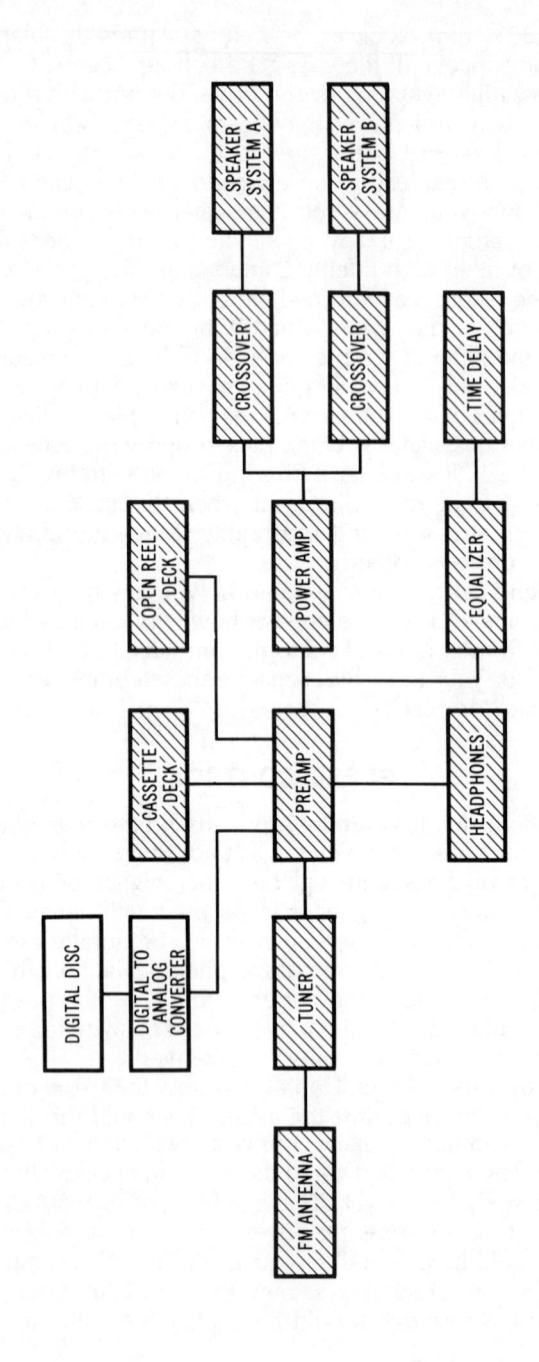

7

Digital Versus Analog Audio

The strength of a sound usually varies from moment to moment. The graph in Fig. 7-1 shows how the sound can change. We can convert the sound to its electrical equivalent by using a microphone and so, instead of having a changing sound we have a changing voltage. But the voltage wave is comparable to or is analogous to the sound wave on which it is based and so we can call it an analog of that wave. All we have done, in converting the sound wave into a voltage wave, is to change the wave into a more measurable form.

We can refer to the strength of the wave at any selected moment in terms of volts, or some submultiple such as millivolts (thousandths of a volt) or microvolts (millionths of a volt). We can convert this voltage wave back to its original sound waveform at any time, for one is an analog of the other.

Another advantage of changing the sound wave into its analog is that we can amplify the voltage wave. But the output voltage wave, although much larger, is still an analog of the original sound wave.

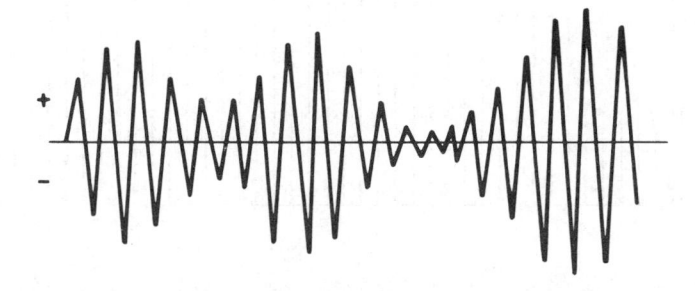

Fig. 7-1. Voltage waveform corresponding to a sound.

ANALOG VERSUS DIGITAL

How do we measure the voltage of a wave whose value keeps changing? As mentioned earlier in Chapter 2 and Chapter 4 we usually work with an rms or root-mean-square method, an averaging technique. Another way would be to determine the instantaneous value, that is, the value of the wave at any selected moment in time. If we did this for a succession of moments we would get a series of numbers, possibly 35, 24, 23, 15, 27, etc. These numbers represent the amount of audio voltage of the wave, moment after moment. If we then composed a listing of these momentary voltages we would have a record of that voltage in number form, that is, digital.

Each of the numbers obtained would be a sampling of the voltage waveform (Fig. 7-2). The greater the number of such samples the more accurate our numeric listing would be. These numbers are part of our decimal system, a system in which there are only ten digits, ranging from 0 through 9. Any other numbers, such as 135, 384, 199, are just combinations of the basic numbers used in the system. And that not only includes whole numbers, but decimals and fractions as well.

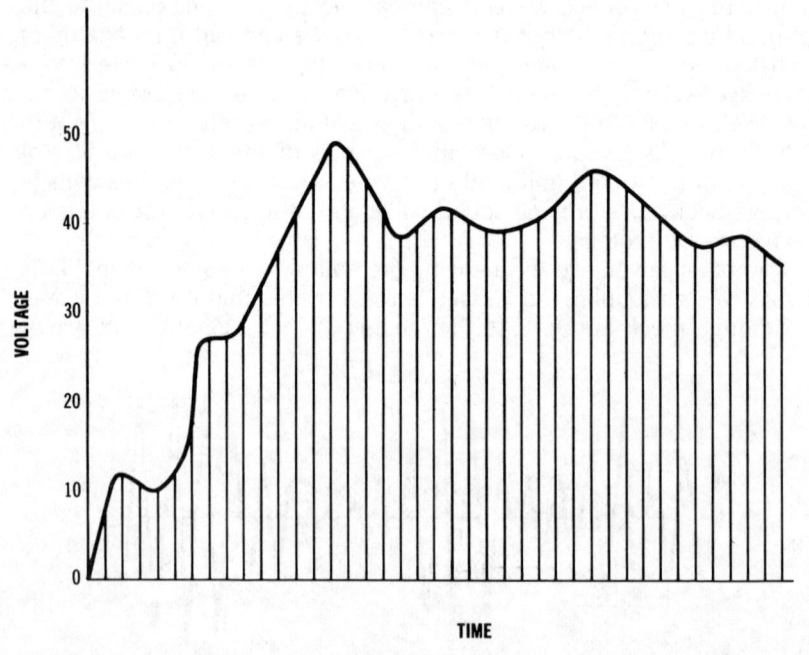

Fig. 7-2. Each vertical line represents the strength of the voltage at a particular moment.

THE BINARY SYSTEM

The decimal system is just one of a group of possible number systems. A quinary system used digits 0 through 4 only; an octal system digits 0 through 7. The simplest of the number systems, though, is the binary for it has just two numbers: 0 and 1. It is this simplicity that makes the binary system so useful in computers, in calculators, and in audio.

Fig. 7–3 shows several groups of binary numbers. No matter how long the number is, or how short, the only two numbers that are used, or that can be used, are zero and one. Because they consist of a succession of zeros and ones such a number listing may not initially make much sense, simply because we are accustomed to thinking only in terms of decimal numbers. And, no matter what number system we select, ultimately we interpret all results in terms of the decimal system. So the binary system is just a tool to help us get results which we cannot achieve directly with the decimal system.

Fig. 7-3. Binary numbers.
```
   10110
  101101
 1010001
11111100
100001001
```

BINARY TO DECIMAL CONVERSION

It is when we first begin to convert binary numbers to decimal numbers that we begin to understand them. Table 7–1 shows the relationship for numbers ranging from zero to twenty in decimal form and their binary equivalents.

The value of a binary digit depends on its position. When placed at the extreme right it has a value of 1. Moving toward the left, the value of the adjacent binary digit is 2, then 4, then 8, followed by 16. This follows a similar procedure used in the decimal system where the value of a number also depends on its position. A number such as 3 has a value based on where it is located. It can be 3, or 30, or 300, or 3,000. All we have done in this example is move the decimal number 3 to the left. But in moving the number the effect is as though we had multiplied by 10. In the binary system moving digits toward the left is equivalent to multiplying by 2. The numbers across the top of the binary digits shown in Table 7–1 give the decimal equivalent value of each of the binary numbers.

In binary, then, 0001 is equal to 1; 0010 is the same as 2; 0100 is the same as decimal 4 and 1000 is equal to 8. Note that binary digit 1 has been moved successively toward the left and each time its value has doubled. In tabular from it would look like this:

239

```
0 0 0 1 = decimal 1
0 0 1 0 = decimal 2
0 1 0 0 = decimal 4
1 0 0 0 = decimal 8
```

Why Binary?

Why should we bother with binary numbers when we are so accustomed to thinking and working with decimal numbers? The answer is that we can use electronic and mechanical devices to represent binary numbers, and then, if we need to, convert those binary numbers to decimal form.

In Fig. 7-4 we have five switches, possibly of the type you have in your home for switching lights on and off. Each switch in its off position can represent the number 0, and when turned on, each switch would then be indicative of the number 1. These switches could be operated mechanically to give us any number in binary from 00000 to 11111. If we put these numbers into decimal form, 00000 in binary is the same as zero in decimal. 11111 in binary is the same as 31 in decimal. So with just five switches we could have the binary equivalent of any number from 0 through 31.

The switches give us our numbers in binary form only. It is then up to us to convert the binary numbers to the more familiar decimal

Table 7-1. Decimal Numbers and Corresponding Binary Numbers

Decimal Value	Binary Value				
	16	8	4	2	1
0	0	0	0	0	0
1	0	0	0	0	1
2	0	0	0	1	0
3	0	0	0	1	1
4	0	0	1	0	0
5	0	0	1	0	1
6	0	0	1	1	0
7	0	0	1	1	1
8	0	1	0	0	0
9	0	1	0	0	1
10	0	1	0	1	0
11	0	1	0	1	1
12	0	1	1	0	0
13	0	1	1	0	1
14	0	1	1	1	0
15	0	1	1	1	1
16	1	0	0	0	0
17	1	0	0	0	1
18	1	0	0	1	0
19	1	0	0	1	1
20	1	0	1	0	0

```
        SW1         SW2         SW3         SW4         SW5
A ──────╱○──────────╱○──────────╱○──────────╱○──────────╱○──────

B ──────╱○──────────╱○──────────╱○──────────╱○──────────○→○──────

C ──────╱○──────────╱○──────────╱○──────────○→○─────────╱○──────

D ──────╱○──────────╱○──────────○→○─────────╱○──────────╱○──────

E ──────○→○─────────○→○─────────╱○──────────╱○──────────○→○──────
```

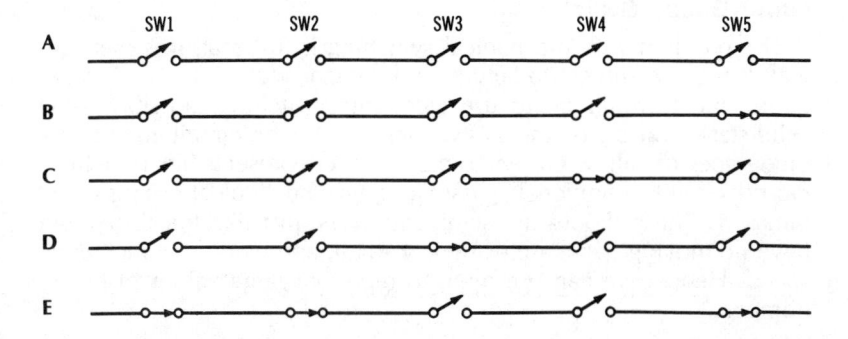

Fig. 7-4. An open switch is zero; a closed switch is 1. In (A) all switches are open and so the sum is 0. In (B) the rightmost switch (SW5) is closed and so the sum is 1. In (C) the second switch (SW4) from the right is closed, representing decimal 2. In (D) the third switch (SW3) is closed, equivalent to decimal 4. In (E) the first, second, and last switches are closed (SW1, SW2, and SW5) and so we have 16 + 8 + 0 + 0 + 1 or decimal 25.

form. Of course we started with just five switches as a simple example, but there is no reason why we couldn't have 50 switches, or 500, or 5,000. The larger the number of switches the larger equivalent number we could produce in binary form, ready for change to decimal.

Fig. 7-5 shows three groups of binary numbers. Above each of these numbers is the equivalent in decimal. By adding the decimal equivalents we get the decimal value of the entire binary. Note that zero in binary does not contribute to the sum of the final decimal number. The first binary number at the right has a value of 1, and then, moving toward the left, values of 2, 4, 8, 16, 32, 64 and so on.

$$16 + 0 + 0 + 2 + 1 = \text{DECIMAL } 19$$
$$1\quad 0\quad 0\quad 1\quad 1$$

Fig. 7-5. Some binary numbers and their equivalent decimal values.

$$32 + 0 + 8 + 0 + 0 + 1 = \text{DECIMAL } 41$$
$$1\quad 0\quad 1\quad 0\quad 0\quad 1$$

$$64 + 0 + 0 + 8 + 4 + 2 + 0 = \text{DECIMAL } 78$$
$$1\quad 0\quad 0\quad 1\quad 1\quad 1\quad 0$$

Other Binary Switches

The problem with mechanical switching is just that, it is mechanical. It is slow, subject to failure, and demands lots of room. A better arrangement would be to use electronic switching, possibly using solid state diodes working as switches. In its open circuit condition a diode does not allow current to pass, hence represents binary 0. In its closed circuit condition it passes current and could be considered binary 1. These diodes are semiconductors, just like transistors, but have no moving parts, are extremely small, and work with incredible speed. Hence, we can use them to represent large values of binary numbers.

Bits

The words "binary digits" are usually contracted to form a new word—bits. If we use three switches we can produce three binary digits or three bits. With three bits the maximum we can count to is decimal 7. That is, we can go from 0 0 0 to 1 1 1. Obviously, then, the greater the number of bits (Fig. 7-6) the higher we can count. An 8-bit binary system cannot give us as large an equivalent decimal number as a 16-bit system. Table 7-2 shows the number of bits and their maximum binary value. The 8-bit and 16-bit systems are the more commonly used.

Fig. 7-6. Each binary number is a bit.

0	ONE BIT
01	TWO BITS
101	THREE BITS
1010	FOUR BITS
01110	FIVE BITS
101011	SIX BITS
1011011	SEVEN BITS
01011101	EIGHT BITS

CONVERTING ANALOG TO DIGITAL

In Fig. 7-7 we have a graph of a commonly used wave known as a sine wave. This wave is in analog form but we can convert it to digital by a succession of measurements along the wave. These measurements are taken at equally spaced intervals of time indicated by the numbers, 1, 2, 3, 4, etc., along the bottom of the graph. These numbers could be in seconds, thousandths of a second (milliseconds), or millionths of a second (microseconds). The numbers arranged vertically along the left represent values of voltage. The wave starts with a value of 20 volts, reaches a peak of 30, drops down to about 10, then comes back up to 20. Fig. 7-7 shows the instantaneous voltages at various equal time periods, starting with zero and ending with ten. Once we have the voltage values in decimal form we can convert

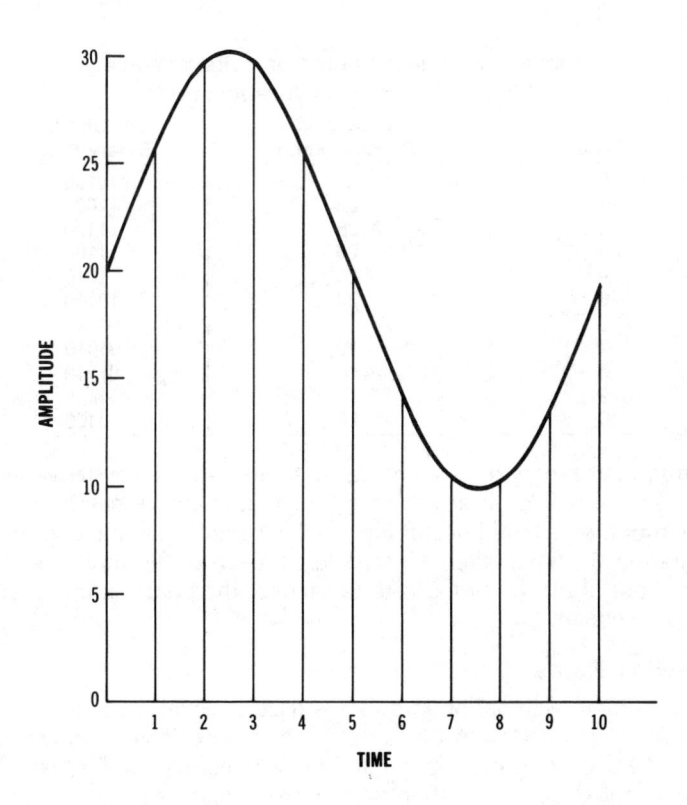

Fig. 7-7. Digitized values of a voltage wave in analog form.

Table 7-2. Maximum Binary Values and Decimal Equivalents

Number of Bits	Maximum Binary Value	Maximum Equivalent Decimal Value
1	1	1
2	11	3
3	111	7
4	1111	15
5	11111	31
6	111111	63
7	1111111	127
8	11111111	255
9	111111111	511
10	1111111111	1023
11	11111111111	2047
12	111111111111	4095
13	1111111111111	8191
14	11111111111111	16383
15	111111111111111	32767
16	1111111111111111	65535

Table 7-3. Digital Values of Voltage Wave (Fig. 7-7) — Five-Bit Arrangement

Time	Voltage in Decimal Form	Voltage in Binary Form
0	20	10100
1	25	11001
2	28	11100
3	28	11100
4	26	11010
5	20	10100
6	14	01110
7	10	01010
8	10	01010
9	14	01110
10	20	10100

them to their binary equivalents as in Table 7-3. The system used in Table 7-3 is a five-bit arrangement for that is all we need.

Each of the vertical lines in Fig. 7-7 is called a sample and in that illustration we have taken 10 samples. If each of the numbers along the bottom of the graph indicates a second, then our sampling rate is one per second.

Sampling Rates

The minimum sampling rate that is used is generally at least twice the highest frequency to be sampled. If, for example, the sampling rate is 40,000 times per second and if the maximum audio frequency to be sampled is 20 kHz, then each wave of that frequency would be sampled twice. The abbreviation 20 kHz (kilohertz) is the same as 20,000 cycles per second. Divide the frequency into the sampling rate to get the number of times each wave of that frequency would be sampled.

Twenty kilohertz is generally considered the upper limit of the audio range. The frequency at the opposite end is 20 Hz or 20 cycles per second. The sampling of the wave at 20 Hz, using the same sampling rate, would be 40,000 divided by 20 or 2,000. This means each wave at the bass end would be sampled 2,000 times. Since the sampling frequency remains constant, the number of samplings depends on the frequency of the audio signal at any particular moment.

The sampling frequency could be a value such as 44.3 kHz or 44,300 times per second; to maintain stability the sampling frequency is derived from a 4.433 MHz crystal.

Quantizing

Representing a signal in terms of numbers is called quantizing. Quantizing an analog signal means taking the instantaneous value of the analog waveform and converting the decimal value of each instantaneous value into binary form. The bits, or smallest possible segments, permit extremely precise description of each sound sample.

THE ANALOG DISC

Phono records or phono discs are cut in analog form. At the time a master record is made (Fig. 7–8) a cutting stylus engraves sound modulations into a groove with the movement of the stylus analogous to a controlling audio voltage. This audio voltage corresponds to the sound in front of a microphone. The movement of a stylus in a phono cartridge follows that of the cutting stylus and it is this movement that generates an audio voltage via the phono cartridge. So what we have here consists of sound, an audio voltage representing that sound, and then mechanical motion. The playback system is simply the recording system, but in reverse.

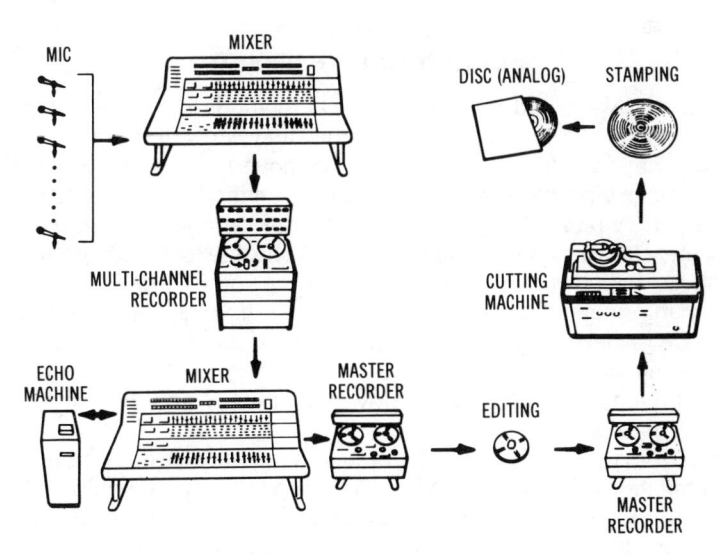

Fig. 7-8. Conventional analog record production system.
(Courtesy Sony Corp. of America)

THE DIGITAL DISC

A completely different approach is used in a digital disc. Instead of varying grooves we have a series of pits and flats as indicated in Fig. 7–9. Seen edgewise they look like an alternate series of raised and depressed surfaces. If a raised surface represents binary digit one, then a depressed surface can be considered as the binary digit zero.

245

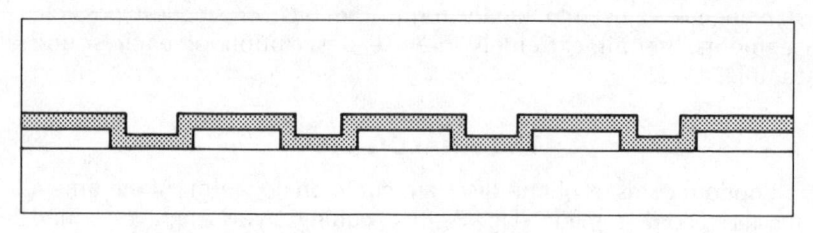

Fig. 7-9. Representation of pits and flats on a disc.

We can, for example, take the binary numbers shown earlier in Table 7-3 (representing the audio voltage waveform of Fig. 7-7) and put these numbers into a phono disc, with groups of pits and flats corresponding to the binary numbers. The problem then becomes one of tracking these pits and flats and then changing the binary numbers into analog form.

MECHANICAL TRACKING

As a start we could use a mechanical method. In this arrangement we would have a stylus tracking the pits and flats cut into the record groove. This would cause the stylus to move up and down and the mechanical vibrations could then be transmitted to a piezoelectric converter. A piezoelectric material has the capability of converting pressure changes into voltage changes. The piezoelectric converter could be called a binary to analog device for its input consists of binary information but its output is a varying voltage similar to that generated when tracking an analog record. The changing output voltage of the piezoelectric converter could then be brought into the usual preamplifier, then to a power amplifier, and finally to the speaker systems.

While this arrangement is theoretically possible it does have serious limitations. In moving from a flat to a pit the stylus would receive repeated blows and we always have a rapid succession of zero signal conditions to maximum. Whether the stylus and its transducer could tolerate such a set of working conditions is doubtful.

VARIABLE CAPACITANCE TRACKING

In the variable capacitance tracking system, illustrated in Fig. 7-10, the stylus doesn't bump up and down but rides along the surface of the flats. As described earlier the digital signals are engraved in the form of tiny pits. Fig. 7-11 shows how small the pits and flats are compared to the grooves in an analog phono record.

In variable capacitance tracking the disc is made of conductive

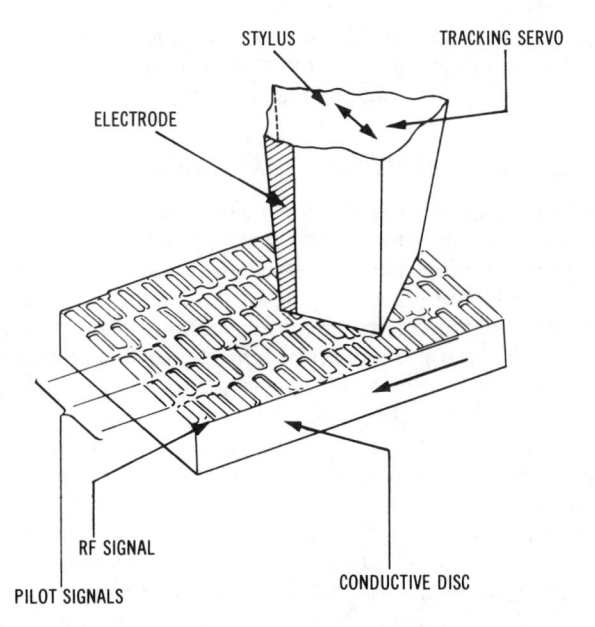

Fig. 7-10. Digital disc system using variable capacitance method.

Fig. 7-11. Pits in the digital disc (left) compared with grooves in an LP phono record.

materials and the signals are detected as changes in electrostatic capacitance. To maintain the necessary accuracy the pits are engraved in guide grooves. Some variable capacitance systems use a dynamic tracking servo system for improved trackability. In this case the disc has additional pilot signals impressed on both sides of the signal pits as shown earlier in Fig. 7–10.

In this system, the capacitance between the disc as one pole and a stylus mounted sensor electrode as the other pole is continuously measured. The variation in the amount of material between the two poles causes changes in capacitance, and this, in turn, causes changes in the electrical signals that are produced.

OPTICAL TRACKING

This digital disc tracking system makes use of a laser beam that reads the presence or absence of signal pits contained in a disc beneath a plastic coating (Fig. 7–12). These readouts take the form of variations in reflected waves which are converted into digital codes and then into analog audio waveforms. There is no direct contact between the disc and the pickup system and so this method requires the use of a dynamic tracking servo system. Fig. 7–13 supplies details of how the scanning is accomplished.

Of course there are advantages and disadvantages in the variable capacitance and optical systems. The variable capacitance system requires mechanical tracking and so the records and the tracking electrode are subject to wear. The optical system is much more complex, is more subject to misadjustment, and is more expensive. However, there is no record wear nor is there any wear in the pickup system.

ANALOG RECORDING

In analog recording the sound is picked up by two or more microphones and the resulting electrical signal is brought into a mixing console. Here the sound engineer regulates, controls, reduces, or increases the signal. Individual instruments or groups of instruments are then recorded on a multitrack recorder. These tracks are subsequently mixed down, possibly altered again by the sound engineer, until there are just two tracks for left and right stereo sound. These tracks, on a master tape, are then used to cut a master disc on a lacquer cutting

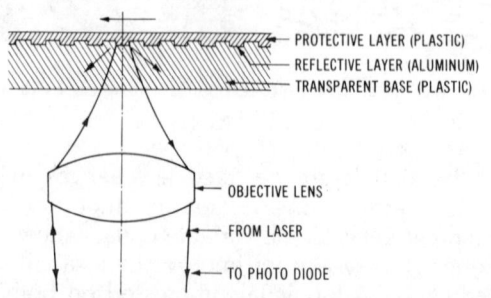

Fig. 7-12. The pulse code modulated disc is scanned from beneath by a laser beam. The flats and pits are coated with a reflective layer of aluminum. A protective layer of plastic is placed above the flats and pits.
(Courtesy Sony Corp. of America)

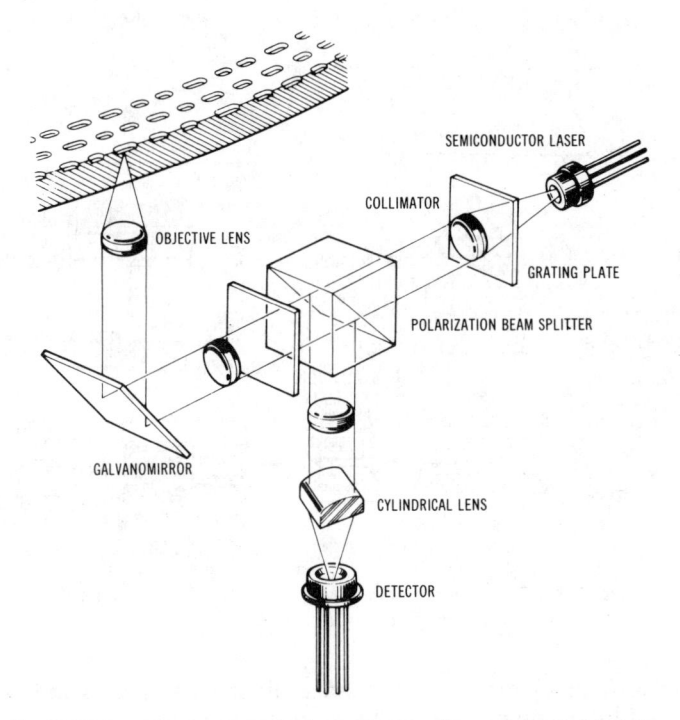

Fig. 7-13. Laser beam method of scanning a pulse code modulated disc.

machine, from which the record stampers are then produced. This process was described earlier in Fig. 6-20 in Chapter 6.

DIGITAL RECORDING

Fig. 7-14 shows the digital recording process. Some of the steps used in analog recording are retained. Sound is picked up by a number of microphones and the electrical output is fed into a mixer. The signal undergoes conversion from analog to digital form in a digital audio multichannel recorder. The signal undergoes editing in a digital audio editor and finally controls a digital cutting machine.

DIGITAL RECORDING ON TAPE

Digital recordings are made by first recording music onto computer tape. An analog-to-digital converter translates the music into 16-bit numbers at a rate of 50,000 numbers per second.

When recording an analog signal on tape the signal has a varying amplitude. When the signal is extremely quiet it may reach down into

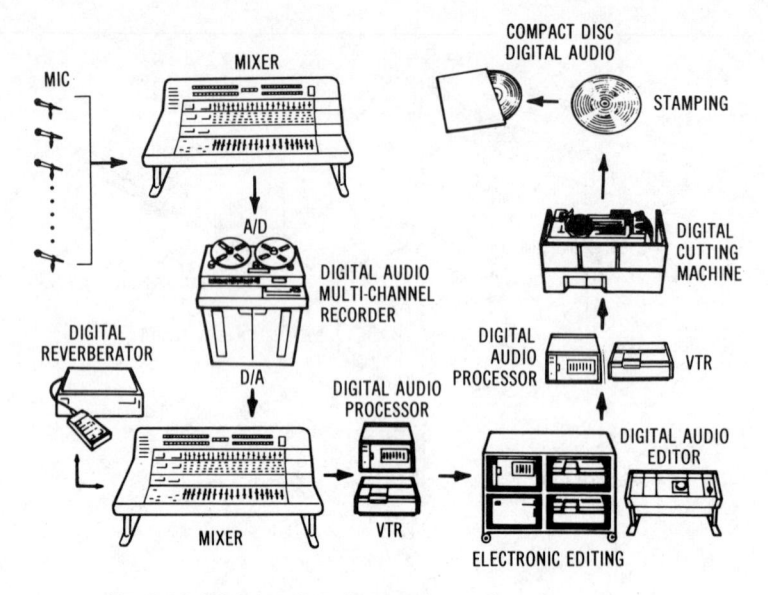

Fig. 7-14. Digital audio editing and record production system.
(Courtesy Sony Corp. of America)

the noise floor of the tape. The noise floor of a tape is the noise produced by magnetic tape in the absence of an input signal. At the opposite end, the signal, if sufficiently strong, will saturate the tape, causing distortion.

This problem doesn't exist when recording tape digitally for all the binary numbers have the same strength and are of equal amplitude. The recording level, then, can be set well above the noise floor, eliminating tape noise, and well enough below the maximum recording level of the tape, eliminating tape saturation distortion. As a result the dynamic range of the tape—the sonic range between the softest and loudest analog sounds that can be registered (usually about 60 dB)—no longer matters. In digital tape recording of 16-bit words, for example, the dynamic range is about 96 dB. In effect, then, using digital recording extends the dynamic range capability of the tape system from about 60 dB to very close to 100 dB. See Fig. 7–15.

In analog recording every effort is made to impress as much signal on the tape as possible, with the result that there is always the possibility of print-through, a situation in which material recorded on one layer of tape "leaks" through to produce an echo signal on a tightly wound adjacent layer of tape. This is eliminated in digital recording since the signal strength is never enough to produce this condition.

At the time of cutting the pits and flats into the phono records (these are cut according to the numbers on the tape), there is no speed variation, distortion, or background hiss. The digital process broadens the dynamic range. Other advantages include an absence of rumble, wow or flutter, and intermodulation. The signal to noise ratio can be more than 90 dB, frequency response essentially flat from 20 Hz to 20 kHz, and harmonic distortion less than 0.05%.

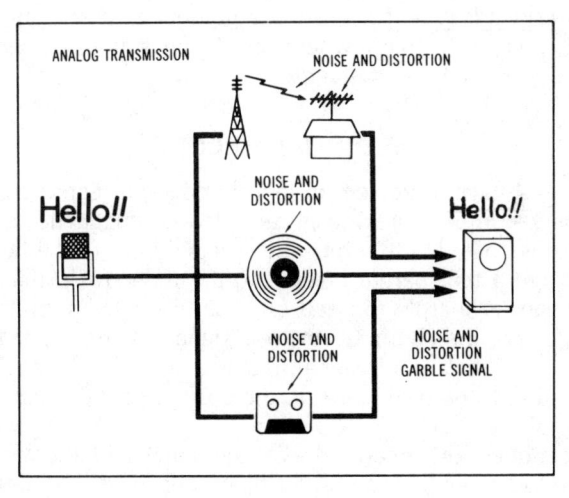

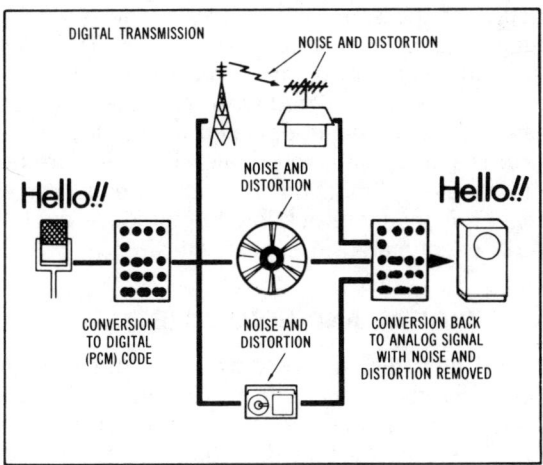

Fig. 7-15. Analog transmission (top) vs digital transmission. *(Courtesy Sony Corp. of America)*

PULSE CODE MODULATION

Pulse code modulation, abbreviated as pcm, is the technique of using the pits and flats technique for impressing binary digits into a phono record.

The binary digits contain all the original musical information. There is also space remaining on the pcm disc for the insertion of other information besides the sound signal. Individual tracks can be selected at random or can be preprogrammed. Error correcting codes can be introduced. A sensor electrode guidance system immediately corrects the electrode path whenever digital signals appear incorrectly positioned beneath the sensor.

THE DIGITAL RECORD

The pits and flats impressed on the phonograph record correspond to the binary numbers previously recorded on magnetic tape. After this the disc is coated with a reflective metallic layer. The laser beam shines through a transparent layer onto the reflective surface. When the laser beam is sharply focused on a flat the maximum amount of light is reflected, indicating a binary 1 signal. When the laser beam strikes a pit the result is a 0 signal. Because the digital signals supply the information to hold the laser in the correct path the sensing stays correct in spite of bumps or vibration.

Dust, scratches, or fingermarks on the surface of the disc are all outside the focus depth of the laser beam and have no influence on the reproduction of the sound. The reverse side of the disc is also protected against damage by a transparent layer. Thus, the disc and optical sensing arrangement supply protection against dirt, scratches, and fingerprints. There is no record wear, regardless of the number of playing times of the laser beam-operated digital disc.

With optical scanning via a solid state laser there are no mechanical or electromagnetic limitations on either frequency response or dynamic range. It is therefore possible to reproduce the full dynamic range of 90 dB encountered in concert halls.

DIGITAL AND SOUND FIDELITY

To understand how digital recording can result in higher fidelity sound, compare the telephone with the telegraph. The telephone is analog; the telegraph is digital and uses a code consisting of dots and dashes. Even if there is noise or distortion the person receiving the code signal only needs to know the difference between a dot and a dash to understand the original message. Not infrequently, a message transmitted in code will get through despite strong electrical noise when a message transmitted by voice will not be understood.

Like the dots and dashes in Morse code, digital audio's pulse code isn't affected by noise and distortion. Only the presence or absence of pulses needs to be detected to retrieve the original music. Besides that, digital audio is recorded along with a synchronizing signal which allows any wow and flutter to be canceled before the music is reproduced.

THE COMPACT DISC

Various noncompatible pulse code modulated discs have been demonstrated, but one that has been extensively publicized is the compact disc (CD) See Fig. 7-16. This is a phono record less than

*Fig. 7-16. Compact Digital Audio Disc. Impervious to dust, dirt, and wear, the CD contains up to one hour of music on one side of the 4.7-inch diameter disc. Sound is reproduced by a lower-power, solid-state laser. The CD was jointly developed by Sony corp. and N.V. Philips of Holland.
(Courtesy North American Philips Corp.)*

5-inches in diameter that can be played for one hour without turning it over, compared with less than a half hour for the average phono record. The CD contains over 6 billion digital sound signal bits without including all the extra bits used for such tasks as speed control or error correction.

Each bit, as indicated earlier, is either a flat surface representing binary digit 1, or a microscopic pit representing binary digit 0. These are laid out in a helical track and one unit of sound information consists of 16 bits. This is known technically among those who work in computer technology as a word, but the description can be misleading. A word is just 16 binary digits or bits and has no relationship to a spoken word. It is a word in a special code referred to earlier as pulse code modulation, or pcm. Analog to binary involves the breaking up of the complex, continuously changing audio signal into discrete bits which are handled as binary ones or zeros, a code represented by the presence or absence of pulses. Recording accuracy therefore depends on how finely the audio signal is divided and how many digits are used in the pulse code.

The CD has no grooves and the surface is smooth. The digitally encoded sound is in the form of microscopic pits, each approximately 0.1 micrometer (μM) thick, 0.16 micron wide and about 0.2 micron deep, along a 2½-7 mile long spiral track. It is sealed with a transparent plastic that protects against dust, dirt, scratches, and other damage. The solid state laser pickup scans this sequence of pits and flats in the form of a concentrated light spot several times thinner than a human hair.

As the pcm disc rotates, it is scanned from behind and from the center outward by a laser beam, detecting the sequence of pits and flats at a rate of approximately 4.3 million bits per second. Response is at the speed of light, immeasurably faster than the conventional stylus in a groove. Each pcm word is read in less than 10 microseconds—10-millionths of one second—at a constant rate.

The speed of the disc is controlled by coded information on the disc itself, eliminating tracking error. The laser beam cannot skate, offers no microphony link, and is completely out of the influence of static electricity. Because dust, dirt, and scratches are sealed out of the digital track and the focal range of the laser beam, nuisances such as clicks, crackles, and hisses are completely eliminated from the output.

The laser pickup scans the digital track at a constant linear velocity. The rotational speed of the disc must therefore be continuously adjusted, from 500-rpm at the inside to 200-rpm at the outside. Tracking, decoding, and rotational speeds are tightly synchronized with a central clock generator which itself is governed by information encoded in the track on the disc.

Stereo Sound Separation

In a CD digital recording the left and right channel sound signals appear alternately in completely separate words. Since they cannot mix in the player, channel separation is more than 90 dB, compared to 30 dB for an analog recording.

Signal-to-Noise Ratio

In digital audio, the signal-to-noise ratio depends on the number of bits in a word, that is, in the accuracy with which the audio signals are expressed. In the CD system, each bit contributes about 6 dB to a signal-to-noise ratio of more than 90 dB. This compares with an average signal-to-noise ratio of 50 dB to 60 dB for the usual LP. See Fig. 7–17 for Compact Disc Specifications.

Integration

The CD player is compatible with car-fi or home-fi. The output from the laser pickup is a stream of electrical pulses in 16-bit pcm code but in the digital/analog converter (Fig. 7–18) this stream is decoded, word by word, and synthesized into the conventional form of stereo audio signals.

The decoder checks each word to see that it is correctly formed and avoids errors by correcting any that are not correct. By reconstructing the decoded sound values the words synthesize an audio signal which represents information on the disc.

HYBRID RECORDINGS

When a recording is made magnetic tape is used to capture the sound. Tape, though, adds noise of its own, notably hiss, and also has dynamic range limitations. To overcome these problems the analog output of a microphone can be converted to digital format by an analog to digital converter. Then the data is recorded on tape in digital form.

The output of the tape is digital but this can be converted back to analog by a reverse process, that is, by a digital to analog converter. The audio signal can then be fed to a record-cutting lathe in analog form.

The advantage of a hybrid recording is that existing record players can continue to be used. Digital recording converts the analog output into pulses identified by the computerized digital recorder as numerical values. This is done by sampling the audio spectrum at a series of defined points at over 50,000 times per second.

Since all that is recorded by the digital tape recorder is numbers, there is perfect accuracy. No matter how poorly the numbers are stored on the recorded medium, when played back they cannot be decoded inaccurately.

Almost ten times as much dynamic energy, before tape saturation, can be captured by a digital tape recorder. The frequency response of a digital recorder is literally from the lowest possible bass notes to the highest possible harmonics in the musical spectrum. Harmonic and intermodulation distortion, wow and flutter, tape hiss, and normal analog tape abnormalities are nonexistent.

Achievable audio performance

Number of channels	2 or 4[1])
Frequency range	20 Hz - 20 kHz
Dynamic range	> 90 dB
S/N ratio	> 90 dB
Channel separation	> 90 dB
Harmonic distortion	< 0.05%
Wow and flutter	Quartz crystal precision

Signal format

Sampling frequency	44.1 kHz
Quantization	16 bits linear/channel
Encoding	2's complement
Error correction system	Cross Interleave Reed Solomon Code (CIRC)[2])
Modulation system	Eight to Fourteen Modulation (EFM)[3]
Bit rate	4.3218 Mbits/sec.

Frame Format

12 data words of 16 bits	: 24 symbols of 8 bits
4 error correction parity words of 16 bits	: 8 symbols of 8 bits
Control and display symbol	: 1 symbol of 8 bits
Frame before modulation	: 33 symbols of 8 bits
Frame after modulation (EFM) (33 symbols of 14 bits)	: 462 channel bits
Symbols for multiplexing and LF suppression (3 bits per symbol of 14 bits)	: 99 channel bits
Synchronisation pattern incl. 3 bits for multiplexing and LF suppression	: 27 channel bits
Total frame	: 588 channel bits

Fig. 7-17. Specs for

HALF-SPEED MASTERED RECORDING

Half-speed mastered recording is a process in which two-track master tapes are run at exactly half of their normal recorded speeds during the mastering (lacquer cutting) process. The cutting lathe is also run at one-half its normal speed, and so the playback ratio remains constant.

Error correction

Maximum correctable burst length	4000 bits ($\approx$ 2.5 mm)
Maximum acceptable burst length (by combined error correction and interpolation)	14000 bits ($\approx$ 8.4 mm)

Disc

Diameter	120 mm
Thickness	1.2 mm[4]
Diameter of centre hole	15 mm
Programme area start diameter	50 mm
Programme area maximum diameter	116 mm
Sense of rotation (seen from reading side)	anti-clockwise
Scanning velocity	1.2 - 1.4 m/sec
Rotation speed	500 - 200 rpm (approx.)
Maximum recording time	60 min. stereo[4]
Track pitch	1.6 µm
Material	Transparent plastic, with aluminiumised reflective coating, sealed with protective lacquer

Optical stylus (laser)

Wave length of AlGaAs laser	0.78µm
Numerical aperture	0.45
Focus depth	Approx. 2µm
Beam diameter at disc surface	Approx. 1.0 mm

[1]) 4 channels with reduced recording time.

[2]) CIRC: new error correction code for protection against scratches, with high error correction capability for random errors and low probability of undetectable errors.

[3]) EFM: new modulation method for increasing packing density and meeting requirements of optical servo systems.

[4]) Single sided disc (double sided disc optional).

a Compact Disc.

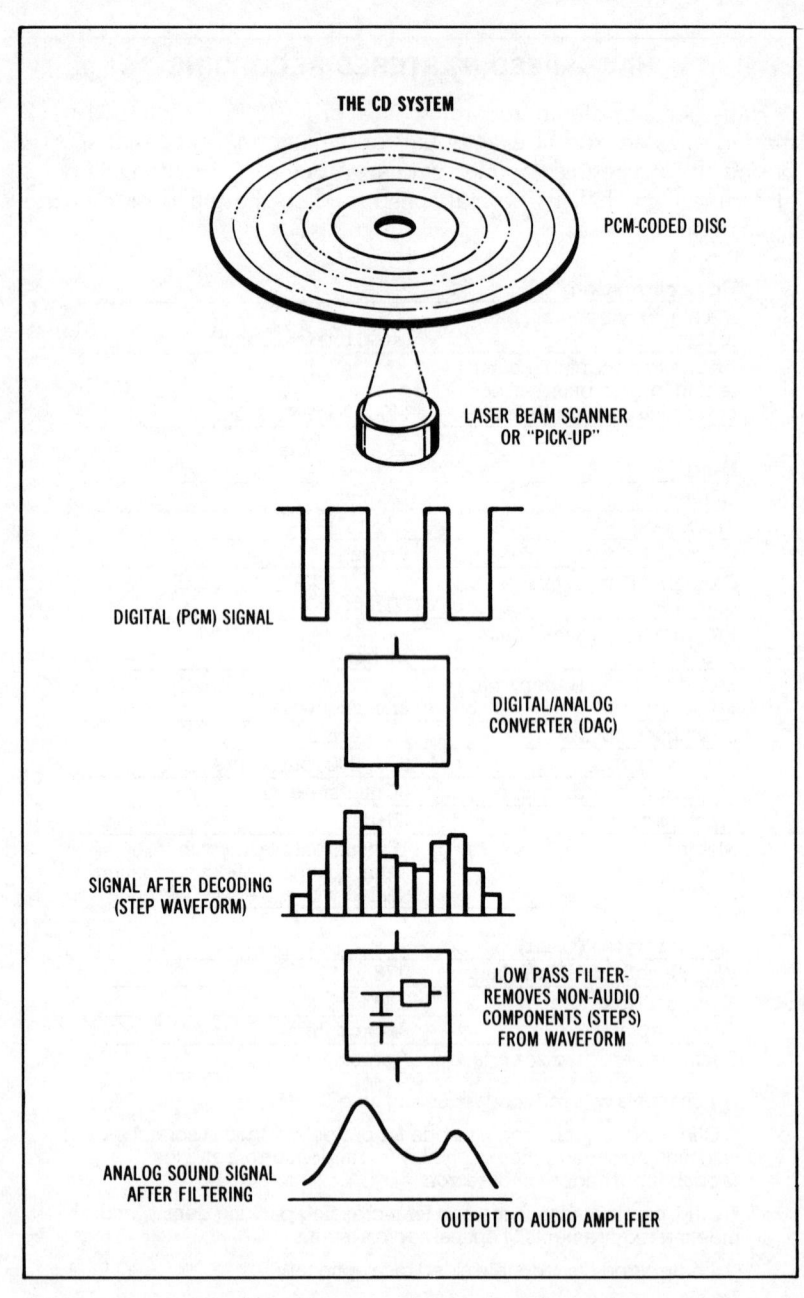

Fig. 7-18. Conversion of pulse code modulation to an analog sound signal.

This slower cutting speed allows the cutting engineer to cut a more perfect groove on the lacquer. The result is a record with greater dynamic range, wider frequency response, and less distortion than records cut at normal speeds. Half-speed mastered records are played at 33⅓ rpm like all other records.

DIGITALLY REMASTERED RECORDINGS

A digitally remastered album is a combination of digital recording techniques and half-speed mastering technology. The two-track analog tapes are played at half their normal speed and recorded onto a digital recorder which is also run at half speed. This transfer process prevents the analog tape deck's electronics from reaching its tolerance limits and significantly reduces analog distortion.

The digital tape is then used to cut the master lacquers at normal speeds. The advantage of this process is the combination of digital accuracy and the transfer of information at half speed. The result is fewer distorted musical signals on the master lacquer.

THE ANALOG PHONO RECORD

While the pulse code modulated disc offers great promise, the analog disc is making steady improvement. In general, discs are plagued by two serious problems. The first of these is the limitation in music range. When an analog record is cut, studio recording engineers compress the music signal to keep record grooves from becoming too wide. The average record plays less than a half hour per side, an achievement that is due to signal compression. Without such compression records would have much shorter playing time. But compression reduces dynamic range. High amplitude sounds are weaker; soft sounds stronger.

Another problem is record noise. In sound reproduction via analog phono records we have an extremely hard surface, a diamond, rubbing against another hard surface, vinyl grooves.

DIRECT TO DISC RECORDING

Direct to disc recording is exactly what its name implies, with the sound cut directly into the phono record without first being recorded on tape. Tape has serious recording limitations. It can stretch, causing distortion; it can pick up sound from other parts of the tape, a characteristic known as print-through; it has a limited range of highs and lows and there is always some background hiss.

Direct to disc isn't new. It was the first method ever used in making phono records since it preceded the invention of the tape recorder.

The difference is that the modern direct-to-disc recording technique makes use of the automated disc cutting lathe and up-to-date pressing and plating processes. With direct-to-disc peak signal headroom is better. But to be able to hear the difference in direct-to-disc compared with recordings via tape it is desirable to use audio components that are top quality. But even with average components direct-to-disc will sound clearer than regular records.

Direct-to-disc records are more expensive than discs using the tape recording process. The higher cost is attributable in part to the fact that these are limited editions. Only a certain number of pressings can be made from each master. With ordinary records the audio signal is on tape which can always be used to cut another master.

DYNAMIC RANGE AND NOISE

Various companies have made efforts to cope with the limited dynamic range of phono records and their noise. One of these companies is dbx, Incorporated.

With the music signal compressed by an exact 2-to-1 ratio dbx encodes the phono record. Then when the record is decoded, that is, during playback, the music signal is expanded again in exact mirror image fashion, a 1-to-2 ratio. This process is called linear decibel companding. Most records have less than 50 dB of dynamic range, with a top of about 60 dB for the best records; dbx encoded discs have a dynamic range of up to 75 dB. Using studio recording, digital mastering the digital dbx discs can supply up to 90 dB or more of dynamic range.

Fig. 7–19 shows what happens. During decoding, when the music signal is expanded, disc surface noise is driven down to virtually inaudible levels. As the illustration shows, the input signal put onto the master tape has an input dynamic range of 90 dB. The encoded signal compressed at a 2-to-1 ratio has a dynamic range of only 37.5 dB. The decoded signal is brought up to 90 dB, but disc surface noise is practically inaudible. Two things are required, though: the purchase of dbx encoded records and a special decoder attached to the hi-fi system.

COMPATIBLE EXPANSION

Compatible expansion (CX), was introduced by CBS; it is compatible since CX discs can be played on any record player without additional equipment.

Like the dbx system, CX records require a decoder to get full benefit of CX encoded discs, but the decoder isn't absolutely essential, since CX records can be played without it.

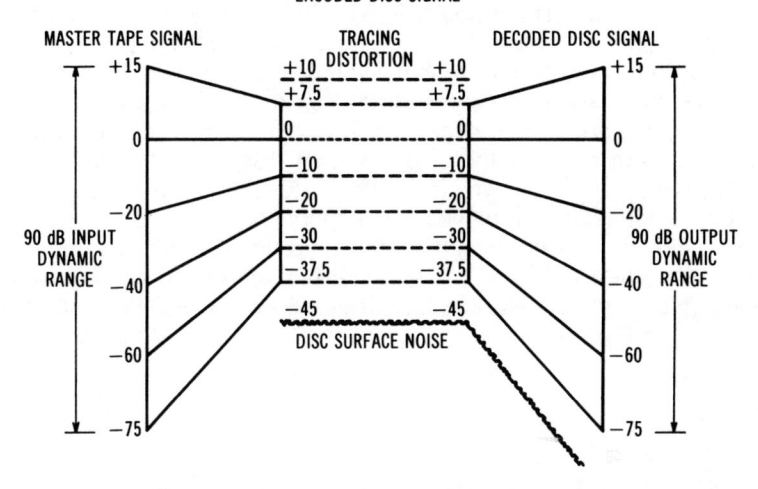

Fig. 7-19. Effect on noise of encoding and decoding. *(Courtesy dbx, Inc.)*

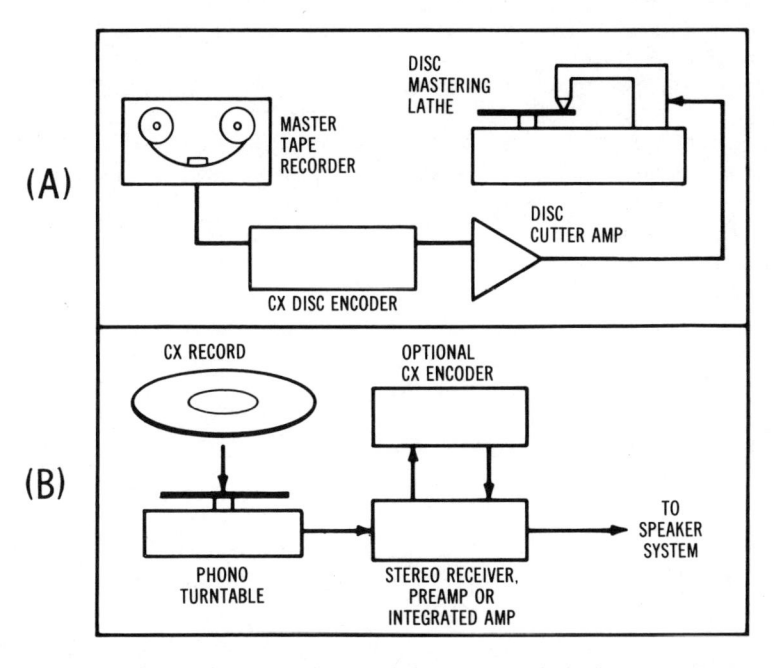

Fig. 7-20. The CX phono record is encoded (A). During playback (B) an optional CX decoder can be used. CX records are compatible and can be played back in a hi-fi system, with or without the decoder.

There is a difference between CX and dbx techniques. In the dbx method loud sounds are compressed, soft sounds are expanded during the encoding process. With CX, the soft passages are strengthened but the loud tones aren't touched. Thus, CX encoded discs do not have as much sound compression as dbx records.

Fig. 7-20 shows the CX method. The sound is recorded on a master tape recorder and is then fed into a CX disc encoder that controls a disc cutter amplifier, which, in turn, feeds the audio signals into a disc mastering lathe. In playback the CX phono record is played on a conventional turntable which supplies the signal to the preamp section in a stereo receiver or a separate preamp or the preamp in an integrated amplifier. The signal is routed to a CX decoder and then returned to the power amp for delivery to the speaker system.

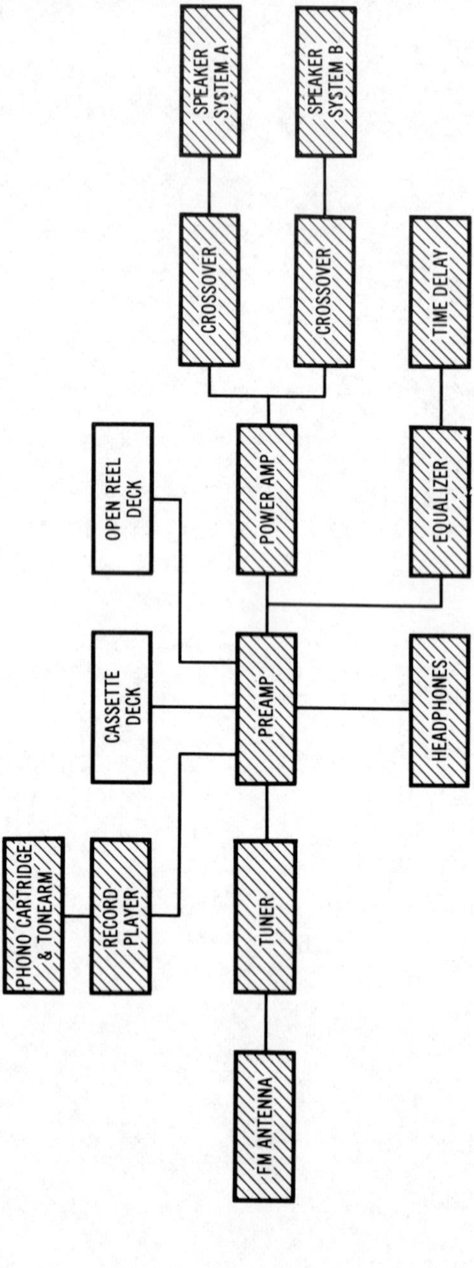
SPEAKER SYSTEM A
SPEAKER SYSTEM B
CROSSOVER
CROSSOVER
TIME DELAY
OPEN REEL DECK
POWER AMP
EQUALIZER
CASSETTE DECK
PREAMP
HEADPHONES
PHONO CARTRIDGE & TONEARM
RECORD PLAYER
TUNER
FM ANTENNA

8

Tape Decks and Magnetic Tape

There are two types of sound sources: passive and active. A receiver is a passive sound source for it is a one-way sonic device. It supplies sound but all we can do with that sound is to listen. The same is true of a turntable, tuner, or microphone, for these are also one-way sound sources.

A tape deck is a completely different kind of component for it is active and sound can flow in two directions—away from the deck during playback; toward the deck during recording. With a tape deck we become participants. Instead of simply listening to the sound, we can transfer music to tape from records or broadcasts, we can dub from one tape to another, we can use one or more microphones and record "live," or we can modify sounds by using various recording techniques.

TAPE PLAYERS VERSUS TAPE DECKS

A tape player is a complete, integrated unit. It contains not only the tape playback equipment, but a preamplifier and power amplifier, and often its own speakers. Tape players are used for playback only, not recording, hence are passive devices. Such units supply music via tape but aren't considered hi-fi.

A tape deck supplies a sound signal and, like a record player, has need of the preamps and power amps and the speakers in the hi-fi system. For hi-fi a tape deck isn't an integrated unit but is regarded as a separate component.

Next, we discuss the different types of tape decks that are available today.

TYPES OF DECKS

There are three basic types of tape decks: 8-track cartridge, open reel, and cassette. There are some 8-track cartridge units that are true decks, but these are quite rare. For the most part the 8-track cartridge unit isn't a deck but is a player only. It isn't part of the usual component in-home hi-fi system and its greatest application has been as a tape player for autos.

The two remaining decks, open reel and cassette, are true decks and can be designed for quality sound reproduction. Of the two, the cassette tape deck is the more popular and the more widely used, although just a few years ago the reverse was true.

Open Reel Decks

The open reel deck, also called reel-to-reel, has its largest application in recording studios, radio stations, and by dedicated audiophiles. Unlike cassette tapes, open reel tapes must be threaded from one reel to the other, from a supply or feed reel to a takeup reel. These range in size from as little as 5 inches to as much as 10½ inches. Fig. 8-1 is the outside front view of a representative open-reel deck.

With other open-reel decks you can record four separate sound tracks individually in synchronization with each other, or you can re-record any track at any time, also in sync with the other tracks.

Some open-reel decks provide synthetic electronic echo and some will let you supply enhancement in the form of acoustic reverberation. In professional recording studios, the "dry" sound—the music directly from the instruments—is recorded separately, and then controlled amounts of reverberation are added later.

Tape Speeds

Most open reel tape decks are two-speed types, giving the user an option of 3¾ or 7½ ips (inches-per-second). But there are also some three-speed decks that include 15 ips and some four speed types that work at 1⅞, 3¾, 7½ or 15 ips. The higher speed means that more tape is presented to the recording head and so more oxide on the tape is available for receiving and storing the signal. There are some open reel decks whose frequency response extends beyond 25 kHz at 15 ips.

Each lower speed in an open-reel deck can be obtained by dividing successively by two. Thus, if we consider 30 ips as the top speed, the next lower speed is 30/2, or 15 ips. Dividing 15 by 2 supplies 7½ ips. Dividing by two again gives the next lower speed or 3¾ ips. Continuing this way we have 1⅞ ips and then ¹⁵/₁₆ ips. This does not mean, though, that every open reel deck can operate at all these speeds.

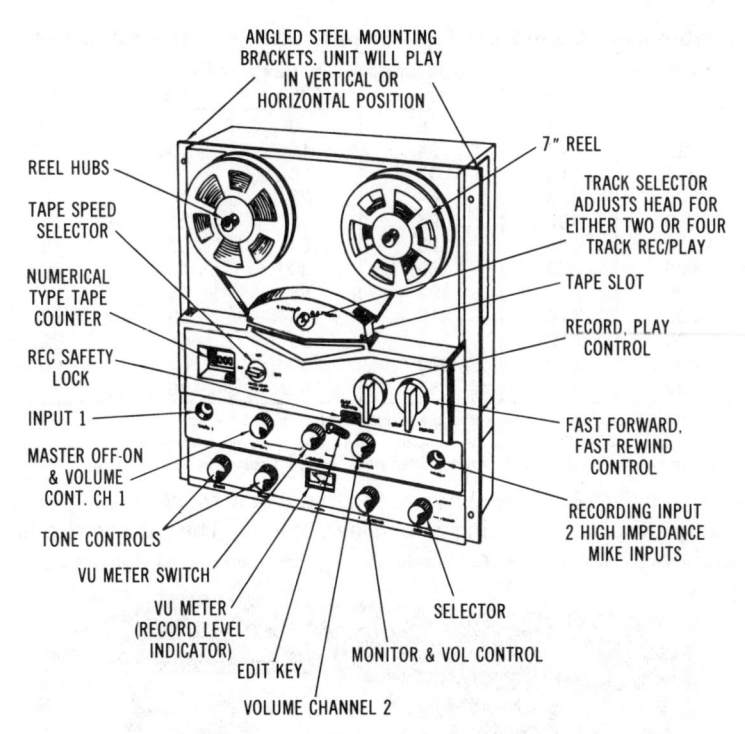

Fig. 8-1. Open-reel tape deck controls.

Table 8–1 supplies the playing time in minutes of various lengths of tape at different tape speeds. Thus 1200 feet of tape played at a speed of 3¾ ips will supply 64 minutes of playing time. Obviously, the maximum playing time is obtained by using tape having the greatest length and played at the slowest speed. The maximum, based on Table 8–1, would be 3600 feet of open reel tape at $^{15}/_{16}$ ips, supplying 768 minutes of playing time or 12 hours and 48 minutes.

It is commonly believed that the faster the speed, the better the quality of sound reproduction. That depends on how you apply this rule. Thus, it is possible for a speed of only 1⅞ ips to give better results on a quality tape deck than 15 ips on a machine much lower in quality. Using the same machine, recording at a higher speed may give better quality than recording at a lower speed if you use the same sound source for both. You cannot improve the quality of poor sound simply by recording at a higher speed.

Cassette Decks

An open reel deck is so-called because the two reels—supply and takeup—are positioned right on the outside of the deck. For a cas-

Table 8-1. Open-Reel Tape Speeds and Tape Length in Feet

Feet	Tape Speed—Inches Per Second				
	15/16	1⅞	3¾	7½	15
150	32	16	8	4	2
225	48	24	12	6	3
300	64	32	16	8	4
600	128	64	32	16	8
900	192	96	48	24	12
1,200	256	128	64	32	16
1,500	320	160	80	40	20
1,800	384	192	96	48	24
2,000	426	213	106	53	27
2,400	512	256	128	64	32
3,000	640	320	160	80	40
3,600	768	384	192	96	48

Time in minutes, calculated to nearest whole minute, 1 direction, 1 track only.
For total mono time on 4-track recorders, multiply time by 4; for stereo, multiply by 2.
For total mono time on 2-track recorders, multiply time by 2; for stereo, use given figures.

sette deck (Fig. 8–2) these two reels are not only much smaller, but are housed in a plastic container known as a cassette. The cassette also has a supply of ⅛-inch wide tape with the speed usually at 1⅞ ips.

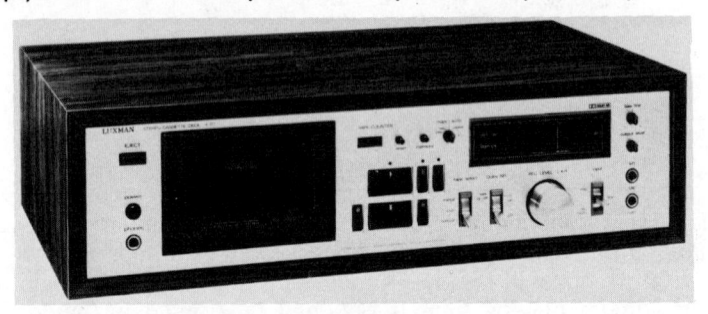

(A) Luxman Model K-117 stereo cassette deck.

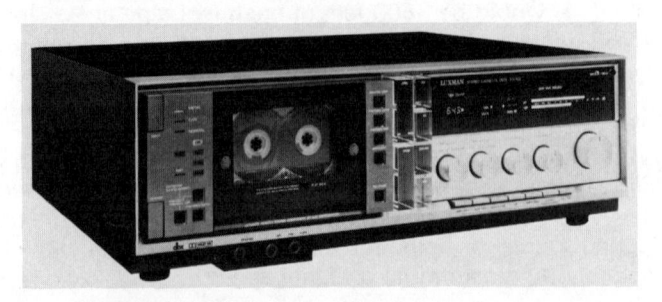

(B) Luxman Model KX-102 three-head stereo cassette deck.

Fig. 8-2. Two modern stereo cassette decks.
(Courtesy Luxman, Div. of Alpine Electronics of America)

Cassettes

Cassettes are commonly available in 30, 45, 60, and 120 minute recording and playing times. These are identified by C numbers. Thus, a C-30 will play for a total of 30 minutes—15 minutes on each side. After 15 minutes of playing time, the cassette must be removed and turned over. In some cassette decks playing the reverse side is done automatically.

Magnetic Tracks

Just as you can have a railroad depot with trains leaving and coming on various tracks, so too do we have tracks on cassette tape. These are invisible magnetic tracks. For stereo recording we need two tracks; one for left channel sound, the other for right channel sound. See Fig. 8-3. For stereo recording we have two tracks "going" and two more tracks "coming." Each track has a width of 0.6 millimeter (mm), with a gap of 0.35 mm between tracks. The speed at which the tape runs past the heads is 4.76 cm/second or 1⅞ inches per second (ips).

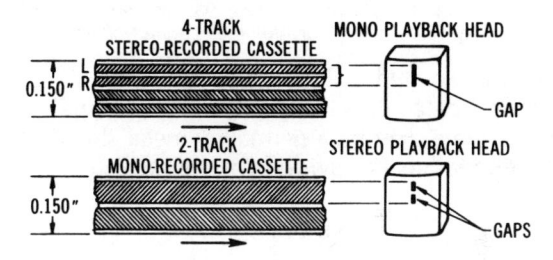

Fig. 8-3. Stereo is recorded as two pairs of left/hand tracks, but either mono or stereo can be played back on cassette deck. *(Courtesy B·I·C-Avnet).*

Inside the Cassette

From the outside all cassettes look alike. This doesn't mean that all cassettes are manufactured the same way but rather that all cassettes are compatible. Physically, they should all be the same size. But while all cassettes are *look alikes*, it's what's inside that counts.

Inside the cassette (Fig. 8-4) is a pair of hubs or reels for holding the tape. These must be made with the utmost precision for if they are not the winding and unwinding of the tape on these hubs will produce wow and flutter, as discussed later in this chapter.

In its movement the tape must be guided by guide rollers that are necessary to make sure the tape can move freely without binding and to ensure that the tape is correctly positioned with respect to the heads in the tape deck. Liner sheets are used inside the cassette to make certain the edges of the tape do not slip from side to side. These

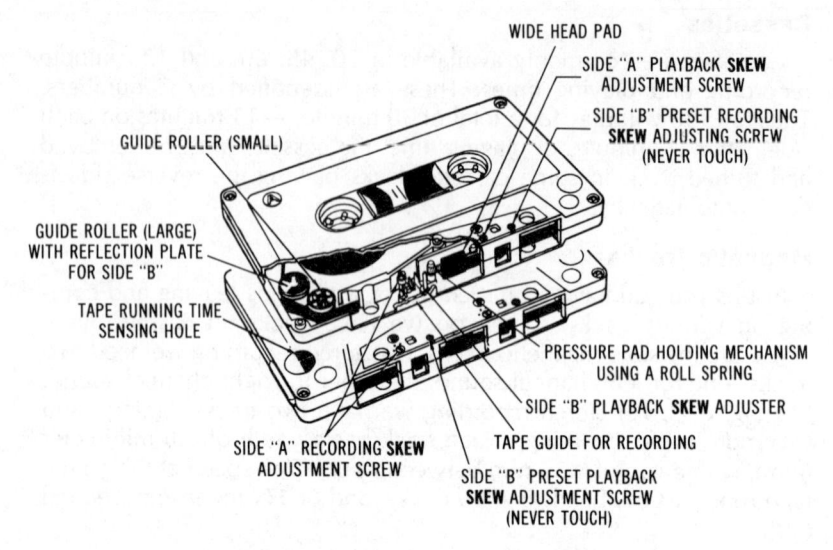

Fig. 8-4. Details of cassette construction.
(Courtesy Luxman, Div. of Alpine Electronics of America)

are generally impregnated with a lubricant, such as silicone, to make them as friction free as possible. And, since the tape is sensitive to magnetic fields there may be a permalloy metal shield plate to keep outside magnetic influences away from the tape.

CASSETTE TAPE

Tape consists of a thin 3.81 mm wide polyester film coated with minute particles of iron oxide, gamma ferric oxide, chromium dioxide, pure metal, or other combinations. Typical particle length is 0.00003937 inch, or about one micron. A micron is the thousandth part of a millimeter. The particles are held to the tape with a glue-like binder.

A C-30 cassette has 150 feet of tape, a C-60 has 300 feet and so on, up the line. A C-120 has 600 feet. But since all cassettes are physically compatible, that is, every cassette made must fit into every cassette recorder, the only way to get a longer playing cassette is to use thinner tape. And so, as you go from a C-30 to a C-60 to a C-120, the tape keeps getting thinner and thinner, but not proportionally. A C-60 is made with the same thickness as a C-30 for the simple reason that a C-30 ends up with room left over. A C-30 tape is about 18 microns thick and so is a C-60. But when we get to a C-90 tape thickness drops to about 13 microns, and a C-120 is just 9 microns.

The C-60 is the most popular cassette length because its tape is thick enough for maximum durability and long enough for most home recording situations. With 45 minutes of recording time per side, C-90s are most popular for dubbing (that is, copying) records and for college class notes. Both sides of most, but not all, LP records will fit onto one side of a C-90, which is more convenient than having to flip a disc in the middle of a playback, and 45 minutes is enough for most college lectures.

Cassette Tape Requirements

Cassette tape must be made so it doesn't break or stretch, its edges mustn't curl, and the base—the long thin ribbon of polyester or acetate film—must retain its strength and flexibility for years. The mixture of iron oxide particles and adhesive binder mustn't separate from the base or wear off because this will foul the playback equipment and cause "drop outs" or holes in the sound.

The magnetic particles on the tape must also be as smooth as possible. The movement of the tape, even at a speed as low as 1⅞-ips, is a wiping action, a constant rubbing or polishing that can wear out recording or playback heads in the cassette deck.

What Happens When You Record?

During recording, each particle on the tape becomes magnetized, some stronger, others weaker. These micro magnets, fixed in position along the length of the tape, form a magnetic record of the sound. And because they form a "record" they can be read, with the help of a tape deck. This record can be erased by the tape deck and so you can record, erase, and re-record as often as you like.

This all sounds simple, and that's the trouble. The polyester film on which the magnetic particles are binder-imbedded can stretch and when it does, distortion results. The metal particles on the tape must be small, but must also be uniform in size; this is a staggering concept when you consider that a reel of tape can contain millions upon millions of particles. Size is just one factor, though, because each particle must have uniform retentivity, another way of saying each of them must be capable of being magnetized to the same degree with the same magnetizing force. In your tape deck, some 35 million particles on the tape move past the record or playback head each minute the tape is running.

The taped music should never fade. Music tapes made in 1946 sound as real and lifelike as they did when they were new. Each particle on tape is only one-sixth of the cross-sectional diameter of a human hair, but it must be perfectly uniform in thickness and magnetic density. Recording tape is virutally indestructible under ideal playback conditions. In one test, a standard tape played 6½ million times in a laboratory suffered no loss of playback qualities.

HOW A TAPE RECORDER WORKS

While open reel and cassette decks are quite different in outward appearance, they work along similar lines. Every tape deck, cassette, or open reel contains a transport consisting of two or more motors for moving the tape from one reel to another.

During the recording or playback process the tape is pulled across the face of two or three heads. Fig. 8-5 shows the arrangement for a three-head deck. The tape comes off a feed or supply reel, and, following a guide, moves across the faces of an erase head, a record head, and a playback head. The tape is moved along by a pressure roller pushed against a capstan (Fig. 8-6), and then over another guide to the takeup reel. While the setup here is for an open-reel deck, the same arrangement is used for a cassette deck. For a cassette deck, the reels and the tape are in an enclosed plastic cassette; for open reel decks they are in the open.

Every tape deck contains a transport consisting of two or more motors for moving the tape from one reel to another. During the recording (Fig. 8-7) or playback process the tape is pulled across the faces of two or more heads. These heads contain coils of wire. In the recording process, audio currents flow through the coil in a head called a recording head (Fig. 8-8). The passage of the audio current is accompanied by a magnetic field which magnetizes the tiny metal-

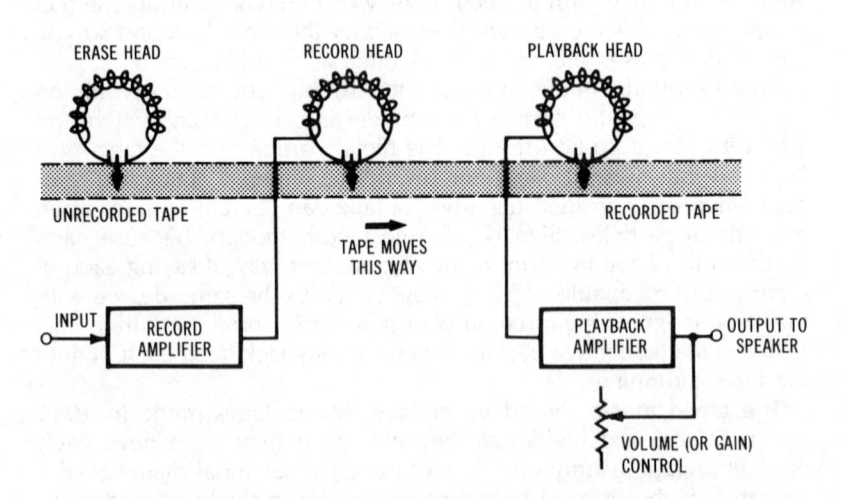

Fig. 8-5. Better types of tape recorders use three separate heads: erase, record, and playback. This illustration shows two separate amplifiers: one for recording, the other for playback. A single amplifier can be used instead. The sound level of the playback amplifier can be adjusted by a volume control.

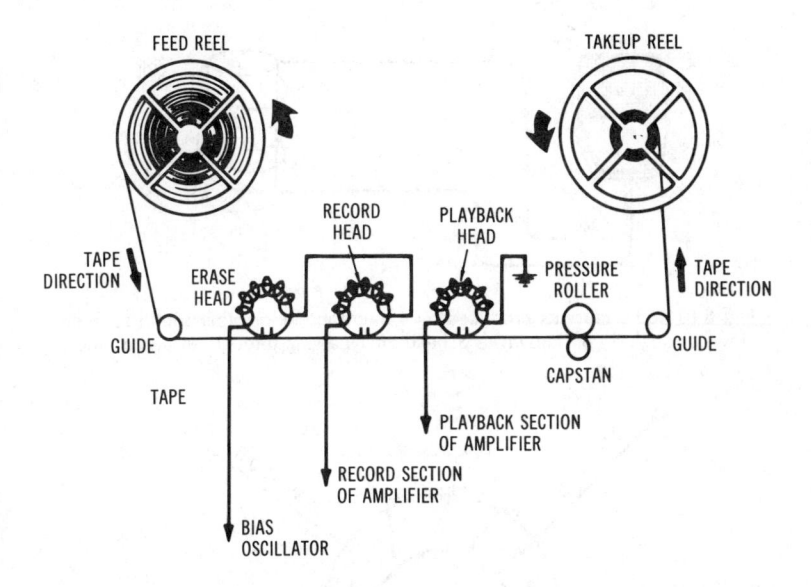

Fig. 8-6. The tape moves from the feed to the takeup reel. The movement of the tape is accomplished by using a rotating metal shaft called a capstan. The capstan turns a pressure roller and it is the rotation of the pressure roller that pulls the tape along.

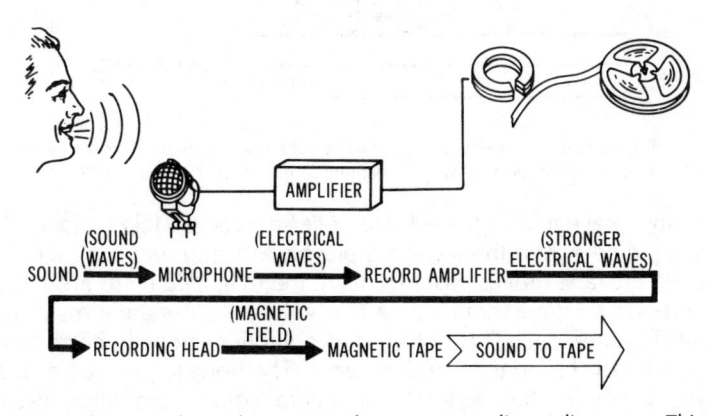

Fig. 8-7. The microphone changes sound to a corresponding audio current. This current is fed into a head coil and its magnetic field is impressed on metal particles on the tape.

lic particles on the tape (Fig. 8–9). Each of the particles becomes an extremely tiny magnet and can retain its magnetism indefinitely. The way in which the particles are magnetized and the direction and extent of that magnetism constitute a "record" of the audio signal.

273

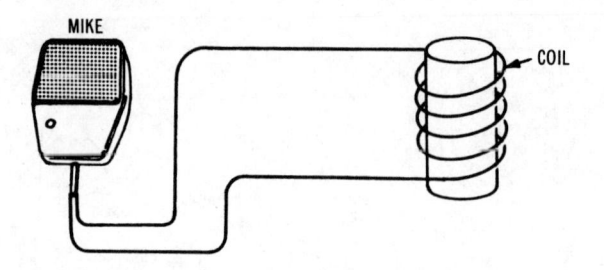

Fig. 8-8. Audio currents produced by a microphone, or other sound source, flows through a coil known as a head after passing through an audio amp.

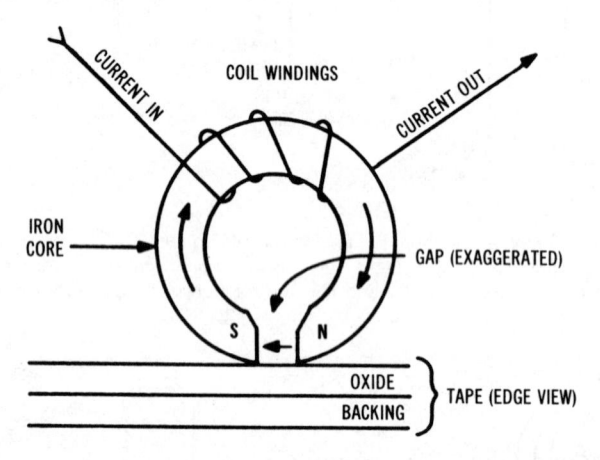

Fig. 8-9. The audio currents flowing through the coil winding are accompanied by lines of magnetic flux which magnetize the metal particles on the tape.

During playback (Fig. 8–10) a reverse process takes place. The tape is pulled across the face of a playback head containing a coil of wire. As the tape moves past this head, the magnetic fields around the particles on the tape induce an audio voltage across the turns of wire in the playback coil. This voltage has a varying strength, corresponding to that of the original audio signal. The voltage generated in the playback coil is then fed into an audio voltage amplifier, usually located inside the tape deck. It is then delivered to an external preamplifier in a receiver, or as a separate unit, or in an integrated amplifier.

In the open-reel deck the feed reel, containing the tape to be played, is mounted on a spindle on the left-hand side of the machine. The take-up reel is slipped onto a spindle on the right-hand side. As the tape moves from the feed reel to the take-up reel it passes the

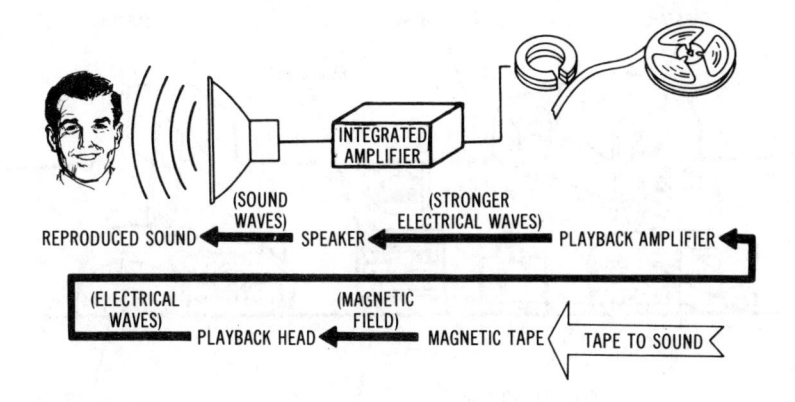

Fig. 8-10. The magnetized tape induces audio currents in the playback head. The audio signal is amplified by a preamp, a power amp, and then is used to drive the speaker system.

erase head first, then the record head, and then the playback head or a combined record/playback head.

THE TRANSPORT

To move the tape requires one or more motors. Less expensive machines use a single motor for moving the tape, while quality machines may use as many as three motors, one for each of the reels, and another as a drive for the capstan.

During its motion, the tape is held snugly between the pressure roller and a capstan (Fig. 8–11). The capstan is a metal shaft driven by a motor. The capstan would be unable to move the tape by itself. The pressure roller is made of hard rubber or some similar substance and there is considerable pressure between the capstan and pressure roller. The revolving capstan turns the pressure roller and it is the pressure roller (sometimes called a pinch roller) which pulls the tape.

Constancy of tape speed is important. The drive speed should not be responsive to voltage or the overall pitch will change as the voltage changes. The capstan shaft must be perfectly round and straight as any variations will result in wow when the variations are slow and flutter when the variations are considerably faster.

The precision capstan becomes the axis and shaft on which a large flywheel is mounted. The inertia of this flywheel stabilizes the rotational speed and helps to smooth short term irregularities caused by variations in friction. A larger flywheel will have greater inertia, resulting in a more stable drive, less wow and flutter, and a more efficient transport.

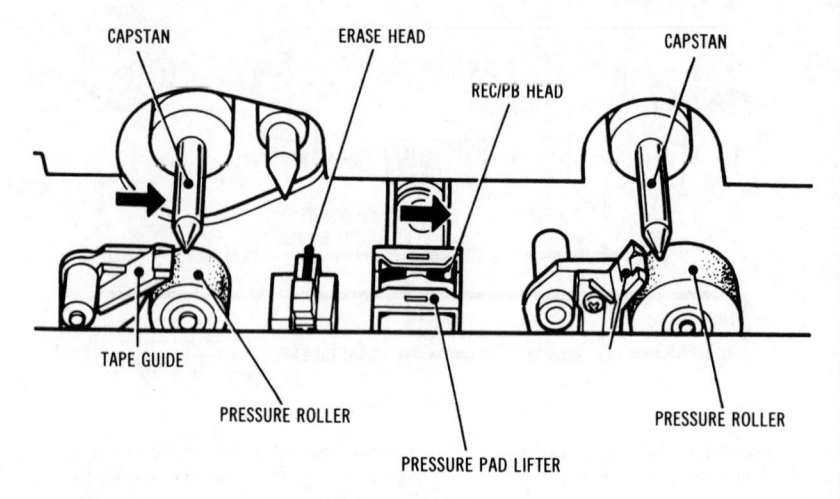

Fig. 8-11. The heavy black arrows indicate direction of travel of the tape. The tape moves between the pressure roller and the capstan. *(Courtesy Nakamichi USA Corp.)*

After a tape has been played, a pushbutton or switch on the front of the tape recorder is operated for rewind. During rewind the pressure roller is disengaged from the capstan, and the motor driving the feed reel for rewind, goes into action.

Although it may seem that way, the tape during play is not pulled along by the rotation of the takeup reel. That is the function of the pressure roller. The takeup reel is motor operated to do two things: first, to wind the tape on the takeup reel; and second, to wind it so that the tape is wound just right—not too loose nor too tight.

The tape, a pliable ribbon, must be *pulled* through its proper path rather than forced or guided into that path by pushing on the side. If it is forced by tape guides, the edge of the tape will become wrinkled and the number of satisfactory replays will be limited. The capstan must be perpendicular to the tape path to aim the tape properly. The axis, as well as the outside diameter of the rubber pinch roller, must be parallel to the capstan axis in all planes.

TESTING THE DECK

Any deck sounds best when recorded and played back on itself. How will it sound with prerecorded tapes or recordings made on a different unit? Work the controls and try to fool it by deliberate misuse of the controls. Does it spill tape? Does it fail to operate? Go through the start/stop cycle in rapid succession? Does it sound the same each time? Listen to a frequency response tape. Is the sound clear or do you hear tremelo (wow and flutter)? The variations you hear on the sine waves will also distort the original sound—whether it is live or prerecorded.

FREQUENCY RESPONSE

High frequency response depends on tape speed and gap dimensions. The narrower the gap (Fig. 8–12) the more extended the response at the treble end. A high frequency signal occupies a very small length of tape. Considered alone, a signal at 15 kHz for a cassette tape traveling at 1⅞-ips requires a playback gap of just 1 micron. But, as usual, there is a trade-off. As the gap is made narrower, efficiency drops and the signal picked up by the tape decreases. This can be overcome by signal amplification, but this adds to the cost of the deck. Further, the amplifier also contributes its own bit to noise, including hiss.

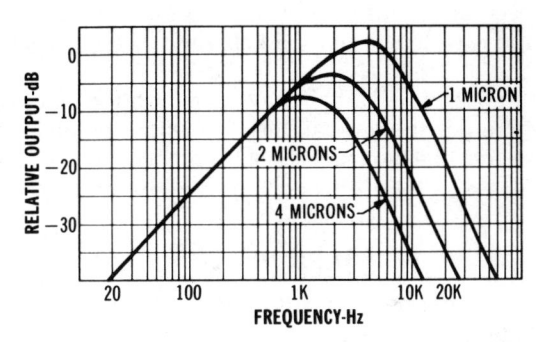

Fig. 8-12. Reducing gap width improves high frequency response. *(Courtesy B·I·C-Avnet)*

Fig. 8–13 shows the response of the playback head equipped with a 4 micron gap. Note that the slower the speed, the earlier the high frequency rolloff, with 1⅞–ips tape starting at 2 kHz and 15–ips tape at 10 kHz. This indicates clearly that tape heads do not have a linear frequency response.

As the audio frequency increases, the magnetic field variations supplied by the tape are more readily detected by the head, up to a point. This is shown if Fig. 8–13 by the sharply rising slope starting at 20 Hz.

Frequency is the inverse of wavelength. As frequency increases, the length of a sound wave becomes shorter. A 15 kHz sound wave recorded on tape moving along at 15–ips will occupy a space of 0.001 inch, or 25 microns, for each cycle of the wave. If the tape moves more slowly, the space occupied becomes less: for 7½-ips it be-

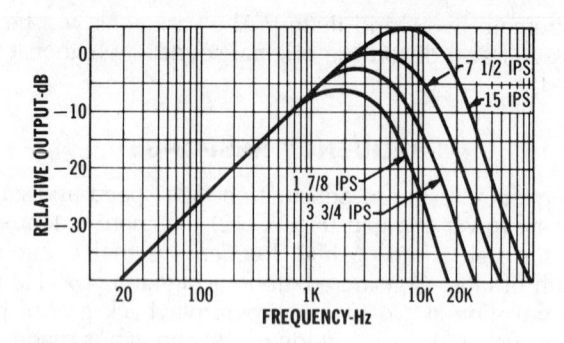

Fig. 8-13. Response of 4 micron gap playback head at speeds ranging from 1⅞ ips to 15 ips. *(Courtesy B·I·C-Avnet).*

comes 12 microns; for 3¾-ips it drops to 6.25 microns and for cassette tape speed at 1⅞-ips becomes 3.125 microns. Looked at in another way, the wavelength of a 15 kHz tone at 1⅞-ips is the same as a 30 kHz tone at 3¾-ips.

A problem arises when a half wavelength on the tape decreases to the point where it is shorter than the dimensions of the head gap. It is at this point that the voltage level from the playback head drops severely.

TAPE HEADS

Every tape recorder needs heads to perform these functions: record, playback, and erase. In the less expensive types of decks, heads may be made to serve a double purpose. Thus, a single head can be made to work for both playback and record. An erase head, as its name implies, is a head which is used to demagnetize tape, that is, to restore it to its unrecorded condition. The erase head can also be used to remove electrical noise from the tape.

A three-head deck uses individual heads for erase, recording, and playback and while this results in a deck that costs more, the extra expense does mean that each head is designed for its specific job. Further, a three-head system offers the only accurate recording control possibility—that of monitoring the tape.

A gap (Fig. 8–14) is used in the heads to permit the magnetic flux lines to reach the metal particles on the tape. Gaps are extremely fine slits in the head. Ideally the gap should be 4 microns for the record head and 1 micron for the playback head. Some gaps are as small as 0.9 micron. In decks using only two heads, the record and playback heads are combined, and so there must be some compromise in gap dimensions. This isn't necessary in three-head decks.

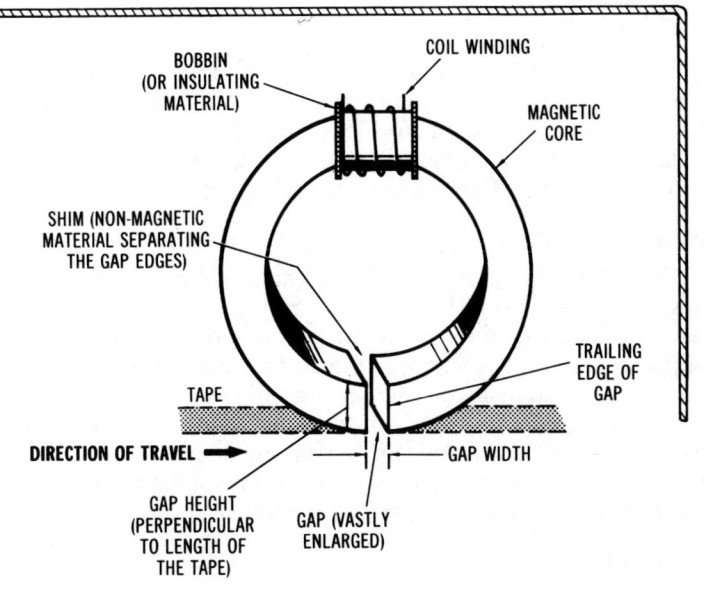

Fig. 8-14. The gap whose dimensions are measured in microns, should be perpendicular to the magnetic tape.

Head Shields

The heads of a tape recorder are shielded to protect the very thin gap of the heads from dust and to shield the heads from any stray magnetic fields. The head shield is a metallic covering for all the heads—playback, erase, and record. In some tape recorders the shield is held in place by a spring lock so it can be removed without the use of tools and replaced as easily. When replacing a shield always make sure that it snaps back into place and that it does so securely and tightly.

With the shield removed, examine the heads of the tape deck. There is an extremely fine, thin vertical slot down the center of each head. It is through these slots that magnetic interlinkages take place between the coil in the head and the metallic particles on the tape.

Head shields are made of various materials but the one more commonly used in better grade decks is Sendust. Sendust is a magnetic alloy of iron, aluminum, and silicon formed by a sintering process in which heat and pressure are used to fuse a magnetic powder. Sendust has several important properties. Physically, it is exceedingly hard, important since it means long head life. Magnetically, it is "soft," that is, it has a fairly low coercivity, important to assure quiet reproduction. A material with high coercivity retains magnetism and thus tends to partially erase the tape and increase the noise level.

The permeability of Sendust is quite high, a necessary attribute to achieve a sensitive playback head. Permeability is a measure of the ability of a material to conduct a magnetic field. Most important, Sendust has a high saturation induction. This means it can carry the high magnetic flux density required to generate a field in the gap sufficient to record on a high coercivity tape.

Head Azimuth

Correct head azimuth means the gap in the head is exactly at right angles to the edge of the tape; if not, there will be poorer high frequency response. The cassette deck is equipped with a screw whose adjustment permits rocking the head to bring the gap into a perpendicular position with respect to the tape edge. You can make this adjustment using an alignment test tape and turning the azimuth adjustment screw until you get maximum output. After adjusting the playback head, do the same with record head azimuth while recording and monitoring a high-frequency tone.

Head Zenith

A head can be perpendicular to the tape but tilted away from it, resulting in what is known as zenith error, causing amplitude changes in the recorded sound. To adjust, turn the zenith adjustment screw (usually located near the azimuth adjustment screw). You can check zenith by sighting along the head using a business card held vertically against the flat surface of the deck for comparison.

PRESSURE PADS

The tape must make firm contact with the heads during record or playback. A commonly used method of ensuring this is to use a pressure pad. A pressure pad placed inside the plastic shell of the cassette may be just a bit of felt which pushes against the tape. The pad holds the tape against the head, but does not interfere with the movement of the tape. The pressure pad is usually mounted in some type of metal holder and then, by spring action, pushes against the tape. In some cases the felt pressure pad uses a beryllium copper spring. Some cassettes rely on a thick piece of foam or a thin felt strip layered over a plastic base to act as both pad and spring.

THE CASSETTE SHELL

Every cassette is made of two plastic halves or shells which must be carefully aligned and joined. In some cassettes the half-shells are welded and in others they are kept together by screws. Small deviations, even those not readily visible, can cause substandard performance and outright failure.

A pair of positioning guide posts are built into the well of every cassette deck. These are intended to hold the cassette in exactly the right position to ensure proper tape drive and alignment with the record/playback head. But the pair of corresponding holes in the cassette itself, which are supposed to fit over the posts, must be properly shaped and aligned. If not, there may be tape slippage resulting in wow and flutter and an inability to drive the tape altogether. In extreme cases, a poorly designed and poorly made cassette case will jam the cassette recorder with possible permanent damage not only to the cassette but to the player itself.

ERASING

Erasing is simply a process of restoring the magnetic particles on the tape to their original unmagnetized condition. The erase current is a high-frequency current passed through an erase coil contained in a housing similar to those used for the record and playback heads. The erase current is accompanied by a varying magnetic field. The effect of this field is to magnetize the particles on the tape but in such a way that no sound will be produced by the tape when it moves past the playback head.

TAPE MONITORING

As the tape moves past the recording head, the signal currents that flow through the coil in the recording head change the magnetization of the tiny particles on the tape. However, as the tape continues on beyond the recording head, the signal which has just been *impregnated* on the tape can be picked up, converted to sound, and listened to for quality. This technique, known as monitoring, gives you an opportunity to make changes in control settings to improve the recorded sound. Because the elapsed time between recording and monitoring is so short, the effect is as though you had listened to the sound as it was being recorded. The only alternative to monitoring is editing—but editing is a mechanical process; monitoring is electronic, is easier, and is faster. This does not mean a monitored tape does not need to be edited, just that the amount of editing subsequently required will be reduced.

SWITCHING

All tape recorders have various switching circuits and Fig. 8–15 is an example of such an arrangement. This particular tape deck has

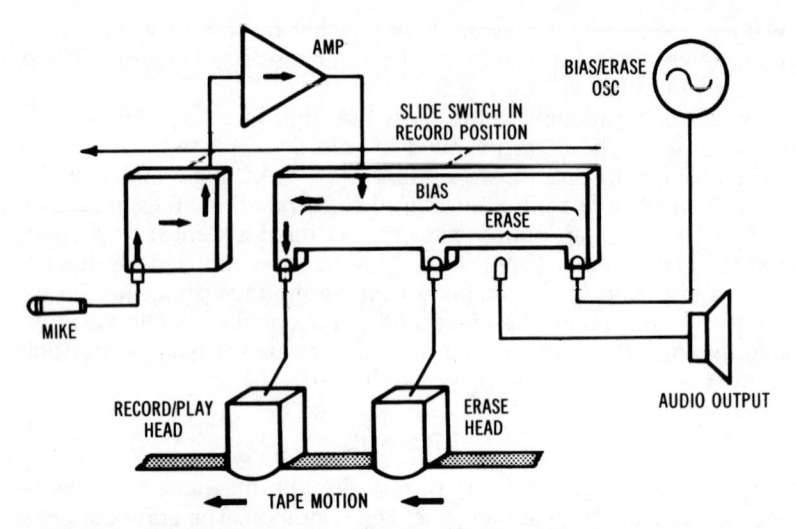

Fig. 8-15. Tape deck with slide switch in record position. Audio output is disconnected.

two heads—one for record/playback, the other for erase—a common combination.

Using a slide switch, the unit is now in the record position. Sound is changed by the microphone to a current that passes through an amplifier, as shown by the arrow, and then down into the record/playback head, via the switch. At the same time an alternating current from a bias oscillator (an ac generator) is brought into an erase head.

The erase head precedes the record/playback head as far as tape motion is concerned. The purpose of the bias current is to demagnetize the tape, to prepare it for recording. Thus, the erase head demagnetizes the tape before it passes under the record head. With such an arrangement we do not need to try to remember if the tape is ready for use, for it always is. The erase head does its work before the record head.

As shown in Fig. 8-15 the speaker is disconnected. This does not mean to imply that the tape deck contains a speaker, for it does not. This is just a way of showing that there is no audio output from the deck when the switching is set in the record mode.

Fig. 8-16 shows that the slide switch has been set in the playback position. This action disconnects the sound input source, a microphone in this instance, but it could be any other sound source as well. However the speaker, simply representing audio signal output, is now connected into the circuit. The bias/erase oscillator is automatically disconnected and the erase head is also out of the circuit.

In the playback position the tape is pulled along the face of the

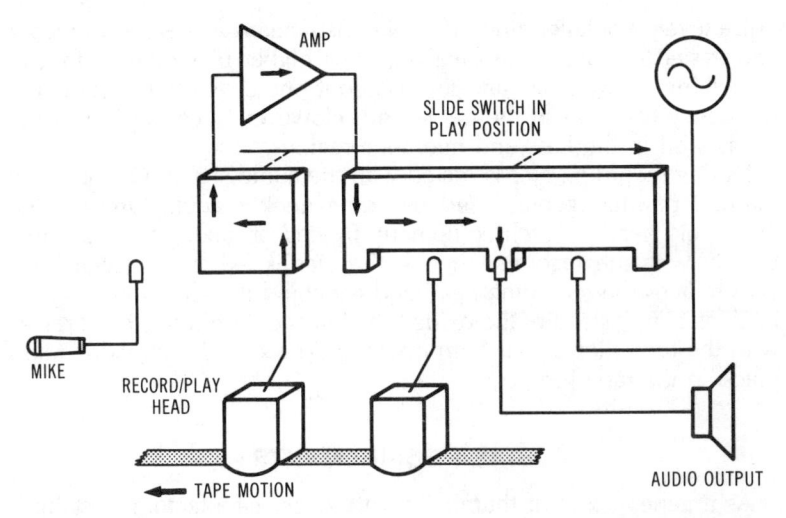

Fig. 8-16. Tape deck with its slide switch in the play position. The tape produces an audio current in the playback head. The speaker symbol represents the audio output and is not intended to indicate the presence of a speaker in the deck.

playback head. This induces an audio current in the playback head which is sent through an amplifier, as indicated by the arrow, and then through the switch to the audio output terminal, as shown by the speaker symbol.

THE BIAS OSCILLATOR

The purpose, in a tape deck, is to have each of the tiny metallic particles on the tape act as a magnetic replica of the sound input. At the same time an alternating current is supplied by an oscillator circuit.

This circuit is an alternating current generator whose frequency is supersonic, that is, inaudible, since the frequency may range from 30,000 Hz to as much as 100 kHz. The bias oscillator does two jobs. The alternating current it supplies during recording helps the recording head do its work of magnetizing the metal particles on the tape. The current supplied by the bias oscillator is also used by the erase head to eliminate any prior recordings.

The amount of bias current that is used is determined by the nature of the magnetic tape. Some tapes require more bias; others less. While some tape decks have just a single bias position, others have switches to control the amount of bias, in two, three, or more steps.

Bias current is essential but is a mixed blessing. Bias produces noise known as bias noise, and the higher the bias the greater the noise level. A smaller amount of bias also means it is easier to record and easier to erase. Tapes that require a higher bias current do give something in exchange and that is about one or two kilohertz better frequency response at the treble end. However, such tapes may not do as well in the bass and midrange regions.

During recording bias is mixed with the audio signal. One unusual method of mixing bias used by some deck manufacturers is the cross-field head. In such equipment, instead of mixing the bias electrically with the incoming signal, the bias is fed to a special head which faces the recording head and is behind the tape, touching the backing side instead of the oxide. The bias field from this head mixes with the recording signal from the record head and the mixing takes place in the tape itself.

BIAS REQUIREMENTS

As a general rule of thumb, cassette deck manufacturers establish the bias frequency at about five times (or more) than the highest recorded audio frequency to avoid beats between harmonics of the audio signal and the bias. The ac bias, for linear output, must not only be a pure sine wave, but must be evenly distributed around its X axis. This means that the positive and negative halves of the bias ac waveform should be identical. There must also be no dc component present in the bias since this would have the effect of changing the operating point of the bias, depending on the dc polarity.

TAPE TYPES

The magnetic particles used on cassette tape are called its formulation, the material of which the magnetic particles are made. These particles are not only composed of selected substances but have a special size, a shape as close to identical as possible, and also an optimum length-to-width ratio.

There are a number of different kinds of tape formulations with each giving the tape certain desired characteristics.

Pure Ferric Oxide

The pure ferric oxide formulation consists of ferric oxide in crystalline shape with a length-to-width ratio of about 10-to-1. The length-to-width ratio of magnetic particles generally varies from 4-to-1 to 10-to-1. Using this particle it is possible to increase the density, the total number of particles per unit area on the tape. Because of the greater density, noise effects caused by open areas between particles are minimized.

Low Noise——High Output

Low noise, high output cassette tapes are so called because the tape formulation consists of magnetic particles of unmodified gamma ferric oxide. The improvement over ordinary ferric oxide is also due to a reduction in particle size and some improvement in shape. These tapes have a higher magnetic particle density, hence a wider dynamic range.

Modified Gamma Ferric Oxide

Modified gamma ferric oxide is yet another category of a cassette tape formulation. To this oxide is added carefully controlled small amounts of either metallic cobalt or magnetite. These substances increase the coercivity and magnetic efficiency of the particles.

Coercive force is the amount of magnetic force required to reduce the magnetism of a particle to zero. Considered from another point of view, coercive force is the signal strength required to record or erase tape and is measured in a unit called the oersted. The coercivity of most ferric oxide tapes is 300 to 350 oersteds, while the coercivity of most chrome tapes is about 500 oersteds, and for metal alloy tapes about 1,000 oersteds. A tape having higher coercivity has increased output in the treble range, but higher coercivity also makes it much more difficult to erase such a tape. Higher coercivity also means higher bias levels, but this, in turn, means a higher noise level. The problem of higher noise has been partly solved by decreasing the particle size of the formulation put on tapes.

Tapes that have a higher coercivity, such as chrome or metal alloy, need a greater amount of bias to obtain maximum sensitivity. Such tapes are also more capable of "holding on" to the signal, especially at the high frequency end. If you use excessive bias on a tape that doesn't require it, the effect will be to deteriorate the high frequency response.

Chromium Dioxide

The chromium dioxide formulation offers better responses at the high-frequency end. Developed by the Du Pont Corporation, these tapes also have a good signal-to-noise ratio. The formulation is designated chemically as CrO_2, and requires a high value of bias current.

Ferrichrome

The ferrichrome tapes represent an attempt to combine the special qualities of chromium and ferric oxides. This is done by putting a thin layer of chromium dioxide over a base coating of ferric oxide, thus obtaining the better high end response of CrO_2 while keeping the superior middle and low end abilities of the oxide tapes. In the ferrichrome tapes there is a layer of about 0.04 mil (four one-hundred-thousandths of an inch) of chromium dioxide and a layer of 0.21 mil (21 one-hundred-thousandths of an inch) ferric oxide.

Metal Alloy

Metal alloy tapes aren't oxide formulations but are pure metal alloy particles. The pure metal tapes have a higher coercivity, an increased maximum output level (mol) at high frequencies and a higher flux density. The advantages of pure metal tapes are improved sensitivity, increased dynamic range particularly at the high frequency end, and lower distortion. While you can use such tapes for playback on existing decks, they cannot be used for recording on any decks except those especially equipped to do so. The high coercivity of metal particle tapes makes it virtually impossible for a conventional erase head, regardless of properties or construction, to do an adequate job.

Tapes in General

All cassette tapes drop off in high frequency response, some more than others. When measured at 16 kHz, the poorest is the conventional, but lowest cost cassette. A typical low-noise/high-output tape is better, while a top rated normal bias tape is still better. In this respect, ferrichrome is superior to the top rated normal bias tapes, but not as good as chromium dioxide tape. The problem with chromium dioxide tape is poor headroom and a higher level of distortion in comparison with gamma ferric oxide tapes.

EQUALIZATION

We cannot record all audio frequencies equally on all tapes. Tapes do not record treble as well as they do bass. This is a function of the formulation used on the tape, whether the magnetic coating is ferric oxide, chromium dioxide, or other material and the speed at which we operate the tapes. The slower the speed, the greater the audio high frequency losses.

A method of equalization is to boost the treble tones before the audio signal reaches the tape. This treble preemphasis is much higher for cassettes running at 1⅞-ips than for open reel at 15 ips. Even more treble preemphasis will be needed if we ever move in the direction of cassettes running at half speed, $^{15}/_{16}$-ips.

Equalization is standardized for each type of tape. This permits tape recorded on one deck to be played back on another.

TIME CONSTANT

The point at which treble preemphasis begins varies with different tapes and is referred to as the tape's time constant. Typically, tapes have time constants of 50 μs (50 microseconds, or 50 millionths of a second), 70 μs and 120 μs. To determine the actual frequency divide 159,154.9431 by the time constant in microseconds. A 50 microsecond time constant means that treble boost begins at 159,154.9431/50 or 3183.0988 Hz. For 70 microseconds treble boost begins at 2273.642 Hz and for 120 microseconds treble boost starts at 1326.291 Hz. The higher the time constant, the lower the point in the audio range at which treble boost starts. Since the signal is preemphasized during recording it must be correspondingly deemphasized during playback.

In cassette tapes, ferric oxide (FeO$_2$) has a 120 microsecond time constant while chromium dioxide and cobalt doped ferric oxide tapes have a 70 microsecond time constant.

LOW FREQUENCY BOOST

In tape recording we get losses at the low frequency end as well as in the treble region. Consequently, the playback amplifier in the tape deck boosts the low frequency end. Treble tones are corrected in recording; low frequency loss in playback. It would be much simpler if we could have a flat frequency response audio signal on tape, but without equalization at both ends of the audio spectrum, the reproduced music would sound weak in its low and high end audio frequency sections.

EQUALIZATION AND BIAS

Equalization goes hand in hand with bias. Many decks now have two or more bias/equalization settings. Normal usually refers to the lowest setting, high to the next position, chromium dioxide to the third or high position, and metal to the fourth or top setting. The purpose of bias/equalization settings is to have the tape provide as flat a frequency response as possible.

SWITCHABLE BIAS

Some tape decks are equipped with front panel switchable bias. The theory is that this will permit the user to set optimum bias for any tape thus meeting the tape's bias requirements closely. However, it is necessary to set both bias and equalization separately, using appropriate instrumentation for all speeds. This requirement is beyond the operating knowledge of most users of tape decks and is an adjustment best left to professional tape technicians.

SENSITIVITY

Sensitivity is a measure of a tape's ability to respond to a fixed level input signal of a given frequency. Ideally, the tape should have the same high degree of sensitivity at all frequencies, an essential factor in getting balanced sound reproduction. Sensitivity is often measured at three points along the audio spectrum: 333 Hz, 8 kHz, and 12.5 kHz.

PRINT-THROUGH

Print-through is a condition in which recorded material impresses itself on several successive layers of tape. This depends on the thickness of the tape, the formulation, and recording level. The thinner the tape, the greater the possibility of print-through, so it is more likely on a C-90 than on a C-30 tape. Vibration and fast rewind, plus long time storage, often produce print-through.

ERASABILITY

If a cassette is to be used repeatedly for recording new material it must be erasable. This poses the question as to *how clean is clean*? It might seem impossible to determine the degree of tape wipeout but if a re-recording contains no annoying residual or *ghost* sounds it can be considered to have good erasability. This is an elementary approach for it does not set up a rigorous standard. A better test would be the ability of a cassette to erase tape as often as required during a single pass through the deck, without resorting to an outside aid such as a bulk eraser.

DYNAMIC RANGE

Every tape, no matter what its quality may be, will deliver some noise on playback, and this noise is variously called the "noise floor" or residual noise. Any music recorded below the residual noise level will be masked or obscured. The separation, measured in decibels, between the noise floor and the point at which saturation of the tape begins is the dynamic range, as indicated in Fig. 8-17. Recording at a lower level brings the signal level closer to the noise floor and reduces dynamic range.

Tape Noise

Noise is an inherent property of the random distribution of the metallic particles that make up the magnetic layer of the tape. These

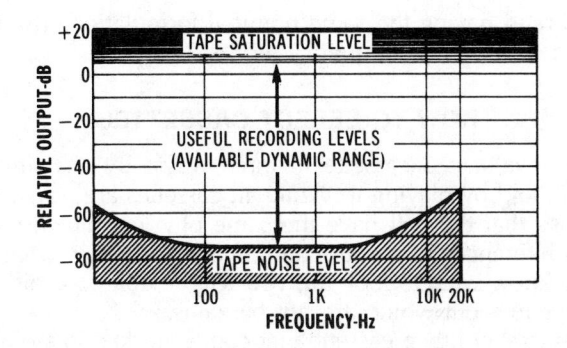

Fig. 8-17. Dynamic range is the separation, measured in decibels, between the noise floor and the tape's saturation level. *(Courtesy B·I·C-Avnet)*

particles do generate signals in the playback head but they are random in frequency and of low level. Because of the low level, only frequencies where the ear is most sensitive are audible, resulting in the familiar sound of tape hiss.

The absolute level of noise is a function of the formulation of the magnetic material, and of the size and distribution of the magnetic particles. It isn't influenced by the recorder or by the level of a recording, and, for all practical purposes, is a property of the tape.

Tape Saturation

There is a limit to the amount of recording signal a tape will accept in a linear manner. At some point the amount of magnetization will no longer correspond to the signal level. As the tape saturates, high frequency rolloff occurs, that is, there is a loss of high frequency response. Recording at a lower level improves the response curve, permitting it to be more "flat" out to its treble limit.

A kind of clipping of the waveform results when excessively high signal levels are put on tape. Known as soft clipping, the tops of the signal peaks are compressed. The degree of compression is measured at the third harmonic distortion of the signal, also known as K3 distortion.

Maximum Output Level

If a signal is recorded at a level giving a third harmonic distortion of 3% then the tape is considered to be saturated, that is, no longer capable of recording higher levels. On playback, the output level relative to a reference level is called the maximum output level, or its *mol*. The *mol* is a property of the formulation of the magnetic layer on the tape and varies not only for different formulations but from brand to brand having the same nominal formulation. The IEC standard for measuring low frequency *mol* is 315 Hz.

HOW TO SELECT CASSETTES

You can evaluate a cassette in three ways: by examining it, by handling it, and by playing it. While all cassettes are physically alike in the sense that they all have the same physical dimensions, there are some differences in construction, and if you know what they are and if you know what to look for, you will be able to avoid cassettes that are cheap—but which are not bargains.

Rule 1 —Never buy a cassette that comes packed in a plain, white cardboard box. Known as *white box* cassettes, these are intended for those who insist on the cheapest cassettes, regardless of recording results or possible damage to the tape deck by such cassettes. They are ordinarily used only in equally low price cassette recorder/players. Top quality cassettes are sold in plastic containers. These not only protect the tape against handling and dropping, but keep the cassette dust free. At the same time the transparency of the plastic enclosure lets you identify the cassette.

Rule 2 —Examine the cassette carefully. Do the two halves of the cassette plastic housing fit together properly? Is the housing warped? Are there any burrs or rough, unfinished edges anywhere on the housing? How are the two halves of the cassette held together—by screws or are they cemented? The screw type is considered a better form of construction, and also lets you disassemble the cassette should you ever need to do so.

Rule 3 —As a further check on the mechanical qualities of a cassette, put it into a deck, turn the unit to its play mode, and then put your ear right up against the part of the deck that holds the cassette. To keep from being distracted, turn down the volume control of your speaker or amp so you get no speaker sound. Listen for rubbing or grinding noises. You should hear very little or none. A noisy cassette means poor mechanical construction, rubbing parts, or loosely assembled parts.

Switch to fast rewind, and with your ear against the part of the deck that holds the cassette, listen again. If you hear a rattling sound, you will know you have a poorly constructed cassette.

Rule 4 —Just the action of putting a cassette into a deck and taking it out again is a test. The cassette should fit in easily, without binding and without being forced. You should not need to exert any pressure to get the cassette into the tape deck. When you depress the eject control on your tape deck, you should be able to remove the cassette tape without difficulty. If you must pry it loose, or if the cassette seems to resist removal, then the physical dimensions of the cassette aren't what they should be or else the cassette housing is warped. You cannot test a cassette by holding it to your ears and shaking it. The mechanism inside the cassette must be loose enough to let the hubs turn.

Rule 5—Check the housing for warpage. You can do this easily by putting the cassette on a flat surface, such as the top of a counter in a store. Put the palm of your hand on the upper flat surface of the cassette. Now try to move your palm. If you get a rocking action the tape housing is warped. You may be able to move the cassette horizontally, but you should not get any vertical movement of the cassette at all.

OPEN REEL VERSUS CASSETTE TAPE

Aside from physical differences, open reel tape and cassette tape are alike for they may both use the same formulations. And they are often afflicted by identical problems. However, the greater length of open reel tape, and the fact that it is out in the open, leads to some special difficulties.

Cupping

The oxide particles on the tape and the base film are two different materials and so have different thermal coefficients of expansion. This means they can expand or contract with changes in temperature, but not at the same rate. The result can be a curvature of the tape along its length. The tape can *cup* inward or outward. You will still get output signal from the tape but it may suffer from high-frequency losses.

A form of cupping can also occur in tape storage after fast foward or fast reverse. This can produce a rough wind with layers of tape protruding from the sides making them subject to creasing, bending, and foldover. If the tape is stored for a long time, normal variations in temperature and humidity will help deform the edges permanently. The result is the edge channels will fade in and out.

Strain Lines

Some open-reel decks wind tape in an excessively tight manner, not necessarily at high speed and without protruding edges. The trouble can be caused by misadjusted takeup tension or by having the tape off center and winding much too close to the reel flanges. After rewind, examine the open reel tape and you may see strain lines, radiating like spokes from the center outward. Other symptoms are noncircular winding or lipped edges. If the same trouble occurs with different brands of open reel tape, then the problem is almost certainly in the deck.

CASSETTE RECORDING—PRECAUTIONS

Before recording on cassette tape there are certain precautions you should take, certain steps to make sure the recording will come up to your expectations. At the start you will be conscious of each step, but in a short time you will be doing them automatically and much more quickly.

1. Make sure the machine on which you will play your cassettes is clean. This means the recording and playback heads, the capstan, and the rollers. You can buy a *head cleaner* cassette which you can use just as you would any other kind of cassette. Or, you can buy special kits made specifically for use on cassette decks.
2. Before inserting the cassette into your deck make sure that the tape inside the cassette is tight by adjusting the takeup hub with your finger. You can also insert a pencil into the hub (Fig. 8-18) to do this. To determine just which reel is the takeup reel, look inside the cassette. The hub without tape or with very little tape is the takeup hub. The other is the supply hub. Each hub works alternately as takeup or supply.
3. Are you using a portable recorder/player? Don't wait for the batteries to die before you replace them. Some recorder/players come equipped with a pushbutton—a *state of charge* switch that will let you know the condition of the batteries. Never keep dead batteries in a portable unit, especially if you do not plan to use it for some time. Batteries can corrode and

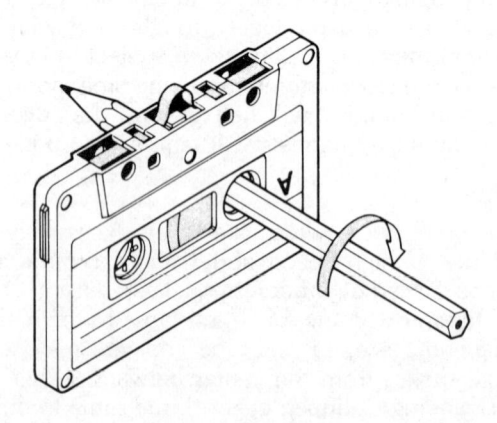

Fig. 8-18. Make sure tape is wound tightly by inserting a pencil into the cassette hub and rotating it. *(Courtesy Luxman, Div. of Alpine Electronics of America)*

when they do they can corrode your playing/recording machine as well. Dead batteries sometimes swell and so may be difficult to remove. All batteries have a certain shelf life. This means a battery consumes itself whether you use it or not. A battery wears out quicker when you use it, but it will also decay when not used, although much more gradually.

4. Cassette decks of the better grade come equipped with a pause control to let you stop the movement of tape. Get into the habit of using it to start and stop the movement of tape. If you do you will find it will reduce possible wow and flutter in your recordings.
5. If you have more than one cassette deck, don't switch tapes back and forth from one deck to the other, unless you are sure that the azimuth of the heads—their alignment—is perfect in both. When you erase on one deck, use that same deck for recording and also for playback. These precautions do not apply to low-cost players, but in high-fidelity units they are of importance. A small difference in head alignment can mean a noticeable difference in erase, recording, and playback.
6. Make sure the bias switch on your cassette deck is set to correspond to the bias requirements of the tape you are using. Most decks have at least a two-position switch, one for normal bias, the other for chromium dioxide.
7. If you are going to record off the air, or are going to record from a sound source such as a microphone, a phono record, or another tape, don't assume everything is going to work just the way it should. Run a test first. Make sure the vu meters or other level indicators on your cassette deck are working and that you have both level controls set the way you want them. Run a recording test and then playback, making a listening check to make sure everything is working properly.
8. If your deck has a noise reduction switch, keep it in the same position for both recording and playback. You can have it on or off, just as you wish. But if you have turned it on for recording, keep it turned on for playback. Mark the outside of the cassette box in some way so you'll always know what you have done. Don't rely on your memory.
9. If you monitor your cassette recordings, do so through headphones. There are several reasons for this. The first is that headphones let you get "closer" to the sound. They will let you hear more of what is going on than speakers. Another is that speaker sound can be picked up by microphones if that is the sound source you are recording.
10. When you make a recording the most important factor is the recording level. If the level is too low, the signal-to-noise ratio

will be poor—just another way of saying that the noise level will be too high when compared to the signal. If the signal is too high, you will get distortion. And so you must steer a course between these two extremes. All decks come equipped with level controls and a pair of volume unit (vu) meters or other level indicators. Most meters have a red area at the right side of the scale, a visual warning that you are in an excessively high recording level situation.

TAPING OFF THE AIR

Taping off the air is an enjoyable hobby and is musically rewarding particularly if you can't get the selections you want in a prerecorded format. There is also a different attitude toward in-home recorded tapes versus the prerecorded variety. No one likes to edit or cut apart prerecorded tapes. The problem with buying prerecorded tapes is that you know nothing of the quality of the tape, but if you do your own recording, whether off the air or from phono records, you have complete control over tape quality.

Before you tape off the air consider that you aren't going to end up with a frequency response of 20 Hz to 20 kHz. Aside from the pipe organ and the bass bassoon, little music is written for frequencies below 50 Hz. Even though a pipe organ may go down to 16 Hz, this is a seldom used note. Broadcast station microphones often are less sensitive to frequencies below 50 Hz. And at the high frequency end, even when your prerecorded tapes and phono discs may call it quits at 15 kHz.

Don't depend on luck. Don't bother to switch on your fm receiver or tuner and tape deck and hope for the best. You'll get a recording but save the adjective "good" for some other time.

You can record off the air with either a cassette deck or an open reel unit. If your open reel deck has a 1⅞-inches-per-second speed then you can record for a little more than 6 hours using a 3600 foot tape. With a 2400 foot tape at 3¾-ips, a commonly used speed, you can record for two hours. Use a 3600 foot tape and recording time becomes 3 hours. With a C-60 cassette tape you can record for an hour, that is, 30 minutes on each side. And a C-120 cassette will give you a total of two hours of recording time. With either cassette or open reel tapes you can get low-cost editing kits that will let you eliminate sounds you don't want to hear.

Before Recording

There are various things you should do before recording broadcasts.

It takes very little dirt to put high-frequency tones out of the running. If you have more than 20 hours of recording and/or playing time on your deck, use a head cleaner cassette or a cotton covered tip and some alcohol to clean the tape heads. Use denatured alcohol, not the drinking stuff.

Don't go running up and down your fm dial in an effort to find some music to your liking. Consult the pages of your newspaper for fm listings. If you have no idea just how long it will take to play the music you want to record, give yourself plenty of tape margin.

Use the pause control on your tape deck whenever you start and stop recording. This will help cut down on wow and flutter at the beginning and end of your recording and will eliminate wasted tape.

Make sure the bias/equalization control on your deck is set to correspond to the tape you intend using.

If you are planning to use a tape that was previously recorded, you may not get complete erasure if the original recording was made on another tape deck. Try using a bulk eraser. If you don't have one run the tape through your deck with the controls set in the record position and with the input signal level control turned completely down. Then rewind the tape, put your tape deck in its playback mode, turn up the volume control, and listen. In this way you should be able to determine just how clean the tape is. As a standard of comparison, run a new blank tape through your machine with the controls set the same way, and listen once again.

If possible, monitor through headphones rather than through speakers. Headphones bring you closer to the sound. If you have no headphones and must use speakers, keep the volume turned down.

Before you go ahead with your off-the-air recording session, run a short test to make sure your equipment is working properly. During the actual recording, your erase head on the deck will eliminate the test you made.

Decide in advance whether you want Dolby® or other type of noise reduction system turned on or off. But if you record with noise reduction, you should play the tape back with the noise-reduction circuit *on*. It's a good idea to mark your tape container with some sort of symbol to indicate whether you used noise reduction or not.

With some decks you can go from one mode, such as fast forward or reverse, without going through the stop position. If your deck isn't so equipped, be sure to use the stop control. This may not be so important with a cassette such as a C-46, but when you get to a C-120 the tape is thinner and is more subject to stretching and tearing.

Set your tape counter to zero. You may be interested in knowing the start and stop points of your recording. Also, during the recording, make a note of counter numbers that correspond to musical passages you like most. This will let you find those passages in a hurry during playback.

Always run a tape through your deck at least once, using regular speed if you have the time, or else fast forward and reverse. Set your controls to the record mode and turn both level controls down to zero. This has a double advantage. It gives you the maximum clean surface for recording and also eliminates the possibility of a tape sticking problem during recording.

Recording

Don't start recording the moment you turn on your deck if you are recording at the beginning of a tape. If your tape has a leader you may lose the start of the broadcast. Instead, turn your deck on, let the leader move past the tape heads and then put your deck in its pause position. At this time touch the reset button of your counter to put the counter to zero.

On your deck you will find a pair of recording level controls. Set these for left and right channel sound so both meter pointers are in the same general area. This isn't easy to do, but at least one meter pointer shouldn't be down while the other pointer tries to climb out of the right side of the case. The idea is to have the recording level for both sound channels approximately equal.

Your vu meters will probably be calibrated in decibels, from -20 dB to $+3$ dB. From 0 dB to $+3$ db the meter scale is in red. If you record at too high a level, the meter pointers will remain in the red areas for most of the time. This indicates you are saturating the tape with signal and so you will get distortion. If you try to play it safe and set your level controls so the pointers of the vu meters hover around the -20 dB region, you will get a poor signal-to-noise ratio. Tape noise and hiss will be high compared to the signal. What you should do is to set your level controls so the pointers get as close to the 0 vu point as possible, with few excursions into the red area of the meters.

A lot depends on the kind of tape you use, for some tapes permit a higher recording level than others. If your deck is equipped with a Dolby® noise reduction system, or comparable type, you should be able to back off about 3 dB in recording level.

By recording off the air you can build your own tape library. Unlike prerecorded tapes, you can be highly selective and record only those compositions you like, or even portions of compositions.

After Recording

One of the advantages of discs over tapes is that it is so easy to locate a selection on records. However, if you put a label on your tape boxes, accompanied by the counter setting for each selection, it shouldn't be difficult to locate any recording you want to listen to. Make sure your deck is in its playback, not record, mode, set your counter to zero to correspond to the start of the tape, and then use the fast forward control while watching the counter to reach the recording you want. Some decks are equipped with a memory so you can get back to any desired selection quickly.

TAPE EDITING

You can edit tape in two ways—by using the erase function of your tape deck or by using an editing kit. If you want to use your deck for editing just set it on record and keep the input level controls turned all the way down. The problem with this technique is that it's difficult to erase exactly. And if you erase somewhere along the middle of the tape, or anywhere else except the start and finish, you'll have a blank unless you can manage to fill it in with recorded sound in some way.

The other method is to cut the tape, remove the section of tape you don't want, and then join the cut ends. This makes it sound a lot easier than it really is. Cassette tape can be recorded in two directions. When you cut across the tape you not only remove what you don't want, but you may be taking out some music you like. However, if you've recorded in one direction only, you can edit by cutting.

To edit tape mechanically you must cut so that the start and finish of what you want to remove ends exactly at the gap in the playback head. This means you must be able to mark the tape so you can make precise cuts. If your deck has more than one head, the playback head is the one at the far right.

To mark the part of the tape to be removed, use a grease pencil. Don't mark with the tape resting against the head gap of the playback head. Instead, move the tape a small distance. Measure this distance and use it as your reference for the start and finish cuts.

The best type of cutter is a single edge razor blade. Don't make a vertical cut since this can produce a click every time the repaired tape passes the playback head. Instead, cut at a 45° angle.

The best procedure, if you plan to edit tapes, is to buy a commercially made editing kit. These are inexpensive and come complete with tabs for joining the cut ends of the tape, an editing block with a 45° groove so you can make precise cuts, and an instruction booklet. You'll find that cutting tape isn't difficult, nor is joining the cut ends any problem if you have an editing kit. The difficulty is removing just that section of tape you want out.

There are two ways you can try to become an editing expert. The first is to dub your original tape, the one you want to edit, onto another tape so that you have a duplicate just in case your editing efforts are less than successful. The other—and this is something you should do—is to experiment editing with a recorded tape you are willing to sacrifice. Start with a voice recording since it is easier to learn with such a tape. You will also find your tape deck counter highly useful for letting you find the start and finish of tape to be cut.

Another method of tape editing is similar to dubbing, copying tape from one deck onto another. It does take some practice. You can remove unwanted sections without having blank areas on the tape, or you can add any sounds you want. You'll need to use the pause control on both decks, so this sort of editing is better if you have someone to help you.

The easiest tape to edit is open reel tape. You can edit cassette tape electronically, but cutting and joining cassette tape takes a lot of patience. Most 8-track decks are players only. If you have an 8-track recorder/player, remember you'll be cutting across four pairs of tracks. No one says you cannot edit 8-track cartridge tape. It just has nothing to recommend it as an indoor sport.

THE BULK ERASER

Just because you have just removed an open-reel tape from its shiny, new box does not mean that tape is completely untouched. Magnetic fields are invisible while some are so weak they can be completely ignored. But wherever a magnetic field exists, such as that around a power transformer or a motor, and is sufficiently strong, and if you somehow move the tape through that magnetic field, you will be recording. That recording will be noise, not music.

The inverse of a brand-new tape is one that has been used over and over again. The end result, in some cases, is that the tape has become so magnetized by the record head that it is literally unfair to demand that the erase head do a complete job. Theoretically, the erase head will remove any unwanted signal present on the tape, theoretically because nothing in an imperfect world is perfect. And so you may want to consider using a bulk eraser.

The bulk eraser is an electromagnet that gets its operating power from the ac line. Such erasers are quite efficient and so it is a good idea to give your tapes the bulk treatment once in a while. And if you really want to get a tape recording with absolutely minimum noise, try using the bulk eraser before recording. And this applies to new tapes as well.

The bulk eraser is an all-or-nothing-at-all device. You cannot erase just part of the tape. For that you still need the erase head. And so, before you use the bulk eraser, make doubly sure there are no passages on the tape you want for future use.

SUPERIMPOSING

The flexibility of the tape recorder—instant record, erase, plus ease of splicing for open reel, and editing for open reel or cassette—will enable you to produce unusual sound effects. Thus, you can superimpose sound-on-sound, that is, put one sound on top of another. Some tape decks come equipped with a superimpose control.

However, if your deck does not have such a control, you will need to inactivate the erase head. There is a reason for this. If tape that has been recorded is sent through the tape recorder with the record pushbutton "in action," then all sound on the tape will be erased since the tape will first move past the erase head and then past the record head. The purpose of this arrangement is to give you a clean recording surface—to put the tape into condition for recording. If your tape recorder has a pressure pad to hold the tape against the erase head, hold the pad back during the time you superimpose.

Since superimposing is really double recording you will need to move the tape away from its close contact with the erase head. You can do this with a small piece of cardboard or plastic.

Another method of mixing is to use two microphones simultaneously. You will need to connect the two microphones in parallel. However, unless you have had some experience in making such wired connections, this method isn't recommended. It is better to use a mixer instead.

MIXERS

A mixer (Fig. 8-19) is a device to enable you to connect two microphones, or other sound sources so that you have just a single input to your tape recorder.

To use the mixer, simply plug the mikes into the appropriate jacks on the mixer and connect the cable of the mixer to mike input of the deck.

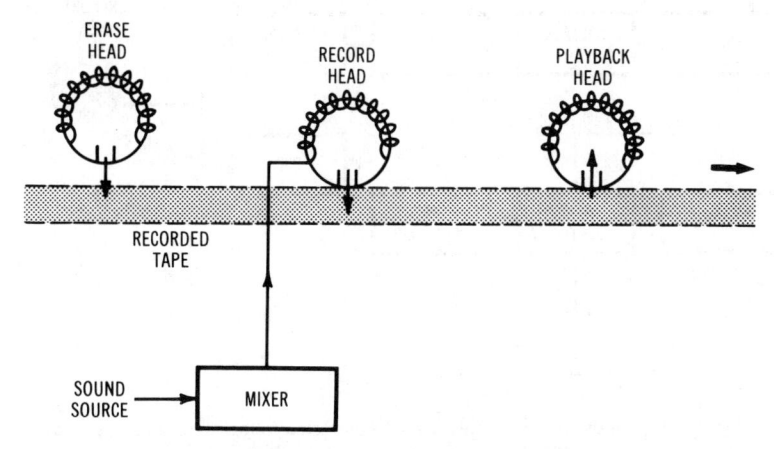

Fig. 8-19. Method of superimposing sound on sound.

There are various types of mixers, some simple, others more elaborate. A simple mixer is just a device for connecting a pair of mikes. A more elaborate mixer will have provision for more microphones and will have a gain control for each. Some come equipped for other sound source inputs, possibly a record player or tape deck (Fig. 8–20).

Mixers are essential for group recordings such as musical ensembles, where more than one microphone is required. The number of microphones to be used will depend on the acoustics of the room and the size of the orchestra. Without the mixer, even a small group would need to huddle around a single microphone. Aside from the awkwardness and discomfort, the microphone would need to be an omnidirectional type so as to pick up sound equally well from all directions.

With a mixer you can announce a record and at the same time have a soft musical background for your voice. Or, by operating the gain controls for voice and music you can have the music gradually take over as your voice level is made lower.

For musicians, amateur or professional, who would like the experience of playing with large orchestral accompaniment, a mixer opens the door to such an achievement. Connect the record player to one jack on the mixer and play to a microphone connected to another jack on the mixer. Gain controls on the mixer will enable you to override the orchestra when you wish, or to merge. With a mixer and the use of the superimpose control on your tape deck you can get unusual sound-on-sound effects.

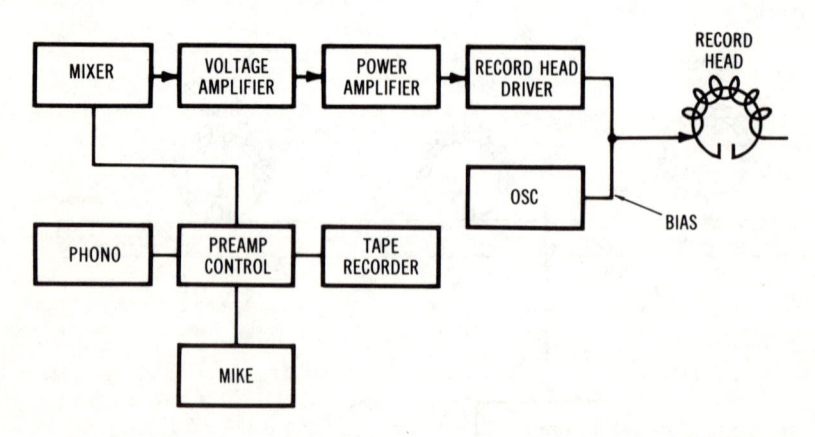

Fig. 8-20. Various sound sources can be connected to the mixer making it possible to have unusual sound combinations.

TAPE DECK FEATURES

Although all tape decks, open reel or cassette, have the same basic functions, that of playback and record, they all do not have the same number of features. However, they all do have certain features in common, such as forward, fast forward, fast reverse, pause, play, record, etc.

Stop Mode

In the stop mode the drive to the reel spindles is disengaged and brakes are applied to both reel spindles. The tape is not in contact with the heads and the amplifier may switch from record to play. In a reel-to-reel transport the pressure pads are withdrawn. In a cassette transport heads are withdrawn. This is the only mode that tape can be threaded in open-reel decks or the cassette removed from a cassette deck.

Start, Play, Run

Normal forward motion of the tape is obtained by pinching it between a rubber roller and a precision rotating shaft or capstan. Brakes and reel spindles are released, and the drive is engaged to rotate the take-up reel to wind up, with the tape being pulled over the heads by the pinch roller. The tape is brought into contact with the heads.

The timing of this sequence is critical. The take-up must start winding up the tape immediately after the tape starts being pulled by the capstan, or a slack loop may be formed. If a loop does form, the take-up reel or hub can gain momentum and possibly jerk the tape causing wow at start up. If the pinch roller contacts the capstan and thus drives the tape before the storage reel brake is released the tape speed will slow down and the tape may be physically stretched, causing permanent distortion. Conversely, when the unit is switched to its stop mode, drive to the tape must cease before brakes are applied to the spindles. If brakes are applied too late, a surplus of tape will be unwound, which can cause the tape to become unthreaded.

Pause or Instant Stop

This feature, pause or instant stop, stops tape motion by moving the pinch roller away from the capstan, but does not withdraw heads from contact with the tape. Brakes are applied to reel- or cassette spindles. Since heads are not withdrawn, pause will stop and start tape more quickly than using run-stop controls. The amplifier is not switched in this mode.

Fast Forward (FF)

The fast forward mode is indicated on the deck control by the letters *ff* or by a pair of arrows pointing toward the right. The spindle brakes are disengaged, and the take up reel is driven at many times play speed so that program material may be searched. The output is muted to eliminate high speed chatter. In fast forward mode the record function is disengaged and the electronics are returned to neutral or play mode.

Record

The record switches the electronics to the record mode and also energizes the erase head. While this control is an electronic function, it is usually placed and interlocked with the tape transport. Due to the fact that "record" erases previous information, several safety interlock concepts are in common usage:

1. A physical latch, key, lever, or button that must be moved to release and allow actuation of the record key. When the transport is switched out of record, the record lock will reset.
2. The switch can be actuated after the transport is in the pause mode. This concept requires a sequence to minimize inadvertent operation and conveniently switches the electronics without moving the tape. This allows the recording level to be adjusted properly before starting the tape.
3. In the cassette, a breakout-tab at the back can be removed to operate a protective interlock lever. This blocks the motion of the record key to prevent any possibility of damage to the recorded program.

Eject

On cassette decks only, the eject feature disengages the cassette from the transport so it can be removed.

Additional Features

While these are the primary features of decks there are others that either improve accuracy of operation or make it more convenient. The number of additional features is often dictated by the cost of the deck, with the lowest price deck units containing only a few, if any, extras.

Level Indicators —While cassette and open reel decks may have a pair of vu meters to indicate recording level in the left and right sound channels, some decks use plasma or fluorescent displays. A plasma display may use fluorescent green tubes with a number of dots per channel, possibly 12. The upper 12 dots permit holding the peak level, from −4 dB to +6 dB in 1 dB increments. Unlike conventional mechanical vu meters instantaneous response is displayed, permitting the user to set up an appropriate recording level. To cope with metal particle tape having sufficient headroom until saturation, a separate meter scale may be provided to read a maximum of +10 dB.

Variable Bias Control —To make the most of basic three head decks, a variable bias system may be used. It is possible to adjust the bias amount by means of a variable resistor and indicator lamp, depending on the tape used.

Azimuth Adjustment —Since cassette tapes do not run between the record and playback heads with an accuracy measured in microns, an azimuth adjustment of the record head may be needed. In some decks there are a pair of azimuth lamps. These turn on when azimuth is precise. Azimuth can be adjusted by turning an azimuth adjustment screw.

Tape Counter —Tape counters can be mechanical or electronic. Most tape counters don't mean much because there is such variation—even among machines of the same model. It is often necessary to do your own calibration. To do this let the tape run for a full minute, starting with zero on the counter. Stop the counter at the end of the minute and make a note of the numbers on the counter. Repeat the test several times and average the numbers to get counter speed per minute. This is done for you in some decks that use a 4 digit 7-segment LED for digital display. The rotation of the tape transport roller is optically sensed and you can read the exact time used for recording or playback in terms of minutes and seconds.

Tape Motion Sensor —An electronic tape-motion sensor stops the transport if a faulty cassette jams or spills.

Separate Tape Compartments —On some cassette decks there are separate compartments for housing a pair of cassettes. Duplication of cassettes is accomplished by inserting an original tape into the playback portion of the deck; recording from it onto a second tape held in the record section of the deck—with the ability to edit the tape in the process. In the separate tape compartments one contains a head designed especially for recording and the other a head especially for playback.

Microphone Input —Some decks have front panel provision for a pair of microphones to provide left and right channel stereo recording.

Recording Mute —The recording mute feature is useful when recording off-the-air fm program sources. While monitoring the program source you can remove unwanted commercials. When transcribing discs to tape, the recording mute feature lets you mute the noise which normally occurs when a phono stylus first makes contact with a disc.

DOLBY®*

There are various types of noise reduction systems used in connection with tape decks but one of the best known and most widely used is Dolby. There are various forms of Dolby noise reduction. Dolby A, for professional use, provides 10 dB of noise reduction from 30 Hz to 5 kHz, rising to 15 dB to 15 kHz. Dolby A is used in professional recording practice to enable studios to produce better master tapes in multitrack recording.

The Dolby B system is designed for home use. It can reduce tape hiss by as much as 10 dB, but it does not reduce low frequency noise, rumble, or wow.

Tape noise, mainly hiss, is most noticeable during the quietest parts of the music. The Dolby system then increases the volume so that the music is recorded at a higher than normal level. The Dolby system can do this because it has the ability to distinguish sounds of different pitch as well as sounds of different loudness.

Dolby B is an encoding-decoding process. These two are complementary, that is, one is the reverse of the other. The first part of Dolby noise reduction takes place immediately before putting the signal on the tape. The second part takes place during playback.

Signals which are sufficiently high in level so as to mask tape noise are recorded unchanged. Lower level musical signals along with noise are recorded with gradually increasing amounts of boost, as indicated in Fig. 8-21. As mentioned in the preceding paragraph, encoding, that is, boosting of the signal takes place prior to recording, so tape noise is not boosted.

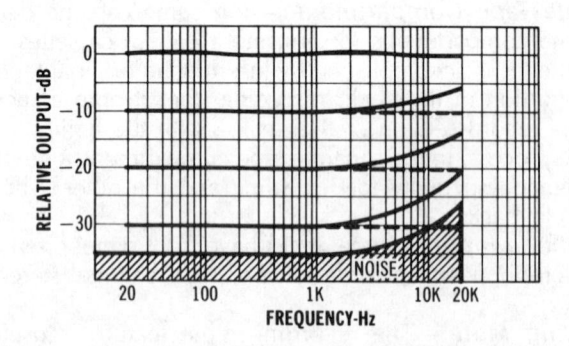

Fig. 8-21. During encoding low level treble tones are boosted. Note—as shown in the graph—the lower the level, the greater the boost.

---

*Dolby is the trademark of Dolby Laboratories, Inc.

Fig. 8–22 shows what happens during playback. Low level signals in the treble region are weakened. The amount of treble cut during playback must be equal to the amount of prior treble boost. Note, as shown in Fig. 8–23, that the noise level is also attenuated. Noise reduction begins at around 1 kHz with the noise being reduced by approximately 10 dB at the top end of the audio band.

When Dolbyized recordings are played back on a deck equipped with Dolby circuitry, the volume is automatically reduced in all the places where it was increased during the recording. At the same time the noise of the tape, now mixed in with the music, is reduced in all the places where the noise would otherwise have been heard.

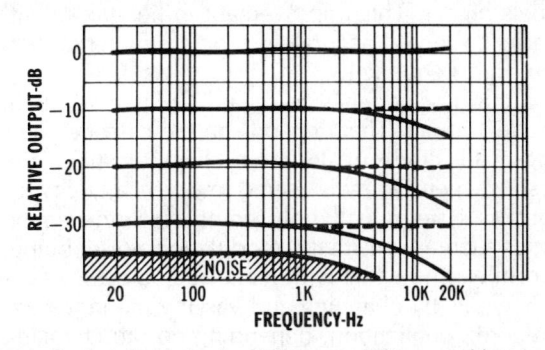

Fig. 8-22. During playback of a Dolby encoded tape, treble tones are weakened, with the amount of attenuation gradually increasing for lower level signals. Note that the noise is also reduced. The dashed lines indicate the frequency response after decoding. *(Courtesy B·I·C-Avnet)*

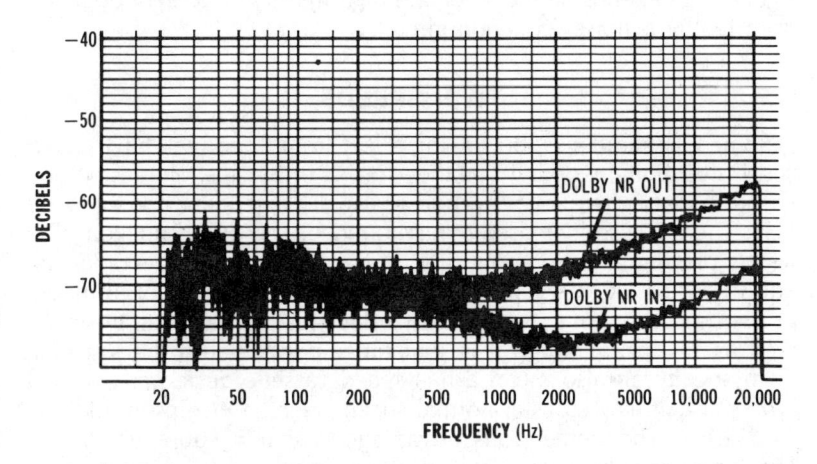

Fig. 8-23. Effect of Dolby noise reduction (nr) on tape noise. The reduction is about 10 dB from about 2 kHz to 20 kHz, but contributes some noise reduction below 2 kHz as well.

Cassettes recorded with the Dolby system can be played on any deck even though it is not equipped with a Dolby decoder. Of course this won't supply 10 dB of noise reduction, but turning down the treble tone control will give normal sound, with some noise reduction as well. The Dolby system will also work with chromium dioxide and high density tapes. Prerecorded tapes that have been Dolby encoded carry the Dolby trademark on the spine.

Dolby HX Headroom Extension system, when combined with Dolby B type noise reduction circuitry, improves the usable dynamic range of any tape formulation, particularly at high frequencies. It permits recording information at 10 kHz, and above, at a level 10 or more decibels higher. This improvement in headroom with any tape makes it easier to record accurately program material that is particularly rich in high frequencies.

There is a curious tendency to regard Dolby noise reduction as a sort of cassette cure-all. That depends on the cassette you use. A cassette having a full circle of desirable characteristics can show outstanding results when Dolby is used. However, a Dolby, or any other noise reduction system, will only emphasize uneven performance characteristics such as sensitivity, modulation noise, output, and sensitivity uniformity. The effect is to make such a cassette even worse.

Dolby HX works by changing and varying the tape recorder's bias level and record equalization, depending on the changing level and high frequency content of the sound being recorded. The effect is to reduce tape saturation, permitting recording at a much higher level. Dolby HX isn't user controlled; instead it operates through a control signal that depends on the level and high-frequency content obtained from Dolby B noise reduction circuitry.

TWIN SPEEDS

Some cassette decks, borrowing a leaf from open reel, have a two speed capability, working at either 1⅞-ips or 3¾-ips. The faster tape speed increases frequency response, usually extending it beyond 20 kHz. There is also increased dynamic range. The disadvantage is that it cuts playing and recording time in half: a C-60 then becomes the equivalent of a C-30, supplying just a half hour instead of an hour.

Twin speed doesn't always mean a deck that can work at 3¾-ips as well as 1⅞-ips, although it is sometimes considered to be just that. Twin speed can also mean a half-speed cassette deck, *i.e.* one that works at half as well as at normal speed, such as $^{15}/_{16}$-ips in addition to 1⅞-ips. The tremendous advantage is that it supplies a really long-time recording and playback deck for the lower speed doubles the time of any cassette. A C-120 instead of being a two-hour tape then becomes a four-hour tape. To be able to achieve a flat frequency response to 15 kHz at $^{15}/_{16}$-ips, such a deck must be capable of response to 30 kHz at 1⅞ ips.

One of the problems of half speed is spontaneous erasure, a condition of partial demagnetization of high frequencies during playback, when the erase head is inactive. This is noticeable at half speed because the recorded wavelengths are half of what they are at normal speed, therefore, such decks require special erase heads.

MEMORY

The advantage of a deck equipped with memory is that it lets you return easily and accurately to any previously determined point on the tape. One method is to set the tape counter to read all zeros at the selected spot; then using either the rewind or replay control. Thus, if you depress the rewind control, you can subsequently touch this control and the tape will automatically return to the spot indicated by the zero settings on the counter.

SPECIFICATIONS

Generally, specifications (Fig. 8–24) for cassette decks include wow and flutter, signal-to-noise ratio, frequency response, and possibly overall distortion.

Wow and Flutter

These are changes in speed resulting in a variation in pitch. Wow is a slow pitch change while flutter is a higher speed change, both caused by the inability of the tape transport to move the tape in a steady, constant motion past the heads in the deck. While these are separate, distinct faults, both are generally combined into one spec, and are expressed as a percentage. The lower the number the better. Better quality decks are rated at about 0.05% or less; figures such as 0.1% to 0.15% should be regarded as the maximum limit.

The frequency of wow is 4 Hz or less, causing a variation in tones at this rate. Flutter is higher and is above 4 Hz, and the resulting variation in tones is higher than wow. Wow and flutter are more noticeable when recording and playing back sustained tones.

Wow and flutter can be caused by a capstan and/or pinch roller which aren't precisely round. It can be due to poorly constructed cassettes, by imperfect hubs or reels, by transport motors which do not work at constant speed. Sometimes wow and flutter can be caused by the devices intended to keep the transport at constant speed. These may "hunt," that is, constantly change motor speed in an effort to maintain constant speed. Friction anywhere along the tape path can also result in tape speed changes.

Heads:	Independent 3 Heads Record Head (ferrite) x 1 Playback Head (sendust) x 1 Erase Head (ferrite) x 1

Drive Motor:
    Capstan drive
    Reel drive
	3 motors
	Quatz Locked D.D. motor x 1
	Coreless motor x 2

Tape Drive:	Dual capstan system
Recording System:	Bridge Recording by Bias Current & Signal Current (BRBS system, Pat. Pend.)
Amplifier:	DC-amp Configuration
Wow & Flutter:	no more than 0.03% (W.R.M.S.)
Signal-to-Noise Ratio:	better than 56dB (Dolby* off) ... CrO_2 tape better than 66dB (Dolby* on) ... CrO_2 tape better than 55dB (Dolby* off) ... LH tape better than 65dB (Dolby* on) ... LH tape
Frequency Response:	30Hz~18,000Hz ±3dB (CrO_2 tape) 30Hz~16,000Hz ±3dB (LH tape)
Overall Distortion:	no more than 1.2% (LH tape, 1kHz, 0dB)
Real Analized Distortion:	no more than 0.3% (LH tape, 1kHz, 0dB)
Separation:	more than 35dB (1kHz, 0dB)
Crosstalk:	more than −60dB (1kHz, 0dB)
Input Sensitivity:	line ; 100 mV mic. ; 0.25 mV (recommended microphone impedance; 600~10k ohms) DIN ; 2mV/1k ohms
Output Level:	580 mV
Headphone Output:	1 mW (8 ohms)
Semiconductors Used:	Transistor (141), Diode (111), IC (48)
Additional Features:	REC. MUTE function, Variable Bias Control, 3-position Equalizer Selector (CrO_2, normal, EX), Azimuth Adjustment Facility with Indicator, Search Function (cue & review), Tape Monitor Circuit, Peak Level Indicator with Peak Hold function; Timer Recording/Playback function, Remote Control (Available with optional remote control box, DOLBY* Noise Reduction System, Oscillator (400Hz, 6kHz), Headphone Jack;
Power Consumption:	75W
Dimensions:	442(W) x 362(D) x 132(H) mm (17-13/32" x 14-1/4" x 5-3/16") (including Rear Protrusions, Knobs & Legs)
Weight:	Net 12.5 kgs (27.5 lbs.) Gross 15.0 kgs (33.0 lbs.)

Specifications and appearance design subject to change without notice.
* Dolby is the trademark of Dolby Laboratories, Inc.

Fig. 8-24. Spec sheet for a cassette deck.

It isn't always possible to make direct comparisons of wow/flutter figures in spec sheets since there are so many ways of making wow/flutter measurements. Five measurement techniques that are used include wrms, an abbreviation for weighted root-mean-square; wrms (JIS); wrms (NAB); Din 45507; and IEEE. The first three measurements, all wrms, are approximately the same and use wow and flutter meters which are damped to average random peak excursions. Din 45507 and IEEE test measurements are also weighted, but these tests result in higher figures of wow and flutter since the specs consist of the highest single figure of flutter and wow obtained during the test. The Din 45507 and IEEE figures for wow and flutter are rarely used in spec sheets since they are invariably higher than the wrms measurements.

Signal-to-Noise Ratio

This spec supplies a comparison between the amount of noise supplied with the inputs shorted versus the output, using the same tape, recorded at a 0 vu level. This spec is supplied in decibels and higher numbers are better. A figure of −55 dB should be a minimum.

Frequency Reponse

This spec has little meaning (Fig. 8–25) unless it is supplied with a plus/minus deviation figure, generally plus/minus 3 dB. The nominal frequency reference point against which the deviation is measured is usually either 400 Hz or 1,000 Hz. It indicates the useful frequency range over which the deck will supply output, but it depends on a number of variables; including the kind of tape used, signal strength at time of recording, and the amount of bias. Generally, frequency response is wider for open reel than for cassette tape.

Open reel decks often supply frequency response figures at their highest speed. At 7¾-ips, using a quality tape, the response can be 30 Hz to 24 kHz, plus/minus 3 dB, although some decks claim a response as high as 28 kHz, at this speed. For a cassette deck the response at 1⅞-ips can be from 30 Hz to 20 kHz, plus/minus 3 dB, using metal tape; from 30 Hz to 18 kHz using chromium dioxide tape and from 30 Hz to 16 kHz using LH tape. For two-speed cassette decks the response is wider at the higher speed.

Distortion

Using too much bias attenuates high frequencies. But if bias is decreased, distortion goes up. In turn this means recording at a lower level, thus reducing dynamic range and decreasing the signal-to-noise ratio.

Distortion, frequency response, and signal-to-noise ratio are interdependent Any one of these parameters can be adjusted at the expense of the other two. That is why, in spec sheets, you may not find information on all three.

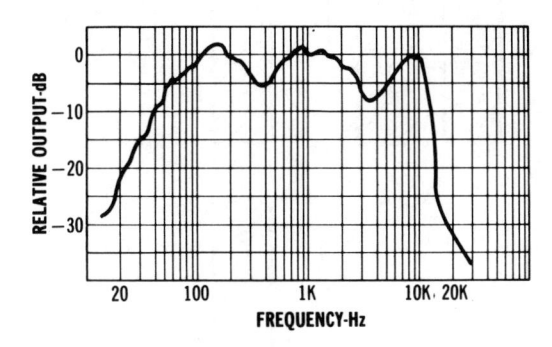

Fig. 8-25. This graph shows a frequency response from 20 Hz to 20 kHz. In this extreme example the deviation would indicate that the response is far from uniform. *(Courtesy B·I·C-Avnet)*

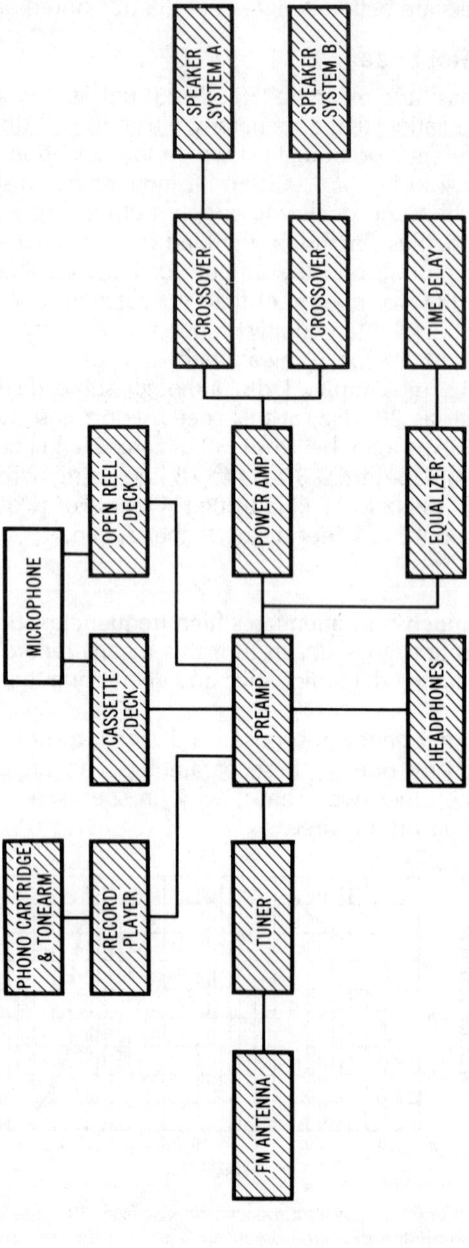
SPEAKER SYSTEM A
SPEAKER SYSTEM B
CROSSOVER
CROSSOVER
TIME DELAY
OPEN REEL DECK
POWER AMP
EQUALIZER
MICROPHONE
CASSETTE DECK
PREAMP
HEADPHONES
PHONO CARTRIDGE & TONEARM
RECORD PLAYER
TUNER
FM ANTENNA

9

Microphones

Although they work at opposite ends of the high-fidelity system, microphones are more closely related to speakers than to any other component. Like the speaker, the microphone (or mic) is a transducer, but while the speakers change electrical energy to sound energy, the microphone has an opposite approach. Its job is to convert sound energy to its electrical equivalent.

BASIC MICROPHONES

There are a number of different kinds of basic microphones: the dynamic, condenser, ribbon, electret, and ceramic. These names are for identification only and are no index to quality. An electret isn't necessarily better or worse than a dynamic. Quality is a relative term and in microphones, as in almost anything else, does not come cheaply.

The Dynamic Microphone

The most widely used of all microphones is the dynamic (Fig. 9–1). Built something like a speaker in reverse, it consists of a coil which can be made to move by sound pressure on a diaphragm to which the coil is attached. As the coil moves between the poles of a permanent magnet, it produces a voltage that corresponds to the sound.

The dynamic microphone is a miniature electrical generator, but instead of supplying power line voltages and currents, its output is made up of a current in the audible range, from somewhere below 100 Hz to somewhere in the region of 15 kHz and higher.

The dynamic microphone is a first cousin to the dynamic speaker. In some intercoms, tiny dynamic units double as speakers and microphones, being used for both purposes.

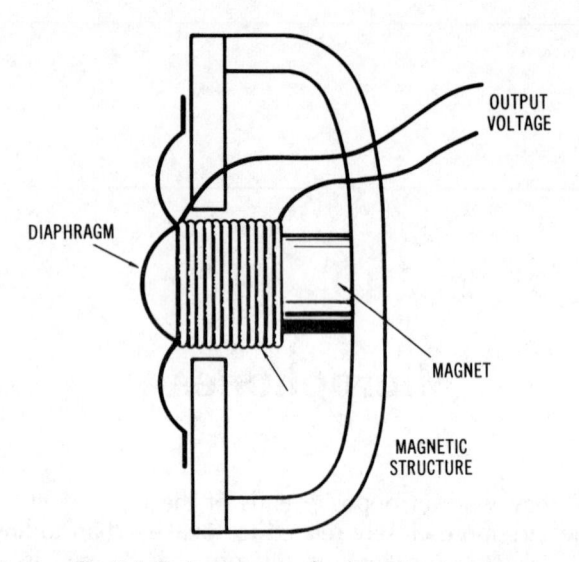

Fig. 9-1. Basic structure of a dynamic microphone.

Dynamic microphones are rugged and dependable and can supply a smooth and extended frequency response. The diaphragm, made of a nonconducting material such as plastic or Mylar*, is supported along its outer edge and can move back and forth freely and easily. The housing of the microphone is sometimes made along lines comparable to that of a vented port speaker for extending and improving bass response. Another technique, comparable to speaker design, is the double element or two-way or coaxial microphone, a dynamic type that has a tweeter and a woofer arrangement, with the sound separated by a crossover network.

Condenser Microphone

The word "condenser" has long been replaced by the word "capacitor" except in the case of microphones. A condenser microphone is basically a capacitor, a pair of metal plates separated by a nonconducting material called a dielectric. When a dc voltage is applied to the capacitor plates the unit takes an electric charge, the amount of charge depending on the area of the plates, their physical closeness (or separation), and the material of which the dielectric is made.

If the plates are separated or brought more closely together the electric charge varies, and it is on this principle that condenser microphones are made.

---
*Trade name of E.I. duPont de Nemours and Co., Inc.

In the condenser microphone one of the metal plates is kept in a fixed position. The other plate, a diaphragm, is free to move and since it moves in accordance with the sound pressure applied, the voltage across the capacitor plates varies at an audio rate.

To be able to reproduce transient tones the diaphragm material must be extremely light. It is usually some kind of plastic material, but since plastic is not an electrical conductor, the diaphragm is given a very fine coating of a metal, sometimes gold; gold because it is extremely stable, is highly conductive, and is unaffected by weather conditions. In this condenser microphone we have a metal backplate, fixed in position with the plastic acting as the dielectric and the metallic plating on the dielectric functioning as the other plate of the condenser.

As they are, these components form a capacitor, but one that is uncharged. A charge, in the form of a dc voltage ranging between 50 and 200 volts, is now put across the two plates. This voltage, known as a polarizing potential, makes one of the capacitor plates negative, the other positive. Because of this polarizing voltage the two plates are attracted to each other, but do not touch.

In the presence of a sound pressure, such as that caused by the voice or musical instruments, the diaphragm will vibrate. As it vibrates, so does the ability of the capacitor to take a charge. When the diaphragm moves toward the fixed back plate the voltage across the two plates will increase. Conversely, when the diaphragm moves away from the back plate the voltage across the plates decreases. This changing voltage is at an audio rate, corresponding to the varying audio pressures put on the diaphragm.

These voltage changes are extremely small and cannot be used directly. Instead they are brought into a solid-state amplifier often built directly into the microphone case.

Electret Microphone

The need for an external polarizing voltage complicates the use of the condenser microphone somewhat. One way of eliminating that polarizing voltage is through the use of the electret microphone, a type of condenser microphone (Fig. 9–2).

The charge on a capacitor doesn't drop to zero when an applied dc potential is removed. Depending on the way the capacitor is constructed the charge can remain for a long time. An electret is a special type of capacitor specifically designed to retain its charge practically indefinitely. In the electret microphone the unit receives its charge at the time the microphone is manufactured. However, like the regular condenser microphone the output voltage is very small and so a solid-state amplifier is included inside the microphone case.

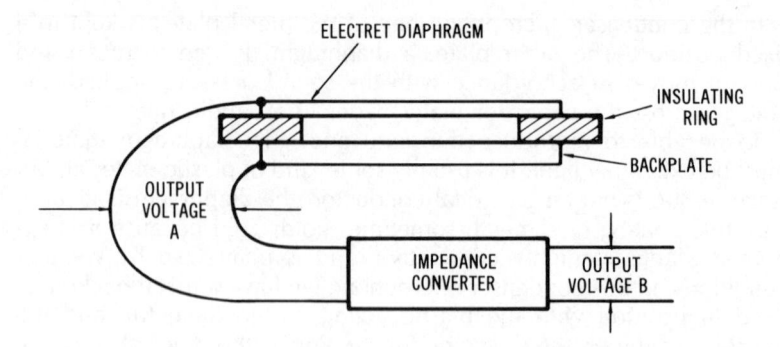

Fig. 9-2. Basic structure of an electret microphone. The output impedance of the microphone is extremely high and so an impedance converter is used to bring that impedance down to somewhere between 50 and 200 ohms.

Ribbon Microphone

The ribbon microphone is similar in concept to the dynamic microphone. It consists of a ribbon made of a metal such as duralumin stretched between the poles of a permanent magnet. The ribbon—the diaphragm of the microphone—moves when subjected to sound pressure. In moving it cuts through the magnetic flux lines supplied by the permanent magnet and so an audio voltage is induced across the ribbon. The voltage generated across the metallic ribbon is an audio ac potential.

Like the condenser and electret microphones, the ribbon has very small signal output and so may contain a step-up transformer built into the case. Because of the very light weight of the ribbon diaphragm the microphone is very sensitive to transient signals.

Both the ribbon and dynamic microphones make use of a conductive element moving in a magnetic field. In the case of the dynamic microphone the conductive element consists of a coil of wire; for the ribbon microphone the element is a single length. The coil of the dynamic microphone is just a convenient way of getting a long length of wire to fit between the poles of the permanent magnet.

The Ceramic Microphone

The ceramic microphone belongs to the family of piezoelectric transducers. Piezoelectric transducers (Fig. 9–3) can be made of crystalline materials such as barium titanate. It is this latter substance that is used in the ceramic microphone.

When a piezoelectric material is flexed it develops a voltage across its faces. This voltage can be a varying one, directly proportional to the amount of pressure. In this way the changing pressure supplied by

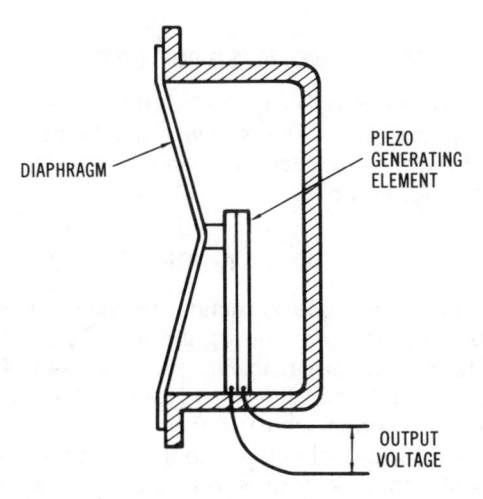

Fig. 9-3. Piezoelectric microphone.

a diaphragm and applied to the crystal can produce a correspondingly changing audio voltage.

The ceramic (or crystal) microphone is about the poorest in terms of quality of sound reproduction. The ceramic microphone does have one advantage: it has high output. This means you can connect such a microphone to a low power amplifier and still get enough amplification to drive a speaker. That is why such microphones are popular with manufacturers who are willing to supply a microphone, amplifier, and speaker at what may seem to be bargain prices. However, in areas of high heat and humidity the crystal microphone may declare a holiday and decline to have any output at all.

MICROPHONE RUGGEDNESS

Of all the components used in a high-fidelity system, only a few require handling. The phono cartridge is one (although cartridge handling is being diminished through automatic operation) and the microphone is another. This means that the microphone must be rugged, not only for indoor, but for outdoor use as well. The microphone must be relatively insensitive to temperature and you shouldn't need to worry about using it in freezing weather or on the hottest summer day. And it should be fairly impervious to humidity.

Shock mounted dynamic microphones should be able to tolerate the rough usage often encountered on stage and during set-up handling. Some microphones feature a built-in filter to control wind noises and verbal popping.

MICROPHONE SENSITIVITY

The sensitivity of a microphone is its ability to respond to sound. The changes in air pressure due to sound reach the microphone from all directions: the front, back, and the sides, as well as from above and from below the microphone.

POLAR PATTERNS

Not all microphones respond to sounds arriving from all directions. A polar pattern is a graph for showing the directional response of a microphone. The graph, shown in Fig. 9-4, consists of a series of concentric circles, one within the other. Each of the circles is identified by a number, starting with 0 on the outermost, then −5 dB, −10 dB, and so on. The circle closest to the center is generally −20 dB. The circles are divided by three diameters cutting the outer circumference into eight equal arcs each having a value of 45°. A small mark at the center of the circles represents the microphone.

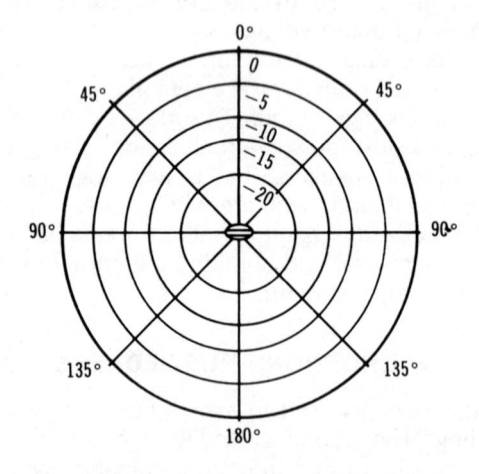

Fig. 9-4. Basic polar diagram. *(Courtesy AKG Acoustics, Inc.)*

Plotting the Polar Pattern

The polar pattern for a microphone is plotted in an anechoic chamber, a completely enclosed room whose walls, floor, and ceiling are covered with sound absorbent materials. A sound generator, set to some desired audio frequency, possibly 1 kHz, is kept in a fixed position while the microphone is moved in a complete circle around it. The distance between the two, the sound source and the microphone, is generally 1 meter or 3.281 feet. The output voltage of the microphone is then plotted on the polar graph.

Omnidirectional Pattern

Some microphones have a 360° sensitivity (Fig. 9–5). This means they can pick up sounds from all directions equally well. Omnidirectional microphones can be any of the types previously mentioned: dynamic, electret, condenser, etc. Omnidirectional microphones designed for hand-held use usually have attenuated bass response below 150 Hz, while those intended for recording and instrumentation use have uniform response throughout the audio range.

Cardioid Pattern

A cardioid microphone is designed to pick up sound entering the front and sides of the microphone while discriminating against sound coming from the rear (Fig. 9–6). Cardioid microphones exhibit a rising bass response as the unit is placed closer than two feet from the sound source.

Cardioid microphones using a single diaphragm lose volume level and high frequency response at 90° from the axis of the microphone. Special two-way or coaxial microphones using separate diaphragms for low and high frequencies are available which have slightly reduced volume level at the side of the microphone but high frequency response remains normal as compared to on-axis response. Two-way microphones are especially suitable for group recording and podium use.

Cardioid microphones usually have attenuated bass response below 150 Hz to minimize handling noise. Some of this type of microphone have a switch in the handle to reduce bass response and to control acoustic feedback.

Cardioid, or heart-shaped response patterns, are superior for controlling acoustic feedback, crowd noise, and other unwanted sounds. This is often referred to as front-to-back discrimination. In practical working terms this means the ability of the microphone to function at higher volume levels without experiencing amplified sound feedback and crowd noise problems.

Cardioid Versus Omni

Just as no two people really hear in the same way, various kinds of microphones can be made to have different *listening* characteristics. An omnidirectional microphone hears well from all directions, while the cardioid pays more attention to sound coming at it from one direction.

Whether you use a cardioid or omnidirectional depends on what you want to do. You can use the cardioid to emphasize pickup from

| 125 Hz | 250 Hz | 500 Hz |

| 1000 Hz | 2000 Hz | 4000 Hz |

| 8000 Hz | 16000 Hz |

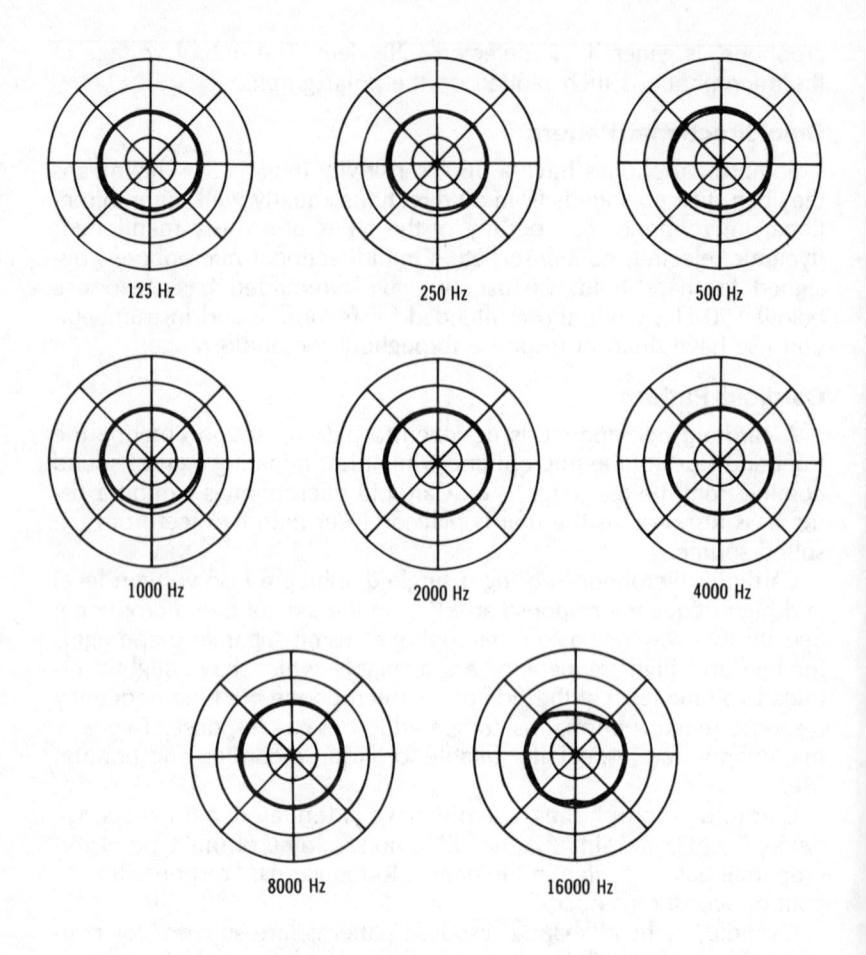

Fig. 9-5. Polar patterns of an omni microphone at different frequencies.
(Courtesy AKG Acoustics, Inc.)

one or a few closely grouped instruments, the omnidirectional from all of them. The omni, though, can pick up background noise while the cardioid discriminates against it. The cardioid, then, is a more sound selective type of microphone.

Supercardioid

There are special versions of the cardioid microphone, called the supercardioid, designed to provide maximum rear cancellation at

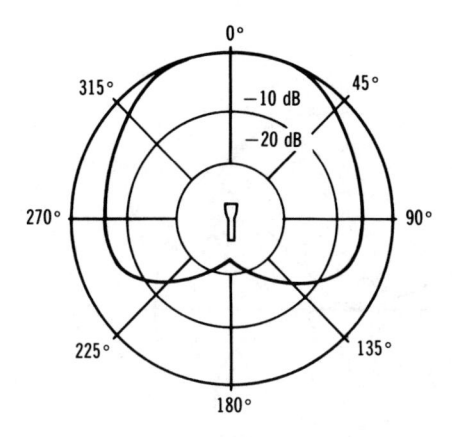

Fig. 9-6. The heart-shaped response of the cardioid microphone plotted on a polar diagram.

150° from either side off axis instead of 180° as in the standard cardioid microphone. The supercardioid is designed for sound reinforcement applications where feedback suppression is of greater concern than audience pickup from the rear of the microphone.

Figure 8 or Bidirectional

Bidirectional microphones are usually of the ribbon or condenser types. These microphones use two diaphragms with out of phase connections for pattern selection (Fig. 9–7). These microphones are very useful in separating close sound sources in recording so that better control can be obtained with instrument sections of an orchestra.

Differential or Noise Canceling

The differential or noise canceling microphone is a special purpose type. This microphone works on the principle that if the sound source, such as a person's mouth, is very close (just a few inches from the front of the microphone diaphragm), the higher sound pressure level on the front of the diaphragm will produce sound—while equal pressure on both sides of the diaphragm will cancel. This type of microphone is most useful for speech communications in noisy environments such as factories, athletic fields, stores, and recording studio control rooms.

Directional or Shotgun Microphones

Shotgun microphones provide concentrated sound pickup only over a narrow acceptance angle. These microphones are usually a

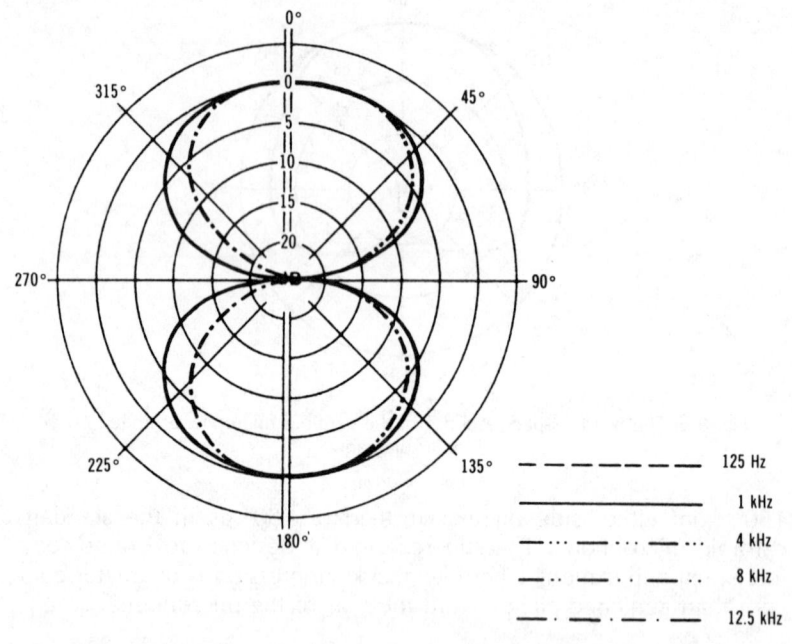

Fig. 9-7. Polar response of a bidirectional microphone. Instead of separate graphs, as shown in Fig. 9-4, the response at various frequencies is plotted on a single polar graph. *(Courtesy AKG Acoustics, Inc.)*

combination of interference tube and cardioid designs designed to reduce sound pickup from the sides and the rear. They find application in the broadcasting of sporting events, question and answer microphones at press conferences, and boom use in film and television studios. The angle of sound acceptance varies with the length of the microphone tube.

PROXIMITY EFFECT

The frequency response of certain types of microphones, such as the cardioid and bidirectional, can change—actually become distorted—when used too close to the mouth. The mouth is capable of producing somewhat explosive spherical sound waves, but as they travel farther from the lips, they tend to flatten and become planar. Hence, the distortion producing effect diminishes as the microphone is moved away. Known as proximity effect, it can produce exaggerated bass response, popping, or booming. If the microphone is held about one inch from the mouth it can result in a bass boost of as much as 16 dB.

Proximity effect is closely related to microphone design. In some microphones proximity effect is sought deliberately. Some performers with a knowledge of proximity effect use this information as part of their recording technique. Most of the bass boost due to proximity effect is below 100 Hz.

MICROPHONE SPECIFICATIONS

Microphones are classified according to type, such as dynamic, ribbon, condenser, electret, with a subdivision based on pattern, such as omni, cardioid, etc. And, like all other components in a hi-fi system, microphones can also be selected based on their electronic characteristics. These include factors such as impedance, frequency response, and output level.

Impedance

Impedance is total opposition to current flow and, like resistance, is measured in ohms. In connection with microphones impedance is understood to mean the *output* impedance of the unit. Impedance matching means that components which are to be connected should have impedance values which are close to each other. Thus, the output impedance of a microphone should be fairly similar to that of the connecting cable. The impedance of that cable, at its other end, should be somewhat the same as the input of the preamplifier to which it is connected. This concept is similar to that in which the output impedance of a power amplifier—usually about 8 ohms—matches the input impedance of the speaker to which it is connected, also about 8 ohms. Sometimes impedances are deliberately mismatched, but when they are matched the purpose is to effect a maximum transfer of signal energy and to minimize signal reflections along any connecting wires or cables.

As a generalization, impedance can be specified simply in terms of "high" or "low," without specific ohmic values attached. High impedance means a high ohmic value; low impedance a low ohmic value. Low impedance indicates small opposition to current flow; high impedance large opposition; consequently low impedance is associated with larger flows of current than with high impedance.

Basically, any microphone can be manufactured as high or low impedance, but low impedance is generally associated with dynamic microphones, high impedance with electret or condenser microphones. Many manufacturers supply combination high/low impedance microphones with connections for user option at the termination end of the connecting cable.

Most consumer tape decks feature unbalanced low impedance microphone inputs specified at 150 to 1,000 ohms. Any low impedance microphone will work properly with these decks as long as the impedance of the input of the tape deck is higher than that of the microphone. Exact impedance matching isn't important in this instance. In some instances it may be necessary to connect two microphones to one input. In such a case, the microphones should be wired in series so that one microphone's internal impedance will not load the other microphone, causing a loss of quality and level.

For high impedance microphones the useful distance between the microphone and the amplifier or tape deck to which it delivers its signal is 20 feet. Beyond this length, cable capacitance causes loss of high frequency response and susceptibility to hum and radio frequency interference pickup is increased. Low impedance microphones can use cable lengths of up to 1,000 feet without interference, hum pickup, or loss of high frequencies.

Frequency Response

Many microphones limit their response to the range between 50 Hz and 15 kHz, but you will also find some that cover 20 Hz to 20 kHz. Like an amplifier, the frequency response should indicate the deviation in decibels with the variation ranging from as little as plus/minus 2 dB to as much as plus/minus 6 dB. See Fig. 9–8.

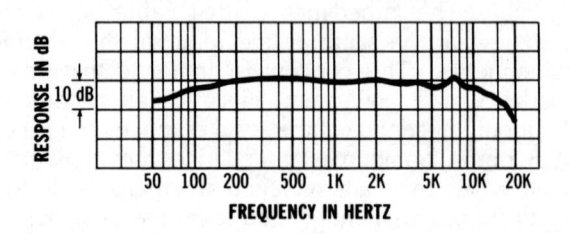

Fig. 9-8. Frequency response of a microphone. (Courtesy Audio-Technica U.S. Inc.)

If you plan on using a microphone for voice only, then the problem of frequency response is somewhat simplified since the frequency range of speech, or singing, is limited. For instrumental music, a wider response is desirable. Some microphones, like speakers, tend to favor a few choice frequencies, and at such frequencies the electrical output may rise sharply. At the same time the microphone may dislike certain tones and for these frequencies the output may drop sharply. The result is an uneven response and the end result is artificial sound. The frequency response of the microphone should be as free of resonances or antiresonances as possible.

The general concept of frequency response for components such as preamplifiers and power amplifiers is that it should be as wide and as flat as possible, that is, having minimum deviation over the entire audio range. However, there is very little music written for frequencies below 50 Hz. A microphone that is responsive to frequencies at 60 Hz or lower is also capable of picking up hum. At the high frequency end not too many people can hear well above 15 kHz. So while a range of 50 Hz to 15 kHz couldn't be called ideal, it is practical. Sometimes microphones are deliberately peaked by the manufacturer to supply boost at a particular frequency, possibly at the treble end to produce *brilliance*.

PHASE

Sound waves or electrical waves can be in step, a condition known as in-phase, in which case they reinforce each other. The other extreme is one in which sound or electrical waves oppose each other. For the most part though, we have a variety of in- and out-of-phase conditions which are partial, and so we get varying degrees of reinforcement or cancellation.

Microphones produce an electrical output corresponding to the sound input. Consequently, if two or more microphones are used, the total output should be the sum of the outputs supplied by the individual microphones. When using a pair of microphones the total output should be about 3 dB higher than when using just a single microphone.

You can assume a pair of microphones to be in phase if you use two supplied by a single manufacturer and follow directions on connections. However, there is no standardization on microphone phasing, so it is possible for a pair of microphones to be out of phase if they are models made by different manufacturers.

You can check the phasing of microphones by using a channel mixer (described in Chapter 8) that can handle two or more microphones and a vu (volume unit) meter. Using voice as a sound source, set the volume level of one microphone for a specific vu meter reading. Now do the same with the other microphone. Work back and forth between the two microphones until you are satisfied that both are producing approximately equal outputs. Make sure that the two microphones are fairly well separated. If they are equipped with an on/off switch, use only one microphone *live* at a time.

Now position both microphones adjacent to each other, and with both microphones on, talk into the area between them, as close to center as you can approximate. If they are in phase, the resulting signal should be about 3 dB higher than that produced by one microphone. You can check to make sure by switching one microphone off or disconnecting it and then talking into the area between the two. Make a note of the vu meter reading. Then switch on or reconnect the unused microphone so you have both microphones working. Talk and then note the vu meter reading. If the microphones are in phase, there should be a 3 dB increase. If, during this test, you get a lower reading when the two microphones are used, then the microphones are out of phase. Don't expect to get exactly 3 dB. This number is mentioned just to give you an approximate idea of what may happen.

It's difficult to take a reading when a meter pointer is moving. Try to use the same voice level at all times and try to aim your voice at the same in-between spot. Your lips should be the same distance from the microphones.

WORKING DISTANCE

The distance between the diaphragm of a microphone and the sound source, whether it is voice or a musical instrument, is called the working distance (Fig. 9–9). For instruments, working distance is usually fixed with the microphone held in position by a stand. For a vocalist the working distance can vary. Some performers practically swallow the microphone, others move it toward or away from the mouth. Some point the microphone directly at the lips; others talk or sing above it.

The maximum sound pressure level (spl) is obtained when the microphone is as close to the sound as possible. If distortion is produced as a result, the fault is generally due to overloading of the following preamplifier rather than the microphone.

The sound pressure level produced by a microphone decreases by 6 dB every time the working distance is doubled. Another way to regard 6 dB is as a 4-to-1 ratio. If, for example, a microphone is positioned one foot away from the sound source and then that distance is doubled, the pressure of the sound level will drop to one-fourth of its original value. If the distance is now increased to four feet, thus doubling the separation distance between microphone and sound source, there will be another 6 dB decrease in sound pressure level.

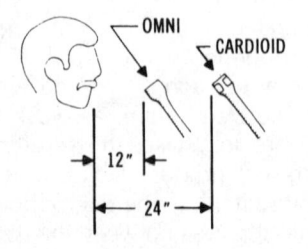

Fig. 9-9. Working distance of the omni versus that of the cardioid.
(Courtesy Electro-Voice, Inc.)

DRY VERSUS REVERBERANT SOUND

In recording there are two basic types of sound: dry and reverberant. Dry sound is that obtained directly from a sound source. Close one ear and put the other within an inch or so from a sound source and what you will hear will be direct sound, or so-called dry sound. Move away from the sound source, possibly a foot or more, and if you are doing this test in a closed room, you will hear reverberant sound in addition to the dry sound. Reverberant sound, also known as wet sound, is sound that is reflected from the floor, ceiling, and the walls, and any objects in the room, including people.

The decrease of spl by 6 dB every time the working distance is doubled applies to dry sound only and doesn't include reverberant sound. It is much more applicable to out-of-doors recording where there is little or no sound reverberation. The amount by which spl will drop as a microphone is moved away from the sound source is greater for a room with little reverberation—a so-called "dead" room, than it is for a "live" room—a room whose surfaces reflect sound quite well.

MICROPHONE SOUND RESPONSE

Microphones are sometimes referred to as *electronic ears,* but there is actually no comparison. We listen in stereo; a single microphone is always monophonic. As a test, try listening with one ear covered and you will not only find it more difficult to distinguish vowels and consonants, but the sound will lose a certain amount of realism.

Like the ears, microphones are designed to pick up sound, but unlike the ears, microphones are nondiscriminatory. The ears, with an assist from the brain, can shut out sounds they don't want to hear. Microphones will "hear" all sounds presented to their diaphragms and will convert those sounds into an electrical voltage equivalent.

THE LAVALIER MICROPHONE

Microphones can be positioned in a number of different ways. They can be hand held, suspended, or mounted on a boom or cross bar. In some instances the microphone can be fastened to an article of clothing using a clip fastener or a tie tack. This latter arrangement, including suspending the microphone around the neck similar to a necklace, is referred to as a lavalier microphone.

The electrical and electronic characteristics of a microphone are not determined by its size, shape, and color finish. The lavalier microphone, although a miniature type, can be any of the types previously described and can have the specs exhibited by its larger counterparts. It can be a dynamic or electret type.

The lavalier is an omni. It is very useful in situations where sound amplification is needed and where the user needs freedom to move around. Usually the lavalier is designed to minimize noise pickup from clothing or noise due to connecting cable flexing.

USING MICROPHONES

There are no absolute rules for the positioning of microphones but there are some general guidelines that can be used as starting points. Microphone positioning depends on whether you are recording a single vocalist or a group, a single instrument or a group, or some combination of voices and instruments. Positioning of microphones also involves other factors such as noise, the need to emphasize one or more instruments, and the possibility of sound feedback. Positioning microphones is also a matter of experience, so experimentation is required. Fortunately, sound recorded on tape supplies instant results so it is possible to move microphones until desired sound results are obtained.

Conference Recording

A single microphone can be used to record a group discussion since the effect of left-right sound displacement isn't needed. This type of recording can be done by having the group seated around a table and, if possible, equidistant from an omnidirectional-type microphone. However, two or more omnis may be needed if the room is highly reflective, with the microphones placed closer to the speakers. If the background noise level is high, two or more cardioid microphones can be used.

Single Vocalist

Hand-held cardioids are best for single vocalists because the close-to-the-mouth use of the microphone will provide a high signal-to-ambient noise capability, providing better separation and control of the vocalist's sound and that of any accompanying instruments.

Vocalists may sometimes indicate a preference for a particular type of microphone, having learned from experience how to use that microphone for the production of desired sound effects. There is no single microphone which can be regarded as a universal type, that is, suitable for every vocalist.

As a general rule, a vocalist should try singing to the side of the microphone or possibly over the top. When running a test tape listen for breath sounds, popping sounds, and sibilants. These interfere with the clarity of the lyrics. Use a windscreen to prevent breath pop, sometimes noticeable with words beginning with the letters *p* or *t*.

Rock Group

The high ambient noise environment will require cardioid microphones with a close talk pickup pattern. Place the microphones as close to the instruments as possible for best separation and control.

Rock groups are characterized by very high sound output levels, sometimes to the point of listening pain. The large output can drive the following amplifier into distortion so microphone attenuator pads may be needed to prevent overload of the preamp or tape deck. An attenuator pad is simply a network of resistors, but is designed so as not to interfere with impedance matching between the microphone and the input to the amplifier or tape deck.

Flute

When played softly the flute can produce a tone that is almost a pure sine wave. For recording this instrument, try a two-way microphone facing the instrument about halfway along its length and spaced about a foot away. If you move the microphone up along the instrument closer to the flautist's mouth you may find you are picking up breath sound; if too far down the instrument you may be missing the sharp attack character of some of the tones. However, the positioning of the microphone suggested here is simply a starting point and the final location depends on the kind of sound you want to get from the instrument.

Reeds and Brass

Reeds include instruments such as the clarinet, sax, and oboe while brass includes the trumpet, trombone, French horn, and tuba. These may be recorded in combination but sometimes are required to produce solo passages, with or without instrumental accompaniment. In the case of reeds, position the microphone near the bell. For a brass group ensemble use cardioids positioned several feet away. As you move the microphone closer in to the instruments you will get more specific detail; farther away gives more blending.

A cardioid has greatest sensitivity along the front line of its axis; less along the sides, 90° off front. You can use this characteristic to pick up maximum sound from instruments facing the microphone; less from those at right angles to the microphone.

Double or String Bass

These instruments supply the low frequency tones giving the rhythm for a group of players. There are various ways of getting the background tones: one technique is to use a lavalier up against the *f* hole. Put some foam around the microphone to keep it from picking up mechanical vibrations. You can also use a contact microphone for this purpose. Alternatively, try a two-way microphone, about a few inches from the f hole on the upper side of the bridge, using a suitable microphone stand.

Electric Bass

For the electric or Fender bass use a two-way cardioid positioned right in front of the bass amplifiers with the microphones facing the speakers directly, that is, on axis. A regular cardioid can be used instead of the two-way type. The output of the instrument, a strong bass sound, can overload the following amplifier. If you note that the bass response is unclear or seems muffled, use an attenuator pad between the microphone and the following amplifier or else try moving the microphone away from the instrument. Since the cardioid will also pick up sound from the sides you may wish to move the instrumentalist farther away from the other performers to minimize their sound.

Guitar

There are two basic types of guitar: the nonamplified or acoustic, and the electric guitar. The acoustic guitar supplies a softer sound and its sound output can sometimes be lost in the presence of other, stronger output instruments.

There are several ways of recording the acoustic guitar. One technique is to put an omni about a foot away from the opening of the instrument. If you move the microphone in closer in an effort to get more of the acoustic guitar effect you may find you are picking up the noise of fingers moving along the strings. Some recordists prefer this as adding realism; others feel it detracts from the sound.

For the acoustic guitar you may want to try contact microphones made especially for this purpose. A condenser microphone is excellent for use with this instrument. Cardioids are preferred to omnis to give the instrument a chance to compete against background sound, but if the microphone is brought too close in toward the instrument you may find that the bass response is too heavy.

For micing the electric guitar follow the same procedure used with the electric bass. Position the microphone in front of and facing the instrument's amplifier, moving the microphone closer in or farther away to get the kind of sound you want.

Drums

Drums are seldom used alone, but rather in sets including a snare drum, splash cymbals, tom tom, hi-hat, and a bass drum. If you want to emphasize the drum beat mount a cardioid facing the skin. The bass drum may be damped or muffled to produce special sound effects. To get the full effect of the drum set without emphasizing any particular one of the instruments mount a condenser omni pointing downward from a boom.

Piano

The piano is a percussive instrument, producing sharp transients so it is essential for the microphone to have good transient response. If the piano is part of a rock, country, and western group there will be a heavy number of transients. The microphone must reproduce these as played; if not, the sound will lose some of its original quality. To determine if a microphone is reproducing transients, listen to a percussive instrument and then listen to the way its sound is reproduced by the microphone. If you can notice the difference, the microphone has poor transient response. Transients determine the character of sound so it is best to get a microphone that interferes as little as possible with the sound's personality.

When recording a grand, whether baby or full size, raise the lid. For an upright, such as a spinet, remove the lid, or, if that isn't practical or possible, raise it. For an upright piano, mount a cardioid on a boom and have it facing down into the piano, about one-third the way removed from the treble end. Experiment with the positioning of the microphone, but it possibly will be somewhere between 6-inches and 12-inches above the opening of the piano. Bringing the microphone too close to the strings may add proximity effect. You will find that the distance of the microphone from the strings will be important in determining the kind of piano sound you want to get, ranging from a sharp attack to overall resonance. You can use a single cardioid or you may want to try two: one close up for distinctive string sound; the other farther away for sound blending.

Bowed String Instruments

Bowed string instruments such as the violin and viola depend on the body of the instrument to add fullness and harmonics to the string sound. A cardioid, boom mounted, and pointing down at the instrument opening, will supply rich sound. If the microphone is brought too close to the strings there may be some raspy pickup as the bow moves across the strings.

Sometimes three microphones are used in recording; quite often more. In a basic three-microphone setup, two of the microphones are used to obtain left/right channel sound. These are cardioids and are mounted so as to supply minimum audience pickup. A third microphone, an omni, can be set some distance back to pick up reverberant sound, adding fullness to the recorded material. In some instances, audience noise is wanted to give the impression of a live performance.

CABLES

Sometimes it's necessary to have a long cable, just a pair of shielded connecting wires, between the microphone and an amplifier or tape deck. In that case, a low impedance microphone is required. Still, long cables do have a tendency to steal treble tones.

Like microphones, cables used for connecting microphones to a following preamplifier can be low impedance or high. A high impedance cable, also called unbalanced line (see Fig. 9–10) consists of a wire conductor passing lengthwise through the center of the cable. This wire, sometimes called the "hot" lead, is surrounded by an insulating, nonconducting, flexible material surrounded on its outside by flexible metal braid.

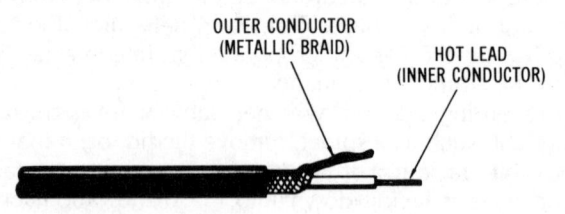

Fig. 9-10. High impedance unbalanced microphone cable.

The metal braid has two functions. Because it is metal it works as a conductor, consequently unbalanced cable is a two-wire line. The signal from the microphone is applied to the center wire and the outer shield braid. The braid also works as a shield to keep out hum or other interfering electrical noises. The braid, sometimes called the "cold" lead, isn't as effective as a separate shield.

A balanced line, as indicated in Fig. 9–11, consists of two separate inner conductors along the full length of the cable and insulated from each other. As in the case of unbalanced cable, the conductors are surrounded by flexible insulating material which is then covered with flexible wire braid. Unlike unbalanced line the braid isn't part of the signal path but is simply used to shield the two conductors from any outside signal interference.

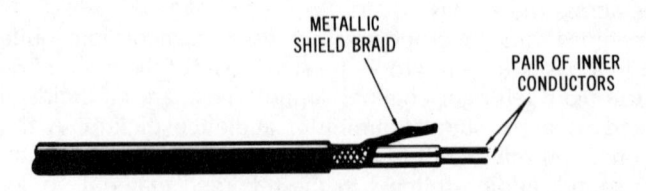

Fig. 9-11. Low impedance balanced microphone cable. The shield braid surrounds both inner conductors.

Glossary

This glossary is not a dictionary and so the words aren't rigorously defined, but are supplied with general descriptions. Not all the terms are contained in the text for sometimes they weren't sufficiently relevant. The language of high fidelity, however, has exploded to such an extent, and the abbreviations so numerous, that a glossary is essential.

A

A-B test Method of evaluating hi-fi components by switching them on and off sequentially.

ac Abbreviation for alternating current. A current that periodically reverses its direction of flow. Ac voltage is also used to describe an alternating voltage.

acoustic feedback The feedback of sound, possibly from a speaker to the sound source, causing ringing, howling. An unwanted sound interaction between the output and the input with the fed-back audio wholly or partially in phase with the input.

acoustic suspension speaker An enclosed speaker system that uses air inside the box to supply restoring force to the cone.

acoustics The study and science of sound. The word is sometimes used to refer to the acoustic characteristics of a listening room, hall, church, or auditorium.

a/d Analog to digital converter.

af Abbreviation for audio frequency. The range of tones whose frequency is considered to extend from 20 Hz to 20 kHz.

afc Abbreviation for automatic frequency control. A circuit used to compensate for frequency drift in tuning or for slight tuning errors.

alternate channel selectivity *See* selectivity.

am Abbreviation for amplitude modulation. A method of imposing an audio signal on a much higher frequency carrier wave by varying its amplitude at an audio rate.

am band A broadcast band extending from 535 kHz to 1605 kHz. A band using amplitude modulation.

am/fm indicator Illuminated callout used to identify the type of broadcast being received, whether am or fm.

am/fm/mpx A tuner or receiver capable of picking up am broadcasts and fm broadcasts in mono and stereo.

am/fm selector Control on the front panel of a tuner or receiver for choosing either am or fm reception.

am suppression This has nothing to do with am broadcasting. Electrical noise signals have the characteristics of am signals. Unless suppressed they can cause a crackling sound. Also known as am rejection.

ambience The acoustical environment of a listening room or concert hall. Ambience is the effect of reverberant sound.

ambient sound External or environmental sound.

ampere Basic unit of current, used for dc and ac. Submultiples are the milliampere, or thousandth of an ampere, and the microampere, or millionth of an ampere.

amplifier A circuit or component used for strengthening an audio signal. See also preamp and power amplifier.

amplitude modulation See am.

Analog Bass Computer Speaker arrangement that anticipates cone movement by reading the output of the power amplifier, controlling cone excursion accordingly.

AND gate A logic circuit consisting of two or more gates connected in series. A circuit in which both gates must be open for signal passage.

anechoic chamber A specially designed room lined with sound absorbing materials. A room that is acoustically "dead" used for testing microphones and speakers. A soundproof room.

ANL Automatic noise limiter.

antenna One or more metallic elements or conductors for receiving broadcast signals. Abbreviated as ant.

antenna element One or more sections of an antenna.

antenna gain Comparison between the amount of signal picked up by an antenna and an antenna used as a measuring standard.

antinode Point at which standing sound waves in a listening room, concert hall, or auditorium reinforce.

antiskating device A mechanism that is part of a phono tonearm that puts a small outward force on the arm to compensate for the inward thrust of the arm when it is in motion. Sometimes called a bias compensator.

antistatic fluid A liquid used for discharging or minimizing the electrostatic force on the faces of a phono record.

ARLL Audible Relative Loudness Level. A method for the measurement of phono rumble.

attenuation Signal weakening or cut. A reduction of signal strength.

attenuator A circuit, usually resistive, for weakening a signal or a band of signal frequencies.

audibility threshold The weakest sound detectable by the human ear. It is approximately 0.0002 microbar measured at 1 kHz.

audio frequency Abbreviated as af. Generally considered to extend from 20 Hz to 20 kHz.

auto blend A circuit that automatically monitors a signal, reducing hiss accompanying weak signals.

automatic eject Also called auto eject. A feature of cassette tape decks in which the cassette is ejected when the tape reaches its end. This is not the same as end of tape play which can occur at any point along the tape.

automatic frequency control Circuit in the fm section of a tuner or receiver which locks in on a selected station. It helps eliminate station drift automatically. Abbreviated as afc.

automatic level control A circuit for keeping the level of a signal within previously determined limits. Abbreviated as alc.

automatic record changer A record player which can be loaded with two or more phono records for automatic sequential play.

automatic replay Also called auto replay. A cassette deck which will automatically replay a cassette.

automatic reverse Also called auto reverse. A cassette deck which will reverse the direction of a tape for replay without the need for removing the cassette.

automatic shutoff Also called auto shutoff. A cassette deck that will stop automatically upon reaching the end of a cassette tape. This term is also used to describe the action of an electronic current sensor in an amplifier that turns the amplifier off if current flow is excessive.

aux input An abbreviation for auxiliary input. The input on an amplifier to which extra signal sources can be connected. Sometimes called an aux terminal.

azimuth Arrangement of a tape head so its gap is exactly at right angles to the edge of the tape.

azimuth alignment The heads in a cassette deck must make a right angle with the longitudinal axis of the tape. Azimuth alignment is the adjustment of heads so they form a right angle with the tape.

B

baffle A flat surface for mounting one or more speakers. A flat surface for supporting a speaker system. A board for separating the rear-produced sound waves of a speaker from front-produced waves.

balance control A variable control for adjusting the sound output level from left and right front speakers. Sometimes called channel balance.

balanced line A cable consisting of two inner conductors covered with a plastic-type insulating material and having an outside flexible metallic shield braid.

bandpass filter A circuit which will pass a selected band of frequencies while attenuating frequencies above and below the passband.

bandwidth The total frequency range of a band of frequencies, from the lowest to the highest frequency.

basket The metal supporting frame of a loudspeaker. It is used to support the surround to which the outer rim of the speaker cone is attached and also the permanent magnet assembly.

bass Pronounced as though it were spelled "base," it is the range of low frequencies, generally extending below 200 Hz.

bass driver A speaker designed for reproducing bass tones.

bass filter *See* low filter.

bass reflex speaker *See* ported speaker.

bass tones Tones generally considered to be in the range of 20 Hz to 200 Hz.

beat The interaction of two waves, usually of different frequencies.

belt drive Platter drive using a belt to connect a pulley on a turntable drive motor with the turntable platter.

biamp system High fidelity sound system using two audio amplifiers, with each designed to handle its own range of audio frequencies. A system in which separate power amplifiers are used for the woofers and midrange/tweeter drivers.

biamplification in/out switch When this switch is in its on position there is separate amplification of high and low frequency audio tones with separate connections to the woofer and midrange/tweeter; when in the off position all audio tones are distributed to all the speakers.

bias Voltage applied to a transistor to determine its class of operation. Also an ac current used in tape decks for erase and to prepare the tape for recording. Sometimes the word is used to describe the sidethrust on a tonearm.

bias compensator See antiskating device.

bias frequency The operating frequency of the bias oscillator in a tape deck. The range of bias frequencies is from 30 kHz to as much as 120 kHz. A commonly used value is 100 kHz.

bias oscillator An ac generator in a tape deck for producing a supersonic current, used as an erase current and to help in preparing the tape for recording.

bidirectional microphone See figure 8 microphone.

bilevel A receiver or a tuner having two levels of audio signal output.

binary system A numbering system that makes use of two digits only, 0 and 1.

biradial stylus See elliptical stylus.

birdies Interference noise caused by an fm station broadcast adjacent to a desired signal.

bit Abbreviation for a binary digit. The binary number 0 or 1.

block diagram A diagram using rectangles to represent circuits or various sections of a component. An arrangement of rectangles to show the movement of a signal from one section of a component to another. A method of showing the relative positions of circuits.

bobbin A form on which the voice coil of a speaker is wound.

boost An increase in the level or amplification of a single frequency or a group of frequencies.

braid Also called shield braid it is used as one of the conductors of unbalanced coaxial cable at the same time working as a shield to prevent the pickup of noise signals. *See also* coaxial cable.

bulk eraser Basically, a large coil of wire connected to the ac power line and used for erasing a complete reel of tape at one time.

C

C-number A number on a cassette that indicates its total playing time. A C-60 plays for 1 hour—30 minutes on each side. Typical C designations are C-30, C-45, C-60, C-90, and C-120.

cantilever Small bar or rod or tubing made of aluminum or beryllium for holding a magnet, coil, or bit of ferrous metal at one end with a stylus attached at the other end.

capacitance The amount of electrical charge a capacitor can hold. An electrical part whose capacitance is measured in farads, a unit that is too large for practical use. Submultiples are the microfarad and the picofarad. A microfarad is a millionth of a farad. A picofarad is a millionth of a microfarad.

capacitive reactance The opposition of a capacitor to the flow of a varying or an alternating current. Capacitive reactance is measured in ohms.

capacitor An electrical or radio part consisting basically of two metal plates separated by an insulating material, the dielectric. Capable of taking and storing an electric charge. Formerly known as a condenser. The word condenser is now used only in connection with microphones.

capstan A tape drive spindle kept in close contact with a pressure roller, with tape between the two. A metal shaft whose rotation turns a pressure roller.

capture ratio The ability of a tuner or receiver to suppress a weak signal while responding to a stronger signal. Measured in decibels.

cardioid microphone A sound-to-electrical-energy transducer having a heart-shaped response. Maximum response is from the front, on axis, with decreasing response at the sides, and minimum response to sounds from the rear.

carrier A radio frequency wave which can be modulated to carry a lower frequency audio wave. A radio wave having constant amplitude and frequency when unmodulated.

cartridge When used in connection with magnetic tape refers to endless loop eight-track tape. The same word is used in connection with record players in referring to a phono cartridge.

cassette Tape made for use with a cassette player or player/recorder. Cassettes have two pairs of tracks, one pair in one direction, the other pair in the reverse direction. A plastic container with two spools or reels containing preloaded tape.

CD Abbreviation for compact disc. A phono record which is pulse code modulated.

ceramic microphone A piezoelectric transducer using barium titanate as the energy converter. Varying pressure on this crystalline material produces voltage changes in proportion to the applied sound.

channel A signal or sound path. Monophonic or mono is known as single channel; stereophonic sound (stereo) as two channel. In stereo sound music from center stage toward the left is left channel sound; from center to the right is right channel sound.

channel balance Same as balance control. A control used to adjust the amount of left and right channel sound from speakers.

channel separation The separation between left and right channel sound, usually specified in decibels (dB).

chromium dioxide tape Also known as chrome or chromium tape. A tape formulation designated as CrO_2 and requiring a high bias current.

circuit breaker Device for interrupting or breaking the connection to a circuit. A protective unit that operates under conditions of excessive current flow.

circumaural headphones Headphones in which the pads completely encircle the ears, excluding outside sound.

Class A An audio amplifier that introduces very little distortion, but is inefficient. An amplifier whose transistors are biased to permit current flow at all times.

Class AB An audio amplifier having some of the operating characteristics of both Class A and Class B amplifiers.

Class B An audio amplifier that requires the use of a pair of transistors, with each transistor working alternately. Operating efficiency is about 60% to 70%.

Class C A heavily biased amplifier, highly efficient, but not suitable for audio amplification. A type of amplifier generally used in connection with the transmission of code signals.

Class D An extremely efficient audio amplifier in which the transistors work as switches.

Class G A system of amplification using two power amplifiers, one for low, the other for higher power.

clipping Cutting of the top and/or bottom of an audio waveform. Clipping can be caused by excessive signal drive to an amplifier. Clipping may sometimes be used deliberately, as in the limiter circuits of fm receivers to eliminate any amplitude changes.

coaxial cable Sometimes called coax. The cable consists of a central conductor, also called the "hot" lead, surrounded by an insulating plastic material. The insulating surround is covered with wire braid which also acts as a conductor while shielding the inner wire from electrical noise. The outer wire braid is known as the "cold" lead. Coaxial cable is sometimes used to connect an antenna to the antenna input of a tuner or receiver. Also called unbalanced cable when used to connect microphones.

coaxial microphone A transducer using separate diaphragms for low and high frequency tones. Also known as a two-way microphone.

coaxial speaker Speaker system with two drivers mounted along the same axis. This may consist of a woofer and a midrange/tweeter. Both speakers are independent units.

coercivity Amount of magnetic force required to reduce the magnetism of a metallic particle to zero.

coil Turns of wire generally wound in circular form. Also called an inductor, or an inductance.

cold lead Sometimes known as a ground lead. A wire connected to the minus or negative terminal of a speaker. The shield braid of unbalanced coaxial cable. These words are sometimes used to describe a wire that is connected to ground.

compact system Audio system in which all the components are supplied completely connected to each other and housed in a single cabinet, often including the speakers.

compatibility Interchangeability of parts or components. All cassettes are compatible and any cassette can fit into any cassette player or cassette player/recorder. Compatibility also refers to the ability of fm mono receivers or tuners to pick up stereo signals and hear them monophonically. It also means the capability of a stereo phono pickup to play mono records.

complementary symmetry Push-pull circuit with each of a pair of transistors handling the audio signal alternately.

compliance The flexibility of the surround in a dynamic speaker. The ability of the stylus assembly of a phono cartridge to move, when a force, measured in dynes, is applied.

component An audio unit, such as a receiver, amplifier, speaker, etc. Radio parts such as coils, capacitors, and resistors are also known as components.

component system A high-fidelity system in which all or most of the components are separate units and which require the use of interconnecting cables.

concentric control A pair of circularly shaped controls with one nested inside the other.

condenser *See* capacitor.

condenser microphone A transducer consisting of a fine coating of metal on a diaphragm facing a fixed metal plate. A separate power supply charges the plates. Sound pressure moves the diaphragm, varying the electric charge at an audio rate.

conductor Any substance which can conduct a current of electricity. A connecting wire, usually made of copper, is a conductor. Other conducting materials are aluminum and silver. Gases and liquids can also be conductors. Conductors vary in their resistance to the flow of current through them. Size for size, silver is a better conductor than copper, but is more expensive.

cone Driving element of a speaker. The element of a speaker that is attached to the voice coil of a dynamic speaker. The outer edge of the cone is supported by a flexible material, the surround, which, in turn, is held in place by a metal frame, a part of the basket.

conical stylus A radial or spherical stylus.

contour control An adjustable loudness control.

counterweight A weight attached to the rear of a tonearm to permit the adjustment of the phono cartridge's tracking force.

cps Abbreviation for cycle or cycles per second. Now replaced by hertz (Hz).

crossover Frequency dividing network used for routing bands of audio frequencies to various drivers. Also known as a crossover filter.

crossover distortion Distortion produced in Class B audio amplifiers during the process of switching from one output transistor to the other.

crossover frequency Frequencies at which various segments of the audio band are routed to different speakers. The borderline frequencies between low, mid, and treble tones.

crossover network A circuit consisting of resistors, capacitors, and coils for dividing the audio spectrum into bands, such as low, midrange, and treble.

crosstalk The leakage of audio signals from one channel to another. The ratio of the wanted signal compared to the unwanted signal, expressed in decibels.

cupping The curvature of magnetic tape, either inward or outward.

cut Signal weakening or attenuation. The reduction of a signal or voltage.

cutting The method used in engraving modulated grooves into a disc. Part of the system in the manufacture of phono records.

cutting stylus A stylus used for cutting grooves into a phono disc.

CX Abbreviation for compatible expansion.

D

d/a Digital to analog converter.

damping An electrical or mechanical method of reducing or minimizing motion. The reduction of resonance effects by the use of resistance. The absorption of energy from a mechanical or electrical device.

damping factor The ratio of the impedance of a speaker system to the internal resistance of an amplifier. It indicates the ability of the amplifier to damp unwanted residual movement of the cone of the speaker.

dB Abbreviation for decibel. A logarithmic unit for giving the ratio between two acoustic, electrical power, voltage, or current levels. The comparison of signal level with a fixed, arbitrary level, such as 0 dB.

dBf A measurement of relative power levels. The reference level is the femtowatt, or the millionth of a billionth of a watt. A unit used to indicate the sensitivity of a receiver.

dc Abbreviation for direct current or type of voltage.

decibel See dB.

deck Component such as a tuner or tape player not equipped with an audio power amplifier.

decoder A circuit, in an fm stereo tuner or receiver, that extracts left and right channel signals from an fm mpx broadcast.

deemphasis In fm receivers, the attenuation of treble tones to correspond to the amount of preemphasis used in broadcasting.

demodulation Sometimes used synonymously with detection. The technique for recovering a modulated signal from its carrier.

demodulator Circuit for separating an audio signal from its associated carrier or intermediate frequency.

detent A click stop for permitting the easy resetting of a control to a previous position. Often used with various kinds of level controls.

differential microphone Transducer in which sound pressures on the sides of the diaphragm tend to cancel.

digital display frequency readout Using microprocessor chips, the exact station frequency is displayed, eliminating the conventional dial pointer and scale. Digital displays are more precise, easier to read, especially at night, and are virtually troublefree since many moving parts are eliminated.

digital pll frequency synthesizer A phase lock loop synthesizer that automatically tunes in the precise frequency being broadcast and displays the station frequency in a fluorescent digital format.

digital time delay An electronic delay of an audio signal measured in milliseconds, resulting in an effect similar to that of a concert hall or auditorium.

digital time delay gain control A control that permits the selection of how much of the whole signal is to be processed or treated with time delay.

DIN (Deutsche Industrie Normenausschuss): German industrial standards.

DIN connectors The German standard for a single connector that handles the functions of two connectors, simplifies installation, and allows quick disconnection.

diode A solid state device, made of silicon or germanium, which permits the flow of current through it in one direction only.

dipole radiator speaker Speaker that radiates sound toward the front and rear with less sound toward the sides.

direct drive Phono player system in which the platter is mounted directly on a motor shaft.

direct to disc The recording of sound directly on a disc without using magnetic tape.

directional microphone A shotgun microphone with sound pickup over a narrow angle.

director An antenna element positioned in front of the active or signal receiving element.

dispersion The way in which sound from a speaker is radiated into a listening area.

distortion Sound which is changed in some way by a component such as a tuner, receiver, amplifier, or speaker system and which is not an exact replica of the original. There are various kinds of distortion. *See also* intermodulation distortion (im); total harmonic distortion (thd); transient intermodulation distortion (tim).

dividing network A single part, such as a capacitor, or a group, such as coils, capacitors, and resistors, used for dividing bands of audio frequencies, separating bass, midrange, and treble tones.

Dolby® A special patented noise reduction system that decreases background hiss by about 10 dB when the source being received has been encoded in Dolby. Dolby encoded tapes have a better signal-to-noise ratio and increased dynamic range. Receivers may be equipped with Dolby decoder circuitry to make maximum use of fm broadcasts encoded with Dolby noise reduction equipment. There are various types of Dolby noise reduction arrangements including Dolby A, Dolby B, Dolby C, and Dolby HX. (Dolby is a registered trademark of Dolby Laboratories, Inc.)

dome radiator driver Device for radiating voice-coil-produced heat and for improving sound dispersion in a tweeter.

double cone speaker Dynamic speaker using a single cone made of two different materials.

drift The detuning or movement off frequency during the warmup period of a tuner or receiver. Drift may be compensated through the use of automatic frequency control (afc).

driver Used synonymously with speaker. Technically, that part of a speaker that produces air movements. So called since a moving membrane or cone pushes air, producing changes in sound pressure that are interpreted by the ear as sound.

drone speaker *See* passive radiator.

dropout Sections of magnetic tape that have lost metallic particles, producing "holes" in the reproduced sound.

dry sound Sound obtained directly from a source. Direct sound from musical instruments or the voice which has not been reflected.

dual cone A speaker that has drivers mounted concentrically, but driven by the same voice coil.

dual if Dual intermediate frequency. A receiver or tuner whose if section is equipped with a double bandpass. An if section that can handle two intermediate frequencies.

dubbing The process of transferring signals from a previously recorded tape to a blank tape.

dummy cassette A nonoperating cassette inserted by manufacturer in tape deck to protect tape unit during shipping.

duo-beta nf Arrangement using low order of negative feedback (nf) for eliminating transient intermodulation distortion and increasing frequency response.

dynamic mass Effective mass of the moving parts of a phono cartridge.

dynamic microphone A transducer using a diaphragm attached to a coil. Sound, actuating the diaphragm, moves the coil in and out of a magnetic field supplied by a permanent magnet.

dynamic range The range in decibels from the softest sound to the loudest, produced by a live sound source. In tape or phono records dynamic range extends from the top level of the noise floor to the loudest sound which can be recorded. The dynamic range of live sound is about 120 dB. For tape and records it is approximately 60 dB, and sometimes lower.

dynamic speaker A transducer using a voice coil mounted between the poles of a permanent magnet and capable of moving a cone.

dynamic test A test made under actual working conditions.

E

echo A sound that is time delayed and is heard after the original sound.

efficiency The ratio of output to input, generally used in connection with speakers. In speakers it is the ratio of audio output power to audio electrical input power. Usually expressed as a percentage, but frequently given in general terms such as "low" or "high."

effective tip mass The stylus, cantilever, and whatever is attached to it comprise the effective tip mass, abbreviated as etm.

electret microphone A condenser type microphone that works without the need for an external polarizing voltage.

electrostatic speaker A speaker in which audio voltage is used to modulate the dc potential put across two large metal plates. A speaker using a pair of large metal plates and a coated film diaphragm. A speaker that works on signal variation of an electrostatic force.

elliptical stylus A biradial stylus having a narrower profile than a conical stylus but supplying better high frequency response.

enclosure A box for housing one or more speakers. The cabinet for a speaker or a speaker system.

equalization A method for making corrections for tape and disc recordings. Playback equalization is the reverse, compensating for equalization changes made during recording. A process of emphasizing and deemphasizing bass and/or treble tones.

equalizer There are two types of equalizers, graphic and parametric. The graphic equalizer is more widely used in home hi-fi systems. Equalizers divide the audio frequency range into a number of sections, five or more. Each control allows a range of change of tone emphasis or deemphasis, generally in the order of plus/minus 12 dB.

equalizer amplifier In a preamplifier a circuit which helps produce a flat frequency response. An arrangement to compensate for the equalization used during the phono recording process.

erase current A high-frequency current passed through the erase coil in a tape deck. A current whose magnetic field restores a recorded tape to its original unrecorded condition.

ETM *See* **Effective Tip Mass.**

F

fader control Control for adjusting sound balance between front and rear speakers.

feather-touch control A microcircuit switch that activates each function rather than mechanical controls.

feedback The transfer of part of the output signal back to the input. If out of phase it is known as negative feedback; if in phase it is called positive feedback. Negative feedback reduces overall amplification and widens frequency response; positive feedback causes ringing, howling, and possibly oscillation.

ferrichrome A tape formulation combining chromium and ferric oxides.

ferrite core antenna A built-in antenna for tuners and receivers for picking up am signals. Consists of a coil of wire wound around a ferrite core. It has good signal pickup sensitivity, is directional, and is compact. It is useful where an outdoor antenna installation isn't practical or possible.

ferrofluid A thick liquid consisting of a colloidal suspension of magnetite used for centering voice coils and as an aid in heat dissipation.

FET Field effect transistor. A transistor consisting of metal oxides and used for voltage amplification. It is used in hi-fi equipment because of its good linearity and stable impedance. A transistor that amplifies voltage, not current. The full name of this transistor is dual-gate MOSFET (metal-oxide semiconductor field effect transistor). A transistor having high sensitivity and high resistance to overloading. Sometimes used in the front end tuning section of a tuner or receiver.

FF In a tape deck an abbreviation for fast foward. Sometimes indicated by a double arrow pointing to the right.

field effect transistor See FET.

figure-8 microphone A microphone with two diaphragms and having a response pattern resembling the digit 8. A bidirectional microphone.

filter A group of parts such as resistors, capacitors and coils used for passing a band of frequencies, or for attenuating the lower end or the upper end of a band of frequencies.

flat plane speaker A driver having a square shape with magnets positioned in each of the four corners.

flutter A relatively rapid wavering in pitch caused by speed fluctuations in the movement of tape. A sound quiver caused by a speed change in a turntable.

flux A line of magnetic force.

flux density The number of lines of magnetic flux per unit area. An indication of the concentration of lines of magnetic flux. The unit of measurement of flux density is the gauss.

flywheel A large, circular, heavy mass used to help maintain stability of rotational velocity. Sometimes attached to the tuning shaft in receivers and tuners, and used in turntables and tape decks to help maintain constant speed.

fm An abbreviation for frequency modulation. The fm band extends from 87.9 MHz to 108 MHz (87.9 megahertz to 108 megahertz).

fm sensitivity *See* sensitivity.

forward/reverse winding time The amount of time to wind tape fully in the fast forward mode or for the complete rewinding of the tape when in reverse. The total amount of time depends on the speed and the length of the tape and is about 24 times faster than the play mode.

frequency The number of vibrations per second or the number of complete cycles per second of an electrical wave.

frequency modulation Abbreviated as fm. A broadcasting technique in which an audio wave is superimposed on a much higher frequency carrier or radio wave by changing the frequency of the carrier.

frequency response A measure of the range of audio frequencies to be passed by a component. Should be accompanied by the amount of deviation expressed in decibels. The range of audio frequencies that can be reproduced by a component with the input signal connected to the auxiliary input and measured with one watt output power from the speakers.

frequency synthesis A circuit in which a free running local oscillator is replaced by a frequency synthesizer that is crystal controlled. *See also* digital pll frequency synthesizer.

front end The tuning section of a tuner or receiver. The input section of these components.

function selector A control used for selecting different sound sources. In components such as a receiver or amplifier it is a control used for selecting various program sources such as am, fm, phono, or tape.

fuse A small section of low melting point metal housed in a cylindrical length of glass and plastic and having a specified current rating.

G

gain The ratio of output to input of a circuit or a component. The amount of signal with reference to a standard. Often expressed in decibels. Gain figures in decibels can indicate a loss as well as an increase.

gap A fine, vertical slit in the head of a tape deck. A slit containing magnetic lines of flux supplied by tape during playback or by coil current during record.

gate Sometimes known as a logic circuit, a gate is a switch capable of making its own decisions whether to turn on or off. There are many different types of gates, such as AND, NAND, OR, NOR. They are all basically alike depending for their operation on the kind and amount of voltage presented to their inputs, and the method and amount of bias.

glide eject An electronic method for ejecting a cassette. Using an electronic microswitch, it eliminates hard to push mechanical eject buttons.

gnd See ground.

grille Cloth or open-mesh metal covering for the front of a speaker. The grille must not interfere with the free movement of sound from the speaker to the listener. To the extent it does so it is called acoustically transparent.

ground Alternative word for minus or negative. Sometimes used synonymously with the phrase "cold lead." A reference point for voltage measurements.

H

harmonic A multiple of an electrical or sound wave. The basic or fundamental frequency of a wave is sometimes called the first harmonic. If the fundamental is 1000 Hz, the second harmonic is 2000 Hz and the third harmonic is 3000 Hz.

harmonic distortion The production of harmonic frequencies in the output of a component such as a receiver, amplifier, or equalizer, not related to the fundamental frequencies of tones at the inputs of these devices. Harmonic distortion is supplied as a percentage. The smaller the percentage figure, the better.

head A metal housing, equipped with a fine vertical gap, enclosing a coil on a tape deck. A shell-cartridge assembly mounted on a tonearm.

head amplifier A pre-preamplifier used with moving coil phono cartridges.

head azimuth See azimuth.

head zenith See zenith.

headphones Sometimes called earphones. A transducer, in pairs, for mounting on or around the ears. Headphones require very little audio driving power.

heat sink One or more metal plates, sometimes in corrugated form, for radiating heat away from a device such as a power transistor. Heat sinks are often used in the output stages of power amplifiers.

hertz Abbreviated as Hz. A measurement of frequency representing the cycle per second. 50 Hz is the same as 50 hertz or 50 cycles per second. Also see kHz and MHz.

high filter A circuit network for attenuating frequencies above some selected frequency such as 8 kHz or 10 kHz. Also known as scratch or high cut filter. A circuit for removing or weakening high pitched noise such as hiss and record surface noise.

high polymer film speaker Speaker using a vapor deposited aluminum coated film for producing sound.

horn speaker Speaker which uses a horn for coupling speaker sound to the air.

hot lead A wire that is connected to the plus terminal of a dc voltage source. The inner wire or conductor of coaxial cable is sometimes called the "hot" lead. The wire connected to the plus terminal of a speaker is also referred to as the "hot" lead.

hub A reel for holding tape.

hum A low frequency tone, often 60 Hz or its second harmonic, 120 Hz. An unwanted sound picked up from the ac power line.

Hz Abbreviation for hertz, the cycle per second; 60 Hz is 60 cycles per second. This abbreviation is replacing cps, the cycle per second. Used in connection with frequency.

I

IC *See* integrated circuit.

IEEE Abbreviation for Institute of Electrical and Electronic Engineers.

if *See* intermediate frequency.

IHF Institute of High Fidelity. Group of American manufacturers of high-fidelity sound equipment for setting up testing standards and for improvements in sound technology.

im *See* intermodulation distortion.

image rejection The reception of a signal at two or more points on the tuning dial, with only one the true station signal. False signals are called images. Tuners or receivers that cannot reject these false signals supply sound with ghosts. Image rejection is in decibels and the higher the better.

IMD *See* intermodulation distortion.

impedance The total opposition of combined resistance and reactance to the flow of current. Measured in ohms.

induced magnet cartridge A phono cartridge with a tiny bit of ferrous metal attached to a cantilever. The movement of the cantilever moves the metal thus changing the inductance of a nearby coil or coils. This causes a variation of the magnetic field around the coil at an audio rate.

inductive reactance The opposition, in ohms, of a coil or inductor to the flow of current through it.

inductor A coil, generally of copper wire, although other metals can be used.

infinite baffle Speaker system using a sealed enclosure. An air tight enclosure.

input sensitivity control A continuously variable control, or a multiposition slide switch for adjusting the input sensitivity of a component.

insulator Any substance that has an extremely high opposition to the flow of current through it. Glass, plastics, and rubber are insulators.

integrated amplifier An audio amplifier that includes a preamplifier and a power amplifier.

integrated circuit Abbreviated as IC. Parts such as resistors, transistors, and diodes etched onto an extremely small chip of silicon.

interference Unwanted signals that may cause changes in the received signal or add noise to it.

intermediate frequency Abbreviated as if. The intermediate frequency of a tuner or receiver is produced by the mixing of the incoming radio signal and the ac voltage generated by a local oscillator. At one time coil/capacitor circuits were used to tune the intermediate frequency, but quality receivers now use ceramic if filters. These are immune to vibration and never need to be realigned. The intermediate frequency used in fm tuners and receivers is 10.7 MHz. In am tuners and receivers it is commonly 456 kHz or 465 kHz.

intermodulation distortion The interaction of tones producing new tones not harmonically related to the original music. Abbreviated as im or imd. Distortion caused by the mixing of tones in a nonlinear stage of an amplifier. Expressed as a percentage.

ips Abbreviation for inches per second. Frequently used with reference to the movement of tape.

isotweeter Speaker using a printed circuit voice coil.

J

jack A socket designed for the easy insertion of a plug. Used in combination with a plug it permits the easy connection and disconnection of wires and cables.

K

k Letter used as a multiplier representing 1,000. 5 k means 5,000. 100 k is 100,000. 5 k ohms is 5,000 ohms.

kHz Abbreviation for kilohertz or a thousand hertz.

kilohm A thousand ohms. Sometimes written as k ohms (kΩ).

L

L signal In stereo, the left channel signal.

labyrinth speaker Speaker that supplies a long path for the rear sound wave produced by the speaker's diaphragm.

laser Beam of coherent light used for tracking pulse code modulated discs.

latching fast forward See locking fast forward.

lavalier microphone Omni microphone of very small size worn around the neck or on some article of clothing.

LED Light-emitting diode.

left/right control An adjustable device for allowing left or right channel sound to come out of left/right located speakers, as desired.

level Strength of a signal or voltage. A level control is used to adjust the strength of a signal or voltage. A volume control is a type of level control. The controls for boost or cut on an equalizer are level controls.

light-emitting diode Abbreviated as LED. A solid-state unit that can emit light. LEDs come in various sizes and are used as illuminated callouts.

limiter Circuits in an fm receiver that clip the tops and/or bottoms of waveforms to produce a signal of constant amplitude. A way of removing unwanted amplitude modulation in an fm signal.

line output The output terminals of a preamplifier or tape player providing a signal voltage for a power amplifier.

load Any device that accepts audio power or uses audio power. A speaker is a load on its preceding power amplifier. Any component that takes power from the ac line is a load on that line.

load impedance The amount of load, measured in ohms. The smaller the impedance, the heavier the load. The load impedance of most speakers used with in-home hi-fi systems is 8 ohms.

locking fast forward and/or rewind A feature that permits the control for fast forward and/or rewind to be locked in position. With this feature these controls do not need to be held down manually. Also known as latching, and called latching fast forward or latching rewind.

logic circuit Decision making or gating circuit.

long throw A designation applied to the voice coils of speakers. A voice coil that is free to make long forward and backward movements. A long throw voice coil in a speaker helps provide good low-frequency response.

loudness control Also known as loudness contour. The human ear does not hear low or treble frequencies well at low sound levels. A loudness control boosts these tones when the volume is turned down. An on/off control used to bring up the level of bass and treble tones when listening to low volume sound.

low filter A filter designed for removing low frequency noises, such as hum and rumble or to weaken such noises. A circuit for attenuating audio signals below some selected low frequency, such as 50 Hz or 100 Hz. Also known as a bass filter, low cut, or rumble filter.

low noise, high output tape A tape whose formulation consists of unmodified gamma ferric oxide.

LSI Abbreviation for large scale integrated circuit.

M

mA A milliampere or thousandth of an ampere. 5 mA is 5 milliamperes of current, whether ac or dc.

μA A microampere or a millionth of an ampere.

magnet materials Any material capable of being magnetized, or an alloy of such materials.

masking effect A loud sound covering or hiding a weaker sound. Music can mask noise if the level of the noise is much lower than the music.

maximum input power When used in connection with speakers is the maximum audio power that can be applied to the voice coil before possible damage.

mc *See* moving coil phono cartridge.

Meg Abbreviation for a million. 10 meg is 10 million. 10 megohms is 10,000,000 ohms.

memory Preset tuning. A method of preselecting radio stations, usually some combination of am and fm, for instant recall. Also used in tape decks for quick selection of wanted passages.

memory set A receiver or tuner feature that lets you lock-in a station on a particular pushbutton. The tuner memorizes this station for a selected pushbutton until you release the memory and reset it.

metal alloy Also known as pure metal. A tape formulation made of pure metal alloys.

mfb *See* motional feedback speaker.

MHz Abbreviation for megahertz, a million cycles per second.

microfarad A submultiple of the farad, basic unit of capacitance. A microfarad, abbreviated as μF, is a millionth of a farad.

micron The thousandth part of a millimeter. Approximately 0.00003937 inch.

microphone Member of the transducer family; it changes sound energy input to a corresponding electrical energy output.

microphone sensitivity Ability of a microphone to respond to changing levels of sound.

microprocessor Logic circuits mounted on a chip made of silicon. A logic circuit is an electronic switch. Microprocessors can handle two numbers only, binary 0 and 1.

microvolt Abbreviated as μV. A millionth of a volt.

midrange The range of audio frequencies between bass and treble tones.

midrange driver A speaker designed for the reproduction of midrange tones.

midrange tones Tones generally considered to be in the frequency range of 200 Hz to 3 kHz.

millivolt Abbreviated as mV. One one-thousandth of a volt.

minimum impedance The smallest amount of impedance presented by a component or a part of a source.

minimum input power The smallest amount of audio power that can be supplied to speakers to produce satisfactory sound output.

MM See moving magnet phono cartridge.

modified gamma ferric oxide Cassette tape formulation to which carefully controlled amounts of metallic cobalt or magnetite have been added.

modular system Prepackaged hi-fi system minus a rack.

modulation The process of superimposing an audio signal on a higher frequency radio wave, a carrier.

monaural An individual having just one ear. Sometimes mistakenly used as a synonym for monophonic.

monitor terminals Terminals for connection to the output of a tape deck.

monitoring Listening to a program in the process of being taped to evaluate recording quality.

mono Abbreviation for monophonic or single channel sound.

mono/stereo switch A front panel control to permit selection of a mono or stereo program.

monophonic Mono or single-channel sound.

MOSFET Metal-oxide-semiconductor field effect transistor. See FET.

motional feedback speaker Speaker using an accelerometer, a piezoelectric transducer, for measuring the acceleration of a cone.

moving coil phono cartridge Phono cartridge in which coils are mounted on the end of the cantilever. The coils are moved in or out or toward and away from a magnetic field. Abbreviated as mc.

moving magnet phono cartridge Abbreviated as mm, it is a phono cartridge in which a magnet on the end of the cantilever moves toward and away from two or more coils.

mpx Abbreviation for multiplex. See multiplex.

μsec Microsecond. A millionth of a second.

multiamp system Hi-fi system using two or more audio power amplifiers. A system with separate amplifiers for bass tones and for midrange/treble tones, or with separate power amplifiers for all three bands of tones.

multipath reception The arrival of fm signals via several paths of different lengths due to signal reflections from buildings, bridges, or hills.

multipath signal A signal that takes one or more reflected paths.

multiplex Technique used in fm broadcasting and reception for the transmission and separation of stereo signals. The transmission of two channels on a signal carrier so they can be independently recovered in the receiver.

multiplex decoder *See* decoder.

music sensor A feature that lets a tape player scan forward automatically to play the next selection on a tape.

μV Microvolt. A millionth of a volt.

mV Millivolt. A thousandth of a volt.

N

NAB National Association of Broadcasters.

NAND Gate A logic circuit consisting of negative AND gates.

neg Negative or minus.

network In audio, a frequency dividing circuit in a speaker system or an electronic crossover network in a multiamp installation. A circuit for performing a special function.

node Point at which standing sound waves cancel.

noise Unwanted sound. Unwanted signals including noise such as hum, hiss, rumble, various kinds of interference, and component generated signals.

noise canceling microphone *See* differential microphone.

nominal voice coil impedance The value of the impedance of the voice coil at a particular frequency, often at 1,000 Hz.

nonlinear A component whose output isn't a true reproduction of its input.

NOR gate A logic circuit consisting of negative OR gates.

nude diamond A stylus made completely of diamond attached to a cantilever.

O

ocl Output capacitor-less. An output circuit which delivers its signal or audio power directly to a speaker system and not via capacitor.

octave A doubling or halving of a frequency. A ratio of 2 to 1 between frequencies. 16 Hz to 32 Hz is sometimes considered a first octave in the audio range. 32 Hz to 64 Hz is the second octave.

ohm Basic unit of resistance or impedance. Multiples consist of the kilohm or thousand ohms; megohm or million ohms.

omni speaker A speaker or driver that radiates sound in all directions.

omnidirectional A component that is equally sensitive to the reception of signals from all directions, or which radiates signals in all directions. Sometimes written as omni.

omnidirectional microphone A microphone which responds to sounds from all directions.

OR gate Two switches connected in shunt or parallel. Either one gate OR the other must be open (switch in a closed position) to permit the passage of a signal.

oscillator An electronic circuit for producing an alternating voltage and current. The frequency of the ac produced by the oscillator depends on the electrical value of its parts. A local oscillator is used in superheterodyne receivers and tuners to generate an ac signal for mixing with the received signal.

output impedance The impedance across the output terminals of a component.

output stage A final stage. In a power amplifier it is the stage which supplies audio power to following speaker systems.

output voltage The amount of signal voltage supplied by a sound source such as a tuner, phono cartridge, or microphone.

P

parallel connection The connection of points of identical polarity to each other: plus to plus, minus to minus. Two components that are wired in shunt. Two components connected across each other instead of being wired sequentially.

parasitic array A group of directors and reflectors used as part of an antenna system.

parasitic element An element of an antenna, such as a reflector or director, having no electrical connection to the active or signal receiving element of an antenna.

passive radiator Drivers generally used as woofer supplements. Passive radiators have no voice coils; simply vibrate in step with an existing speaker.

pc board Printed circuit board. At one time all wiring of electronic circuits was done point to point, using individual wires. The wires or conductors are now etched on a plastic-type board, with electronic parts such as resistors, coils and capacitors mounted on the board. When this is done all connections are automatic by means of the etched wiring.

pcm Abbreviation for pulse code modulation.

peak The maximum amplitude, positive or negative, of a voltage or current or power.

permalloy A type of material, noted for its hardness and durability; used for shielding the coil of a head used in a tape deck.

phase lock loop Abbreviated as pll, this circuit is used in tuners and receivers to improve stereo signal separation and to reduce distortion over a wider frequency range than conventional coil-and-capacitor circuits.

phase shift The advance or lag, with respect to time, as a signal passes through a component. The variance in time displacement of a signal with respect to some other signal.

phasing When used in connection with speakers phasing means that the speakers must be wired so their cones move forward or back in step. The cones or speakers are then said to be in phase.

phono input sensitivity Measured in millivolts it is the minimum amount of signal a preamplifier requires from a phono cartridge.

phono overload Maximum amount of phono signal acceptable at the input of a preamplifier.

photocell A device belonging to the transducer family, capable of changing light energy to electrical energy.

pictorial diagram A diagram consisting of drawings or photographs of components showing how connections are made between them.

piezoelectric speaker Speaker that produces sound through the bending of a ceramic or other crystalline material instead of through the movement of a cone. Used for the reproduction of treble tones. A speaker using a crystalline material having piezoelectric properties.

plasma display A fluorescent display for indicating signal level.

plasma speaker Speaker using ionized air or helium/air mixture for producing sound.

pll *See* phase lock loop.

plug A part that fits into a jack for the easy connection and disconnection of wires and cables.

polar pattern A graph showing the directional response of a microphone.

polarity Used as a reference to components such as batteries or to speaker terminals. Polarity involves two words: plus or positive, minus or negative. Polarity symbols are (+) and (−). Color coding is sometimes used to indicate polarity: red is used for plus; black for minus.

ported speaker Also known as bass reflex or vented port speakers, these drivers channel rear produced sound to the front through a port or duct.

positive Abbreviated as pos. Having plus polarity. *See* polarity.

power amplifier A component which delivers voltage and current to the voice coils of speaker systems at an audio rate.

power bandwidth The bandwidth of an amplifier measured at its half power points. The bandwidth of an amplifier when the component delivers 50% of its rated power output.

power handling ability The maximum amount of power which can be safely handled, used or dissipated by a component. The maximum audio power rating of a speaker.

power output A quality amp has its power output given in terms of rms (root-mean-square) or continuous power. This indicates how much power it supplies with a minimum of unwanted noise or total harmonic distortion.

preamp Abbreviation for a preamplifier. An audio voltage amplifier that supplies a driving audio voltage to one or more following audio power amplifiers. A device which accepts small signals from program sources such as a tuner, tape deck, or record player.

preemphasis In fm broadcasting an increase in the strength of treble tones prior to broadcasting.

pre-preamplifier *See* head amplifier.

pressure pad Pad made of felt or foam for holding tape against heads in a tape deck.

pressure roller Tape roller turned by a revolving capstan.

print-through A condition in which recorded material impresses itself on several successive layers of tape.

programmable set *See* memory set.

proximity effect Exaggerated bass response when a microphone is used too close to the mouth or an instrument.

pulse code modulation When used in reference to phono discs means the construction of pits and flats representing binary digits 0 and 1.

pure ferric oxide A tape formulation consisting of ferric oxide in crystalline shape.

push pull *See* complementary symmetry.

pxe Abbreviation for piezoelectric transducer.

Q

quantizing Representing a signal in terms of numbers.

R

R or r Symbols for resistance, the opposition to the flow of either alternating or direct current. The basic unit of resistance is the ohm. Multiples are the kilohm (thousand ohms) and the megohm (million ohms).

R signal In stereo, the right channel signal.

radio frequency Abbreviated as rf. A wave whose frequency is above the audio range, generally 50 kHz or 100 kHz or more.

rated output power The maximum amount of audio power that can be delivered by a power amplifier without exceeding its specified distortion rating.

ratio detector A demodulator used in an fm receiver or tuner for separating audio frequencies from the intermediate frequency.

reactance The opposition of a coil or capacitor to the flow of an alternating current. Measured in ohms. When combined algebraically with resistance it results in impedance. The opposition of a coil to current flow is called inductive reactance and in a capacitor, capacitive reactance.

receiver An integrated component consisting of a tuner, preamplifier and power amplifier using a common power supply and mounted on a single chassis.

record warp Any deviation from flatness in a phono record.

rectifier A device, generally a solid-state diode, which changes ac to pulsating dc. Any device which only permits the unilateral flow of current.

reflector An antenna element positioned behind the active or signal receiving element.

relay An electromagnetic type of switch consisting of a coil which becomes magnetized when current flows through it, and mechanical switching elements actuated by the coil. Relays are used for remote operation of other components. They are sometimes used as protective elements in power amplifier output circuits.

resistor A radio part, generally made of wire or carbon, used to help control the flow of currents. Value is measured in ohms. The greater the ohmic resistance the greater the opposition to current flow. Resistors can be variable, fixed, or tapped.

resonance May be mechanical or electrical. The tendency of some part to resonate at a particular frequency. An increase in signal level at some frequency due to resonance.

response *See* frequency response.

reverberant sound Sound reflected from surfaces such as floor, ceiling, and walls. Sometimes called wet sound.

reverberation time The amount of time it takes for sound to decrease by 60 dB.

rf Abbreviation for radio frequency.

RIAA Recording Industry Association of America. The standards of disc recording and frequency curves established by the RIAA.

ribbon microphone A transducer using a metallic ribbon diaphragm stretched between the poles of a permanent magnet.

ribbon speaker Speaker using a narrow element of aluminum foil or aluminum coated mylar as the sound producing element.

rim drive Platter driving technique using an idler wheel mounted on a motor shaft, making contact with another wheel on the inside rim of the platter.

rms *See* root-mean-square.

rolloff A decrease in frequency response at the bass or treble end of a response curve, or at both ends. Indicated in decibels per octave.

root-mean-square A method of averaging audio voltage, current, or power. In this method of measuring a varying quantity such as an ac voltage, current or power, a number of instantaneous values are taken over a complete cycle of the wave. These values are then squared, that is, multiplied by themselves. The results are added and then an average figure is calculated. The final answer is the square root of this average.

rumble Low frequency noise resulting from vibrations in the platter and motor of a turntable and also from record warp.

rumble filter *See* low filter.

S

sampling rate The number of times the instantaneous value of a wave is sampled, per second.

scan sense Used during automatic scanning. A feature that lets the tuner or receiver lock in only on stations that will produce adequate sound output from the speakers.

scanning A tuning process, possibly automated, moving up and down a broadcast band. Scanning is tuning the am and fm bands and can be done manually or automatically.

scratch filter *See* high filter.

selectivity Also known as alternate channel selectivity. The ability of a receiver to select the weaker of two stations broadcasting on adjoining frequencies. Designated in decibels, higher amounts are preferable.

semiconductor A solid-state component made of germanium or silicon.

sendust A magnetic alloy of iron, aluminum, and silicon, formed by a sintering process. Used as a head shield in tape decks.

sensitivity Sometimes called usable fm sensitivity. It is the smallest signal provided by an antenna which the tuner or receiver can convert into a satisfactory sound signal. It is expressed in microvolts (μV) or in dBf and is defined by the Institute of High Fidelity as the signal strength at which the tuner or receiver will suppress noise by 50 dB. Thus, a tuner with 2.0 μV of usable fm sensitivity can convert a signal from the antenna of two one-millionths of a volt and suppress the background noise by at least 50 dB.

separation *See* stereo separation.

series connection The connection of parts in sequence so that the same current flows through them.

Shibata stylus A modified form of the elliptical stylus.

shield braid *See* braid.

shotgun microphone *See* directional microphone.

signal-to-noise ratio A comparison between signal and noise levels, measured in decibels. A ratio of wanted signal voltage to unwanted noise. Abbreviated as snr. This figure indicates how quiet and hiss-free the background will be in relation to the signal. A signal-to-noise ratio of 60 dB means that the signal is 1,000 times stronger than the noise. The higher the figure the better.

slew rate   Rate of voltage change per microsecond.

slide control   A control whose motion is horizontal or vertical, as opposed to rotary.

snr   *See* signal-to-noise ratio.

solenoid   A coil having a moving metallic element. A solenoid can be considered as a motor having restricted movement.

solid-state   Diodes and transistors or circuits using hese parts. Diodes, transistors and related devices made of semiconductor materials such as germanium or silicon.

speaker impedance   The total opposition of the voice coil of a dynamic speaker to the flow of audio currents through it.

speaker phasing   *See* phasing.

speaker system   A group of two or more speakers housed in an enclosure. The speaker system generally includes a crossover network—a frequency divider system for routing bands of audio frequencies to speakers.

speaker wire   Wire pairs that are coded by color or identified in some other way for connecting speakers to the speaker output terminals of an amplifier or receiver.

spec   Abbreviation for specification. A spec sheet details the electrical and electronic characteristics of a component.

spider   Highly flexible material made of cloth or plastic impregnated cloth for holding the voice coil in position. A spring-like device made of some fibrous material.

spl   Abbreviation for sound pressure level.

spurious response   The reaction of a tuner or receiver to self-generated signals. Measured in decibels. Self-generated false signals produced by a tuner or receiver. These signals can be radiated, causing interference in other tuners or receivers. They can also affect the tuner or receiver producing such signals. The self-generated signal can interfere with a wanted station signal. Expressed in decibels, higher values are better. Sometimes also known as spurious rejection.

station preset   *See* memory set.

stereo   Abbreviation for stereophonic or two-channel sound.

stereo indicator   An indicator such as a lamp which glows or a readout which becomes illuminated on receipt of a stereo signal. Light-emitting diodes (LEDs) that indicate fm stereo reception rather than monophonic (mono).

stereo/mono switch   Fm stations broadcast in stereo and/or mono. A tuner or receiver that can switch to the correct mode automatically or manually. A useful feature in fringe reception areas.

stereo separation A measure of the ability of the multiplex decoder in an fm tuner or receiver to separate left and right sound channels in stereo broadcasts. Measured in decibels. For best stereo effect channel separation over the frequency range of 20 Hz to 15 kHz is important. A good tuner or receiver should have at least 25 dB separation, with higher figures better.

stereophonic A sound system using two channels—left channel sound and right channel sound.

strain lines Tape that is wound excessively tight with strain lines radiating outward from the center of the reel. A term applied to open reel tape.

strapping Technique for combining left and right stereo amplifiers to supply higher power output in mono.

stylus Formerly known as a needle. A small diamond used for tracking modulations in record grooves.

subsonic filter Filter circuit for minimizing or eliminating sound produced by record warp or by turntable rumble.

subwoofer A speaker designed to handle low-frequency tones, possibly as low as 16 Hz.

supercardioid microphone A variation of the cardioid microphone providing maximum rear cancellation of sound at 150 degrees from either side of the axis instead of 180 degrees as in a standard cardioid.

superheterodyne A receiver consisting of a tuner section, a mixer, and a local oscillator. These are followed by intermediate frequency (if) stages feeding into a demodulator. The output of the demodulator is the recovered audio signal.

superimposing In tape recording a method of putting one sound on another.

supra-aural headphones Open air type of headphones which permit listener to hear extraneous sounds.

surround Highly flexible material used for supporting the outer edge of the cone in a dynamic speaker. The surround receives support from the metal basket holding the speaker assembly. The surround is highly compliant, sometimes in the form of a half roll.

switch Sometimes abbreviated as sw, it can be purely mechanical, electronic, or some combination of the two. Mechanical types can be pushbutton, slide, lever, toggle, or any other arrangement for turning the switch on and off. Unlike mechanical switches electronic switches have no moving parts and should theoretically last forever.

switched outlet An outlet on the rear of a component that becomes "live" when the component's power switch is turned on.

synchronous motor A motor whose rotational speed is locked into the frequency of the ac power line.

synthesizer Refer to digital pll frequency synthesizer.

T

tangential tonearm A linear tracking tonearm. A tonearm that moves across the phono record in a straight line.

tape bias An alternating magnetic field having a frequency from about 30 kHz to 120 kHz, impressed on magnetic tape to make it easier for an audio signal to magnetize the tape.

tape deck A tape player/recorder which does not contain an audio power amplifier. A program source for use with a high-fidelity system.

tape formulation Types of magnetic particles put on tape, such as ferric oxide and chromium dioxide.

tape head Refer to head.

tape hiss Random noise produced by tape in the treble region.

tape memory Circuit in a tape deck that locates the start of a musical program.

tape monitor A technique for checking the quality of recordings in the process of being made. Possible only with tape decks using separate playback and record heads.

tape monitor switch In its on position this switch permits listening to a recording in the process of being made.

tape sensitivity A measure of a tape's ability to respond to a fixed level input signal of a given frequency.

test tone switch A switch used for adjusting the recording level of a tape deck.

thd See total harmonic distortion.

three-way speaker system A speaker system in which the audio spectrum is divided into three bands: bass, midrange, and treble, with each of these bands handled by separate speakers.

tim See transient intermodulation distortion.

time constant When used with reference to magnetic tape is the point at which treble preemphasis begins.

time delay A technique in which rear speaker sound is delayed so it does not reach the listener's ears simultaneously with front speaker sound.

time delay distortion The time delay of different parts of the audio spectrum by a speaker.

time delay muting Circuit for eliminating switching thumps coming from speakers when the hi-fi system is turned off or on.

tone control A variable control for adjusting bass or treble level. There are usually only two tone controls on a receiver, but some do have a separate tone control for midrange tones.

tone defeat A switch that defeats the tone control, producing a response that is independent of any tone control setting. Tone defeat may be made part of a tone control.

tonearm shape Tonearm shapes include I, J, S, and straight.

total harmonic distortion Unwanted changes of the original sound caused by nonlinear circuit design in the tuner, receiver, or amplifier of a sound system. The sum of all harmonic distortion products. Generally expressed as a percentage. Abbreviated as thd.

track Narrow segment along the length of tape. This narrow, invisible path is magnetized and is used to induce an audio voltage in the playback head in a tape player. Stereo reproduction requires the use of a pair of tracks. The tracks are that part of the tape along which magnetization takes place during recording.

tracking A measure of the velocity of a stylus tip in cm/sec (centimeters per second) while maintaining good groove contact.

tracking error Deviation of the cartridge axis from the record tangent, measured in degrees.

tracking force See vertical tracking force.

transducer A device for changing one form of energy to another. A microphone is a transducer since it converts sound energy to electrical energy. A speaker is a transducer, converting electrical energy to sound energy. A battery, electric light bulb, and an electric generator are all transducers.

transformer A device, generally consisting of a pair (or more) of coils wound on a coil form. The transformer may have an air core or some ferrous material such as iron. The transformer can be a step up or step down type, or may have a turns ratio of 1 to 1.

transient Voltage or current wave having sharp, short-lived peaks. Certain musical instruments such as the banjo, guitar, mandolin, and piano produce numerous transients.

transient intermodulation distortion Abbreviated as tim. Caused by excessive negative feedback and a slew rate that is too low.

transient response The ability of a component to respond to rapid changes in the signal waveform.

transistor A tiny bit of silicon or germanium, highly purified, to which an extremely small impurity, such as boron, is added. Also known as solid state in contrast to tubes whose elements operate in a glass or metal enclosed vacuum. Transistors are more efficient than vacuum tubes and with the exception of certain types, such as power output transistors, produce no heat.

transmission line speaker Speaker using a metallic cone positioned vertically with each conic section producing sound at a precise time.

transport Motorized system for moving magnetic tape from one reel to another.

treble driver Speaker used for reproducing treble tones.

treble tones Tones generally considered to be in the range of 3000 Hz to 20 kHz.

triamplification System in which the audio band is separated into three segments with each handled by its own audio amplifier.

tuner Section of a receiver for tuning in broadcast stations. May be a separate component or integrated into a receiver.

tuner-amplifier A receiver containing a tuner, preamplifier, and power amplifier.

tweeter Speaker designed for the reproduction of treble tones.

twin speed tape deck A cassette deck capable of working at two speeds: 1⅞-ips and 3¾-ips; or 1⅞-ips and ¹⁵/₁₆-ips.

two-way microphone See coaxial microphone.

two-way speaker system A speaker system in which the audio spectrum is divided into two bands: bass and midrange/treble. See also three-way speaker system.

U

unbalanced line A cable having an inner, central conductor and a shield braid covering. Both are used for carrying the signal. The center conductor is called the "hot" lead; the outer braid the "cold" lead.

unswitched outlet An ac outlet on the rear of a component. An outlet that is "live" at all times.

usable fm sensitivity See sensitivity.

V

V Abbreviation for voltage, whether ac or dc. Often used in connection with a number, such as 6 V, 120 V, etc.

variable reluctance phono cartridge A phono cartridge in which a bit of ferrous metal is mounted at the end of a cantilever and is moved closer to or away from a group of coils, thus affecting their inductance at an audio rate.

vented port speaker See ported speaker.

vertical tracking force The amount of downward pressure applied to a stylus. Abbreviated as vtf.

virgin vinyl Vinyl that is 100% pure and does not contain reprocessed vinyl.

voice coil A coil of wire in a dynamic speaker. A coil that is attached to a speaker cone.

voice coil impedance Opposition of the voice coil in a dynamic speaker to the flow of audio currents. Measured in ohms, the value is most often 8 ohms, but can be 4 or 16 ohms.

volt Basic unit of electrical pressure. Also known as voltage, electrical pressure, electromotive force (emf), and potential difference (pd). Abbreviated as V.

volume control A control for adjusting the overall output level of a sound system.

vu meter A volume unit meter, calibrated in decibels.

W

watt The basic unit of electrical power. Submultiples are the milliwatt, or thousandth of a watt, and the microwatt, or millionth of a watt. A kilowatt is a thousand watts.

wavelength The distance from the start of a single cycle of a wave to its end. Usually measured in meters or some submultiple of the meter.

wet sound *See* reverberant sound.

whizzer Speaker with two cones fastened to the same voice coil, with one larger than the other. A tweeter cone mounted on the same voice coil as a larger midrange. The whizzer is not supported by a surround.

wide-band Term applied to a filter. A wide-band filter is one capable of passing a broad range of frequencies.

wiring diagram A diagram showing how the components of a sound system are connected. *See also* block diagram and pictorial diagram.

woofer Speaker for reproducing bass tones.

working distance The distance between a microphone and its sound source.

wow A slow change in pitch caused by speed changes in a tape deck or record player. *See also* flutter.

wrms Abbreviation for weighted root-mean-square.

Z

zenith Positioning of a tape head so that it has no angle of tilt away from the head.

Index

A

Ac line power, 125-126
Acoustic suspension, 167-169
Acoustics, 17, 187-189
After recording, 296-297
Air vents, 45
Am signal, 56
Amplifier (amp), 28
  classes, 130-131
  controls
    integrated, 142-145
    power, 142-146
  efficiency, 130
  integrated, 28-29, 123-154
  power, 31, 147-153
Amplitude, 56
Analog, 174-176, 238, 245, 249, 251, 259
AND gate, 72-73
  /OR, 74
Antenna(s), 26-27, 58, 60-65, 95
Anti-birdie filter, 83
Antiresonance, 322
Audio
  and hi-fi systems, 18-40
  digital versus analog, 237-262
  output power, 135-137
  signal, 56
Auto
  blend control, 90
  dx/local, 90
Automatic record player, 200
Aux (auxiliary) terminals, 94
Azimuth, head, 280

B

Balance control(s), 87-88, 109, 145
Bandpass, 82
Bandwidth, 81-83, 152-153

Basic
  gating circuits, 72-75
  microphone, 311-315
Basket, 159
Bass, 327-328
  reflex enclosure, 184
  tones, 39-40
Before recording, 294-296
Bi-amp system, 41
Biamplification, 138-139, 141
Bias, 284-286
Bidirectional microphone, 319
"Binary digits," 242
Binary
  numbers, 240
  switches, 242
  system, 239
Bits, 242
Boom boxes, 166
Boost, 153
  low frequency, 287
Brilliance, 323
Bulk eraser, 298
Buzz, 235
Buzzing sound, 235

C

Cables, 40, 330
Cantilever, 216-217
Capacitance tracking, variable, 246-248
Capacitor, 312
  -less circuits, output, 129
Capstan, 272-273, 275
Cardioid, 170, 317-319
Carrier, 57, 59
Cartridge, 211, 217-218, 222-226
Casseiver, 55
Cassette(s), 16-17, 267-271, 280-281, 291-294

Ceramic, 182
  microphone, 314-315
Channel separation, 102
Circuits
  basic gating, 72-75
  output capacitor-less, 129
  protector, 127-128
  short, 127
Circumaural headphones, 196
Clipping, 129
Closed box enclosure, 184
Coaxial, 163
Cogging, 202
"Cold" lead, 330
Compatible expansion, 260-262
Complementary symmetry, 139, 142
Complex waveforms, 49
Compliance, 223-224
Computer, Analog Bass, 174-176
Condenser microphone, 312-313
Cone, 158, 162-163, 180-181, 185, 193
Conference recording, 326
Conical styles, 219-220
Connecting
  antenna, 63-64
  headphones, 195
  speakers, 192-193
Connections, tuner versus receiver, 91-95
Contour control, 111
Control(s), 84-91, 109-113
  amplifier, power, 145-146
  integrated amplifier, 142-145
Conversion, binary to decimal, 239-240
Converting analog to digital, 242-245
Cooling, 128
Corona, 171
Cost difference, 54-55
Cover, dust, 233-234
Crossover(s), 39, 40, 139, 141, 160, 192
Crystal oscillator, 79-81
Cupping, 291
Cut, 153
Cycles per second, 47

D

Damping, 152-153, 218
dBf, 99-100
Deck, cassette, 16-17
Decibels (dB), 95-97
Decimal conversion, binary to, 239-240
Decoding, 261
Deemphasis, 83, 108
Demodulator, 56-57, 69
Detector circuit, 57
Deviation figure, 309
Diaphragm, 174
Differential microphone, 319
Diffuser lens, 164-165
Digital
  analog versus, 237-262
  and sound fidelity, 252-253

Digital—cont
  audio editing, 250
  disc, 245-246
  display, 78-79
  recording, 249-250
  transmission, 251
Digitized tuner, 78-79
Dipole, 61
  antenna, 27
  radiator, 170-171
Direct
  drive, 204, 205
  to disc recording, 259-260
Directional microphone(s), 319-320
Directors, 62
Dirt, 232-233, 252-253
Disc, 201, 245-246
Distortions, 66, 116, 118-119, 147-150,
  186-187, 256, 309
Dodecahedron, 170
Dolby, 304-306
Drive
  motors, 201-202
  system, 202-204
Driver(s), 39, 181-182, 190-191
Drones, 172-173
Drums, 328-329
"Dry" sound, 266
  versus reverberant sound, 325
Dub, 33, 36, 38-39, 265
Dubbing, 38-39, 108
  switch, 94
Duo-beta negative feedback, 120-122
Dust, 252-253
  cover, 233-234
  jacket, 231
  particles, 231
Dynamic
  microphone, 311-312
  range, 260, 288
    expander, 45
  speaker, 158-160
  system, 161

E

Editing, tape, 297-298
Efficiency, 191
  speaker, 183-184
Eject, 302
Electret microphone, 313-314
Electrical noise, 45, 57-58, 83
Electronic crossover, 39, 40
Electrostatic speakers, 179-180
Elliptical stylus, 220-221
Enclosures, 184-185
Encoding, 261
Erasability, 288
Eraser, bulk, 298
Erasing, 281
Equalization, 106-107, 286, 287
Equalizer, 18, 30-31, 55, 153-154, 188

F

Fader control, 109
Fading, 66
Feedback, negative, 119-122, 142
FETs, 95, 178
Figure-eight, 170-171, 319
Filter(s), 83, 112-113, 147
Flat plane, 172
"Flats," 203, 246
Flutter, 207-208, 307-308
Flywheel, 89, 204, 275
Fm
   antenna(s), 27, 60-64
   /mpx, 20
   muting, switchable, 90
Folded horn, 164
Four speaker system, 22
Frame, 158
Frequencies, subsonic, 121
Frequency, 47-49, 59, 78-80, 101, 117-118, 150, 191, 192, 226, 309, 322-323
Fuse, speaker protection, 93-94

G

Gain, antenna, 64-65
Gap, 277
Gates, logic, 71-75
Gating circuits, basic, 72-75
Graphic equalizer, 18, 188
Ground, 58

H

Harmonic(s), 48-50, 116, 147-149, 256
Head
   amplifiers, 113
   shields, 279-280
Headphone(s), 17, 24-25, 29-30, 84, 91, 157-198
Heil driver, 181-182
Hertz (Hz), 47-48
Heterodyning, 68
Hi-fi
   systems, 13-52
   tuners and receivers, 53-103
Hiss, 230
Horn speaker, 163-164
"Hot" lead, 330
Howl, 235
Hum problem, 234
"Hint," 308
Hybrid recordings, 255-256
Hysteresis motor, 202

I

Image rejection, 101-102
Imaging, 40, 43
Impedance, 119, 137-138, 151-152, 191, 225, 321-322
Induced voltage, 62
Induction motor, 202
Infinite baffle, 167

Input(s), 105-108, 145, 151-152
Inside the cassette, 269-270
Installing your hi-fi system, 43-44
Instant stop, 301
Integrated amplifiers, 28-29, 103, 123-154
Intermediate frequency (if), 56-57
Intermodulation distortion (im), 149-150, 256
   transient, 118-119
Ionization, 171
Ionized air, 171
Ions, 172
Isotweeter, 178-179

J

Jack, headphone, 84
Jacket, dust, 231
Jerk, 301

K

Kilohertz (KHz), 47-48, 244

L

Lacquer, 227, 257
Laser beam scanning, 248
Lavalier microphone, 325-326
LEDs, 76
Level indicators, 302-303
Limiters, 68-69
"Line-of-light" approach, 76
Line power, ac, 125-126
Load impedance, 225
Local oscillator, 57
Logic gates, 71-75
Loopstick, 58
Loudness control, 88, 110-111
LSIs (large scale integrated circuits), 103

M

Magnetic
   field, 62
   fluid, 169
   tape, 265-309
Main balance control, 88
Matching, impedance, 119, 137-138
Mechanical tracking, 246
Memory, 81, 307
Meters, tuning, 75-78
Microphone(s), 35, 311-330
Midrange, 160, 181
   driver, 19, 39
Mirror-image speaker construction, 183
Mixer(s), 57, 299-300
Modulation, 59, 252
Monitoring, 95, 109, 281
Mono and stereo, 15
MOSFETs, 95
Mother mold, 228-229
Motional feedback (MFB), 175, 177
Motor, phono, 201-202

Mpx (multiplex), 20, 69-71
Multipath reception, 65-66
Multispeaker system, 162
Muted tone, 50
Muting, 89, 90, 145
Mylar film, 178

N

Nand gate, 74
Negative feedback (nf), 119-122, 142
Nickel alloy mold, 228-229
Noise, 45, 57-58, 83, 119, 154, 205, 260, 288-289, 319
Nonlinear versus linear inputs, 107-108
Nor gate, 74-75
Numbers, 240-241
Nylon, 233

O

Obstacles, 64
Octaves, 113
Omni radiator, 170-171
Omnidirectional, 64, 317
Open reel, 17-18, 266-267, 291
Operating controls, 86, 144
Optical tracking, 248-249
Or gate, 73-74
Oscillator, 57, 283-284
"Out of round," 203
Outdoor Fm antenna, 27
Outlets, 85
Output, 89, 102-103, 129, 135-137, 147, 226, 321
Overload, 115-116, 129
Overtones, 48-49, 50

P

Parametric equalizer, 18, 188
Parasitic elements, 62
Passband, 82-83
Passive
    crossover, 40
    drivers, 196
    radiators, 172-173
    sound source, 265
Pattern(s), 316-319
Phase, 323-324
    lock loop (pll), 69-70
Phono cartridges, 199-235, 213-217
    equalization, 106-107
    input sensitivity, 114-115
    overload, 115-116
Phonograph (phono), 16
Piezoelectric, 178, 182, 246, 315
Piano, 329
Pink noise generator, 154
Pits, 246-247
Plasma, 171
Playback, 272-275, 278
Polar patterns, 316-320
Polarization, 192

Ported, 166-167
Power, 125, 136, 191
    amplifier, 31, 53-55, 123-154
    output, 135-137, 146-147
Preamp, 28, 53-54, 108, 113-119
Preamplifier(s), 28, 105-122
Precautions, cassette recording, 292-294
Preemphasis, 83, 108
Programmable tuners, 81
Protector circuits, 127-128
Proximity effect, 320-321
Pulse code modulation, 252

Q

Q, 186
"Quack," 185
Quality, 13
Quantizing, 244-245
Quartz, 80, 182
    phase locked loop (pll) motor, 202

R

Radio frequency amplifier, 56
Radiator, 169-171
Ratio, 100, 119
Receiver(s), 15-16, 53-103
Reception, 64, 65-66
Recording, 94, 249-251, 294-296, 303
Record(s), 199-235, 208-209, 302
    changer, 200-201
    player(s), 16, 199-235
        specs, 204-205
Reflectors, 62
Regeneration, 120
Resonance, 322
Response, 117-118, 150, 151
Reverb (reverberation) unit, 188-189
Rf (radio frequency amplifier), 56
Ribbon, 173-174, 314
Rim drive, 203-204
Root-mean-square (rms), 97-98, 136-137, 147, 238
    weighted (wrms), 98
Rumble, 201, 206-207

S

Sampling rates, 244
Scanning, 80-81
Second harmonic, 49
Segue, 35
Selectivity, 59, 101
Selector switch, 88-89, 111-112, 145
Sendust, 279-280
Sensitivity, 59, 98-100, 114-115, 151, 288, 316
Separation, stereo, 102
Shibata stylus, 219, 221
Short circuits, 127
Shotgun microphone(s), 319-320
Signal
    am, 56

Signal—cont
  audio, 56
  separation, 59
  strength
    input, 105-106
    meter, 75
Signal-to-noise ratio (s/n), 61, 75, 100, 119, 150-151, 255, 308-309
Sine waves, 48-50
"Slaves," 196
Slew rate, 118, 151
Sound
  pressure level (spl), 197, 324
  response, microphone, 325
Sound-on-sound, 300
"Sound processor," 31, 32, 37
Speaker
  acoustic suspension speaker, 168-169
  dynamic, 158-160, 161-179
  efficiency, 183-184
  protection fuse, 93-94
  specifications, 189-192
  switching, 147
  terminals, 91
  wire, 93
Speakers
  adding more, 127
  and headphones, 157-198
  electrostatic, 179-180
  labyrinth, 180-181
Specifications, 97-103, 189-192, 307, 321-324
Specs, 13, 46-47, 113-119, 147-153, 189, 192, 204-208, 222-226, 256-257
Speed control, 235
Spider, 159, 169
Spurious response, 102
Stabilizer, vacuum disc, 208-210
Stamper, 230
Standing waves, 189
Stereo, 15-16, 90, 102, 110, 223, 255
Stereophonic sound, 15
Strain lines, 291
Strapping, 136
String bass, 327-328
Strobe, 235
Styli care, 234
Stylus, 217, 218-222
  conical, 219-220
  elliptical, 220-221
  Shibata, 219, 221
  van den Hul, 222
Subsonic
  filter, 147
  frequencies, 121
Subwoofer, 162
Supercardioid microphone, 319
Superheterodyne, 66-67
Suppression, am, 102
Supra-aural headphones, 196
Switchable bias, 287

Switchable fm muting, 90
Switch, 85, 88-89, 90, 94, 110-112, 145
Switches, 83-84, 242
Switching, 281-283
  speakers, 147
Synchronous motor, 202
  hysteresis, 202
Synthesizer, frequency, 79-80
System(s), 22, 39, 41-43
  audio and hi-fi, 18-40
  basic, 15
  compact, 13
  component, 13-14
  dealer, 14
  hi-fi, 13-52
  modular, 14
  prepackaged, 14

T

Take-up, 301
Tape
  decks, 265-309
  monitor switch, 88, 145
  monitoring, 109
  reels, 17
Taping off the air, 294-296
Test tone switch, 90
Testing the deck, 276-277
Timbre, 50
Time constant, 286-287
Time delay, 145, 186-187
Tone(s), 39, 40, 50
  controls, 85, 110
Tonearms, 210-212
Total harmonic distortion, 147-149
Tracking, 211, 224-225, 246-248
Transient, 118-119, 151
Transmission, 251
  line, 60, 180-181
Treble Tones, 39
Tremelo, 277
Triamp, 43
Triaxial, 163
"Tuned," 75
  circuit, 57
Tuner, 28, 53-55, 78-79, 81, 91-95, 99
Tuning
  control, 89-90
  tuner or receiver, 75-84
Turntable(s), 199, 200-201
Tweeter, 19, 39, 160, 181-182

U

Ultra bass tones, 162
Unbalanced line, 330
"Underdrive," 184
Unit of efficiency, 183
Upper-order harmonic, 49
Using microphones, 326-329

V

Vacuum disc stabilizer, 208-210
van den Hul stylus, 222
Variable
  bias control, 303
  capacitance tracking, 246-248
  reluctance, 216, 217
Vented-port speakers, 166
Ventilation, 46
Vents, air, 45
Vibration, 43, 164
Voice, 50
  coil, 158-159
Voltage, 102-103, 125, 128, 189, 237
Volume control, 88, 109, 143
Vu (volume unit) meter, 90

W

"Wandering," 40
Warp, record, 208-209
Warranty, 45
Watts, rms, 147
Waveform, 47, 49, 237
Waves, standing, 189
Wear, 253
Weighted root-mean-square (wrms), 98
Whizzer speaker, 163
Wind noise, 315
Wire, speaker, 93
Woofer(s) 19, 39, 159-160, 181
Wow, 201, 307-308
  and flutter, 207-208

Z

Zenith, head, 280
Zero, 239
  center turning meter, 75-77
  -feedback stereo amplifier, 124